Discrete-Event Simulation and System Dynamics for Management Decision Making

Wiley Series in
Operations Research and Management Science

Operations Research and Management Science (ORMS) is a broad, interdisciplinary branch of applied mathematics concerned with improving the quality of decisions and processes and is a major component of the global modern movement towards the use of advanced analytics in industry and scientific research. *The Wiley Series in Operations Research and Management Science* features a broad collection of books that meet the varied needs of researchers, practitioners, policy makers, and students who use or need to improve their use of analytics. Reflecting the wide range of current research within the ORMS community, the Series encompasses application, methodology, and theory and provides coverage of both classical and cutting edge ORMS concepts and developments. Written by recognized international experts in the field, this collection is appropriate for students as well as professionals from private and public sectors including industry, government, and nonprofit organization who are interested in ORMS at a technical level. The Series is comprised of three sections: Decision and Risk Analysis; Optimization Models; and Stochastic Models.

Advisory Editor • Decision and Risk Analysis

Gregory S. Parnell, United States Air Force Academy

Founding Series Editor
James J. Cochran, Louisiana Tech University

Discrete-Event Simulation and System Dynamics for Management Decision Making

Editors

Sally Brailsford

Southampton Business School, University of Southampton, UK

Leonid Churilov

*Florey Institute of Neuroscience and
Mental Health, Melbourne, Australia
RMIT University, Melbourne, Victoria, Australia*

Brian Dangerfield

Salford Business School, University of Salford, UK

WILEY

This edition first published 2014
© 2014 John Wiley & Sons, Ltd

Registered office
John Wiley &
Sons Ltd, The Atrium, Southern Gate, Chichester, West Sussex, PO19 8SQ, United Kingdom

For details of our global editorial offices, for customer services and for information about how to apply for permission to reuse the copyright material in this book please see our website at www.wiley.com.

Library of Congress Cataloging-in-Publication Data

Discrete-event simulation and system dynamics for management decision making / [edited by] Sally Brailsford, Leonid Churilov, Brian Dangerfield.
    pages cm
  Includes bibliographical references and index.
  ISBN 978-1-118-34902-1 (hardback)
  1. Discrete-time systems–Simulation methods. 2. System analysis. 3. Decision making.
4. Management science. I. Brailsford, Sally. II. Churilov, Leonid. III. Dangerfield, Brian Thornley.
  T57.62.D495 2014
  658.4'0352–dc23

                        2013047217

A catalogue record for this book is available from the British Library.

ISBN: 978-1-118-34902-1

Set in 10/12pt TimesLTStd-Roman by Thomson Digital, Noida, India.
Printed and bound in Singapore by Markono Print Media Pte Ltd

1   2014

We dedicate this book to Ruth Davies, Emeritus Professor of Operational Research at the University of Warwick, a great academic and a great friend.

Contents

Preface

The genesis of operational research (OR) in the Second World War was largely characterised by deterministic techniques with a nod to risk evaluations such as in establishing the optimum balance of merchant ships and naval protection vessels in Atlantic convoy sizes. But it was part of the promulgation of OR techniques in large nationalised industries in the late 1940s and early 1950s that simulation came to the fore. This was particularly evident in the British steel industry. The emerging power of digital computers helped enormously and, under the guidance of luminaries such as Keith Tocher, discrete-event simulation (DES) (and the three-phase system) emerged from what had previously been Monte Carlo simulation.

Later in the 1950s in the United States another luminary, Jay Forrester, was settling into a new role at MIT and he saw the possibilities of applying the ideas and concepts from control engineering to the simulation of economic and social systems. Like Tocher, he relied on the growing power of computers. In fact he had been closely involved on the hardware side even to the extent of holding a US patent for random-access magnetic core memory. Forrester launched the field of what was to be system dynamics (SD), then known as industrial dynamics, in a paper in the *Harvard Business Review* in 1958.

Two powerful intellects were responsible for setting in train two separate methodologies in the domain of management science (MS) that, over the subsequent decades, have come to be employed on an enormous variety of applications. But although the simulation landscape has been enriched by their respective capabilities (which were rendered all the more impressive by the advent of icon-based computing) there has been almost no significant attempt to set out their respective merits in a comparative sense, still less to illustrate how they may be used in concert. This book, we believe, is the first volume to address these issues, while also describing agent-based (AB) modelling, another methodology that has recently emerged.

SCB
LC
BCD

List of contributors

Steffen Bayer, Program in Health Services & Systems Research, Duke-NUS Graduate Medical School Singapore

Tim Bolt, Faculty of Health Sciences, University of Southampton, UK

Andrei Borshchev, Managing Director and CEO, The AnyLogic Company, St Petersburg, Russia

Sally Brailsford, Southampton Business School, University of Southampton, UK

Leonid Churilov, Florey Institute of Neuroscience and Mental Health, Melbourne, Australia; RMIT University, Melbourne, Victoria, Australia

Brian Dangerfield, Salford Business School, University of Salford, UK

Shivam M. Desai, Southampton Business School, University of Southampton, UK

Mark Elder, SIMUL8 Corporation, Glasgow, UK

Andrew Flitman, Florey Institute of Neuroscience and Mental Health, Melbourne, Victoria, Australia

Paul Harper, School of Mathematics, Cardiff University, UK

Lee Jones, Director, Ventana Systems UK, Oxton, Merseyside, UK

Maria Kapsali, Umea School of Business and Economics, Umea University, Sweden

Kathy Kotiadis, Kent Business School, University of Kent, Canterbury, UK

Geoff McDonnell, Centre for Health Informatics, Australian Institute of Health Innovation, University of New South Wales, Sydney, New South Wales, Australia; Adaptive Care Systems, Sydney, New South Wales, Australia

John Mingers, Kent Business School, University of Kent, Canterbury, UK

John Morecroft, London Business School, London, UK

Michael Pidd, Lancaster Business School, University of Lancaster, UK

Stewart Robinson, School of Business and Economics, Loughborough University, UK

Kristian Rotaru, Department of Accounting, Monash University, Melbourne, Victoria, Australia

Rosemarie Sadsad, Centre for Infectious Diseases and Microbiology – Public Health, Westmead Hospital, Sydney, New South Wales, Australia; Sydney Medical School, Westmead, The University of Sydney, New South Wales, Australia; Centre for Health Informatics, Australian Institute of Health Innovation, University of New South Wales, Sydney, New South Wales, Australia

Antuela Tako, School of Business and Economics, Loughborough University, UK

Joe Viana, Southampton Business School, University of Southampton, UK

1

Introduction

Sally Brailsford,[1] Leonid Churilov[2] and Brian Dangerfield[3]

[1]Southampton Business School, University of Southampton, UK
[2]Florey Institute of Neuroscience and Mental Health, Melbourne, Australia;
RMIT University, Melbourne, Victoria, Australia
[3]Salford Business School, University of Salford, UK

1.1 How this book came about

To begin at the end . . . the final chapter in this book, by Michael Pidd, contains both a backwards and a forwards look at system dynamics and discrete-event simulation. Historically, both modelling approaches originate from around the same time, the late 1950s and early 1960s. However, over the intervening decades they developed into separate scientific and practitioner communities, each with its own learned societies, academic journals and conferences. Discrete-event simulation (DES) has been a core subject on MSc programmes in operational research (OR) or management science (MS) from the 1970s onwards, and is a standard technique in the 'OR/MS toolkit'. For many operational researchers, 'simulation' is synonymous with DES, and indeed the aim of the UK OR Society's own *Journal of Simulation* (quoting directly from the journal's web site) is to provide 'a single source of accessible research and practice in the fast developing field of discrete-event simulation' (http://www.palgrave-journals .com/jos/index.html). However, this is not true of system dynamics (SD): the SD community was, and still is to some extent, distinct from the OR community. While there is obviously some overlap in membership, there are many members of the international SD Society who are not members of their national OR Society (and vice versa). SD was certainly not taught on the MSc in OR at the University of Southampton in the 1980s and 1990s.

In 2000, the Simulation Special Interest Group of the UK OR Society held a joint meeting with the UK Chapter of the SD Society, entitled 'Never the Twain Shall Meet'. At this meeting David Lane presented a paper (Lane, 2000) in which he discussed the differences between SD and DES and posed the question about whether they were 'chalk and cheese' or were actually two sides of the same coin. This meeting led to the foundation of a new OR Society Special Interest Group called 'SD+' whose aim was to bring the SD and OR communities together. The '+' in SD+ was broader than just DES: it included many other OR techniques and approaches with which SD could interface. The 'Never the Twain' meeting also led to a number of academic papers exploring the similarities and differences between DES and SD, including the well-known study by Robinson and Morecroft which forms Chapter 9 of this book. Indirectly, it led to this book itself!

In some application areas the use of SD has expanded rapidly since 2000, and healthcare – the specialist application field of all three editors of this book – is one such area. However, despite initiatives like SD+, it is still true to say even today that SD is less well known in the mainstream OR community than DES. The number of DES papers at the annual Winter Simulation Conference, the major US conference on simulation, always greatly exceeds the number of SD papers. The main aim of this book is to begin to address this disparity. The book provides an integrated overview of SD and DES, a detailed comparison of the two approaches from a variety of perspectives, and a practical guide to how both may be used, either separately or together.

1.2   The editors

As editors, we should declare our own personal interests. Having started out in the early 1990s as a dedicated user of DES, Sally Brailsford became interested in SD as a result of the 'Never the Twain' meeting, and joined David Lane in co-founding the SD+ group. Subsequently she became a zealous convert, using SD for several modelling studies. Like many other researchers, she was fascinated by the relationship between DES and SD, and in particular in the domain of healthcare (Brailsford and Hilton, 2001), but first used SD in practice in a project to model demand for emergency healthcare in Nottingham, England (Brailsford et al., 2004) which is described in detail in Chapter 6.

Leonid Churilov has a firm belief that real management problems do not come cleanly separated by disciplinary lines and, as a result, can rarely be comprehensively addressed using a single given modelling method. This basic premise is the source of continuous motivation for his keen interest in combining and contrasting different OR/MS techniques for management decision support. His research, in particular, included combining DES and clustering/classification techniques for decision support in hospital emergency departments, the use of both DES and SD for process systems modelling, and the original value-focused process engineering methodology that integrates the approaches from both the decision sciences and business process modelling domains. His work on philosophical underpinnings of both DES and SD worldviews from the critical realist perspective is featured in Chapter 5.

Brian Dangerfield discovered SD (industrial dynamics as it then was) over 40 years ago after a period in an OR unit in industry. Working at the (then) University of Liverpool School of Business Studies, he was a researcher on a project looking at the role of stocks in the UK economy. Rather than take the obvious econometric route he realised that an understanding of the macro role that stocks played in the workings of the economy would be much enhanced if, instead, macro-economic models were SD simulation models. Variables representing stocks had to be divorced from the flows which changed them. He was impressed with the way (i) SD models were forged at the policy level, (ii) separated out resource flows from the information flows driving changes in those resources and (iii) were also able to embrace relevant 'soft' variables which, using another methodology, might be excluded altogether. In sum he concluded that, given his overall knowledge and real-world experience with OR (including traditional simulation techniques), the SD methodology was just about the most promising in the landscape of OR, and his subsequent research career has concentrated on its use and development.

1.3   Navigating the book

To our knowledge this book is unique – there is no similar coverage in one single volume. Books which cover both methodologies (e.g. Pidd's *Computer Simulation in Management Science*, 2009) merely split the page coverage – there is no attempt at integration, or a detailed comparison and description of how both approaches may be used in the same project. This book provides a seamless treatment of a variety of topics: theory, philosophy, detailed mechanics and practical implementation, all written by experts in the field. While some chapters are aimed at beginners, others are more advanced; the book also includes three software chapters which are very practical in nature.

The book is structured in seven unequal sections, three of which only contain one chapter. This Introduction forms the first section, and Pidd's concluding chapter the seventh. The second section, 'Primers', contains two chapters which provide a basic introduction to DES and SD respectively. These chapters are both written by experts with many years' experience of teaching and using each technique: Chapter 2 on DES (Stewart Robinson) and Chapter 3 on SD (Brian Dangerfield). The authors assume no prior knowledge of either technique, or an academic background in mathematics, statistics or OR. They are aimed at students and practitioners alike. The aim of both primers is to provide sufficient understanding to enable the average reader to get a basic grasp of the topic and be able to appreciate the subsequent chapters. The primers do not attempt to provide the breadth and depth of material offered in a typical MSc course in SD or DES, and they do not contain a great deal of technical detail. References are provided for anyone who does wish to delve deeper into the technicalities!

The third section also consists of a single chapter. By way of contrast with Chapters 2 and 3, the authors of Chapter 4 (Kathy Kotiadis and John Mingers) take a strongly academic stance. This chapter provides a theoretical background for much of the discussion in later chapters about combining different modelling paradigms. This chapter, which was originally published as a research article in the *Journal of the Operational Research Society*, discusses the combination of problem structuring methods with hard OR methodologies. Kotiadis and Mingers reflect on the barriers to such combinations that can be seen at the philosophical level – paradigm incommensurability – and cognitive level – type of personality and difficulty of switching paradigm. They then examine the combination of soft systems methodology and DES within a healthcare case study. They argue, by way of the practical application, that these problems are not insurmountable and that the result can be seen as the interplay of the soft and hard paradigms. The idea of yin and yang is proposed as a metaphor for this process.

The fourth section, 'Comparisons', is by far the largest and represents the heart of the book – its *raison d'être*. It contains five chapters, all of which consider contrasting aspects of DES and SD, ranging from methodological and philosophical comparisons using different lenses or frameworks, through to practical aspects and software implementations. In Chapter 5, Leonid Churilov, Kristian Rotaru and Andrew Flitman investigate how the critical realist philosophy of science facilitates explicit articulation of the fundamental philosophical assumptions underlying the SD and DES worldviews, using a practical illustration of simulating process systems. The ultimate aim is to achieve more effective use of simulation for intelligent thinking about, and management decision support in, process systems. The novelty and original contribution of this research is in applying the stratified ontology of critical realism, and the abductive mode of knowledge generation, to examine explicitly the philosophical bases of the DES and SD simulation worldviews. The outcomes of this research are targeted at both the manager, who is the contributor to, as well as the end user of, a simulation model of a real-world process system and, as such, could benefit from a clear understanding of how management knowledge is generated through the modelling process, and the management scientist who chooses to use simulation modelling to support management decision making in real-world process systems, and requires in-depth understanding of the scientific bases of the respective modelling methodologies to apply them in a truly scientific manner.

It is an oft-quoted cliché that if all you have is a hammer, then every problem is a nail. The aim of Chapter 6 is to discuss whether the choice of simulation methodology – DES or SD – is purely down to the personal preference and expertise of the modeller, or whether there are identifiable features of certain problems that make one approach intrinsically preferable to the other. Although from a methodological standpoint the overall comments are generic and applicable to any setting, the chapter has a bias towards healthcare applications as this is the area of domain expertise of the author, Sally Brailsford. A case study in emergency care is presented, where both DES and SD were used to tackle different aspects of the overall problem. The chapter concludes with some general guidelines to assist the modeller in making the choice of technique.

One commonly held stereotype is that SD naturally lends itself to problems where a group of people with potentially conflicting objectives need to be engaged, on an ongoing basis, in the process of developing and running a simulation model, whereas DES is more naturally suited to problems where there is an agreed objective and, after an initial meeting with the client, the modeller goes away, locks him- or herself in a darkened room and spends a month writing code, requesting data, running the model and obtaining results, which the modeller then presents to the client at the next meeting. This is clearly rather an absurd exaggeration, but like all stereotypes it is worth examining to see whether it contains a grain of truth. The use of DES and SD models in group model building projects, where a group of domain experts or other stakeholders come together to build a model together with a modelling expert, is examined in Chapter 7. Steffen Bayer, Timothy Bolt, Sally Brailsford and Maria Kapsali show how both SD and DES models have a social function and can act as an 'interface' between participants in a modelling project. An interface can be understood as a point of interaction which allows two systems to communicate across a boundary. In everyday use 'interface' is, however, also often used as relating to what affords and enables the process of communication across the boundary: interface as the means or mechanism that allows the boundary between subsystems to be permeable. The metaphor 'interface' highlights the potential of models (and of the modelling process) to support information transmission in the widest sense between the participants in a group modelling project but also across the boundary between the model and those engaging with the model. The aim of this chapter is to explore and illustrate how the metaphor of interface can shed light on the variety of social functions models can have, especially in a group modelling context.

The actual model building process is explored in further detail in the following chapter, which focuses less on the social aspects of model building and more on the technical aspects, providing a detailed comparison of DES and SD by analysing the thought processes of the modellers themselves. In Chapter 8 Antuela Tako and Stewart Robinson describe an empirical study which used verbal protocol analysis to study the modelling process followed by ten expert modellers, five using SD and five using DES. The participants were asked to build simulation models based on a case study and to think aloud while modelling. The generated verbal protocols are divided into seven modelling topics (problem structuring, conceptual modelling, data inputs, model coding, validation and verification, results and experimentation, and implementation) and are then analysed. Quantitative analysis of the verbal protocols shows that all modellers switch between modelling topics; however, DES modellers follow a more linear progression. They focus significantly more on model coding and verification and validation, whereas SD modellers focus on conceptual modelling. Observations are also made, revealing some interesting differences in the way the two groups of modellers tackle the case. This chapter contributes to the comparison of DES and SD modelling in management decision making by providing empirical evidence with regards to the model development process.

The 'Comparisons' section concludes with Chapter 9, a contribution from John Morecroft and Stewart Robinson. Morecroft (an expert in SD) and Robinson (an expert in DES) compare SD and DES through an examination of the way each method analyses the puzzling dynamics inherent in a fisheries system. They begin by recounting comparisons between SD and DES made by other authors over the past 20 years or so. They lament that none of these comparisons are offered by neutral parties – all have a main affiliation with either DES or SD.

Morecroft and Robinson start their own comparison by presenting data charts from two well-known fisheries. In one a serious drop in annual fish 'landings' recovered eventually, but in the other the new rate of substantially lower catches has continued for over 40 years. A series of experiments are then undertaken involving building dual models of a stylised fishery – one in SD and the other in DES. The authors compare, step by step, the two emerging models, the equation formulations and the resultant simulated behaviour. They begin with a 'natural' fishery where there is no human involvement and then progress to a harvested fishery where ships are involved and the fish are landed. Initially the harvested fishery is modelled in equilibrium but this assumption is ultimately relaxed – in the SD model by adding a nonlinearity and in the DES model by adding randomness. The SD model exhibits growth initially but ultimately the fishery collapses. Turning to the DES model, there is still a collapse but over a different timescale: the greater the spread of the normal distribution generating the randomness, the faster the eventual collapse. The authors conclude with a final experiment where the harvested fishery now includes endogenous investment in ships. The authors demonstrate that, in both models, the policies for deciding on ship investments are defective. In conclusion they express the paradigmatic differences between the two approaches by stating that DES illuminates interconnected randomness whereas SD illuminates deterministic complexity.

Underpinning the developments in both DES and SD modelling, the improvements in software functionality stand supreme. As Pidd describes in Chapter 15, probably until well into the 1980s both these methodologies were employed in a manner which today's user would find almost unbelievable. Text-based interfaces were the norm and the mainframe-supported software inevitably involved significant delays in delivering (usually printed) output. Common in DES at the time were such packages as GPSS and ECSL; in SD, DYNAMO and DYSMAP. All this changed by the mid-1990s as icon-based software became the norm, deployed on a PC and, latterly, on powerful laptop computers. Now users could experience almost instantaneous output together with the ability to display animated dynamic visual representations of their systems in real time.

This volume would be deficient if it did not afford some space to the consideration of contemporary software offerings. Accordingly, in the fifth section three authors have been invited to write about the software systems with which they are associated. Two were selected as being examples of commonly used software products for the methodologies covered by this volume and which had been in existence for at least 10 years to date; the third is a relative newcomer. The purpose of this book is such that an excessively lengthy documentation of relevant current software would be counterproductive. There are certainly other available systems for both DES and SD simulation and there is no intention to imply that those described here are in any sense superior; they are purely representative. And, going forward from here, Pidd ponders on possible future software developments in Chapter 15.

A word about the inclusion of AnyLogic is necessary as its emergence is relatively recent. Developments in the two principal methodologies covered here (and those related to them) are continuing apace. However, agent-based modelling (ABM) has emerged as a viable alternative to DES or SD simulation. Its use is primarily for certain types of problem, such as where a high level of (network) interaction detail and granularity is desirable and is capable of being represented as a feedback system. ABM would not exist if it were not for the astonishing developments in computer power in recent years. The AnyLogic system claims to allow for coding a problem as a DES, SD or ABM model. Whether one would wish to replicate a problem system in this way is open to debate. Usually the purpose of the investigation will dictate the approach adopted in coding the model. However, the availability of software such as AnyLogic means that, in theory, one would not need to utilise any other software system in order to create a methodologically diverse variety of systems models and, furthermore, it is possible to combine any of the three methods when using this software.

In Chapter 10 Mark Elder introduces a popular DES software platform – SIMUL8. He begins by emphasising that the model can aid understanding for the client such that he or she can have the confidence to change the process being managed for the better. Even rudimentary models can generate insight and a high-level model can produce the spark which spurs further investigation using more detailed models. While the thrust of this centres upon SIMUL8's dynamic 'floor plan' capabilities, and the way that this on-screen animation hooks in the client group, some of Elder's ideas about the benefits derived from the contemporary modelling process, rather than the model itself, would apply equally to SD.

He continues by taking the reader through an example model using the SIMUL8 software, depicting the processes involved in manufacturing metal tables. Some data can be automatically inserted by the software, thereby speeding up model creation. The need for replicated runs (trials) is emphasised together with the concomitant statistical analysis of the data generated. Some attention is given to the importance of verification and validation of the model as well as how to optimise the resources in the model. Finally, he concludes by considering the future for DES software.

Lee Jones continues the discussion of software matters by describing the Vensim SD simulation platform in Chapter 11. He starts by recounting how complexity in the modern business and social environment might be conducive to the growing use of SD. But this has not happened (particularly in business) and he conjectures that at its heart the issue may be bound up with SD software. Tracing the development of SD software platforms since the text-based environments of the 1980s, he recounts how Vensim's functionality has been improved over the years and how software user forums can now dictate the direction of additions to software functionality.

The importance of model units checks is underlined through the story of how NASA's Mars Climate Orbiter was destroyed on entry to the Martian atmosphere, merely because of a mismatch between metric and imperial units in the software driving the craft's orbital trajectory. Jones further considers calibration of a model to past (and indeed future) data using Vensim's optimisation feature. Other developments mentioned include the 'Reality Check' facility, together with 'Story Telling', the often problematic phase of SD modelling where the reasons for model behaviour have to be effectively communicated to the client; reduced form model diagrams, causes trees and strip graphs can all be usefully deployed to this end. Sensitivity graphs and policy optimisations can require many thousands of repeated runs and modern SD software harnesses the raw computing power now available in order to provide features of this sort. The chapter concludes by recounting how innovations in technology are helping the enhanced exposure of SD models: from the promulgation of models running live over the Internet, to the Facebook interactive game about how best to develop our planet ('Game Change Rio'); and, for the immediate future, the possible utilisation of tablet computers and smart phones for model execution and presentation.

In the final chapter in the 'Software' section (Chapter 12) Andrei Borshchev provides an account of the relatively recent offering: AnyLogic. His underlying theme is to illustrate how this platform can be employed to mix three simulation modelling methodologies – DES, SD and ABM – within the same overall problem. By introducing the concept of the statechart, Borshchev instances seven theoretical occurrences of how agents, discrete elements and SD stocks and flows can interact. Going into much more of the model formulation details, he then describes three specific examples: a supply chain where the market is handled using SD concepts while the supply chain itself involves the discrete-event methodology; an epidemic model exploring a clinic's capacity to cope, where the combination here is ABM and DES; and finally a product life cycle and new product investment policy where the formulation involves a mix of ABM and SD. However, it would be wrong to imply that only combinations of two methodologies are possible: the AnyLogic platform can be employed to develop simulation models where all three co-exist.

The sixth section, which contains two chapters, focuses on models and applications and begins with a practical guide to using different levels or scales of modelling. Chapter 13, by Geoff McDonnell and Rosemarie Sadsad, is dedicated to various aspects of multiscale modelling in the context of public health and health services management. Simulation models of complex and multilevel organisational systems like healthcare are often abstracted at one level of interest. There are many challenges with developing simulation models of systems that span multiple organisational levels and physical scales. Several theoretical and conceptual frameworks for multilevel system analysis are analysed and an approach for developing multiscale and multi-method simulation models to aid management decisions is presented. Simulation models are used to illustrate how management actions, informed by patterns in stock levels, govern discrete events and entities, which, collectively, change the flow mechanism that controls stock levels. The proposed approach explicitly considers the role of context when designing and evaluating, in particular, public health actions. It extends current analytical and experimental methods and has the potential to encourage more collaborative and multidisciplinary effort towards effective public health management.

Chapter 14 focuses on applications of combined simulation approaches, and its objective is to demonstrate the use of hybrid modelling for management decision support. The chapter presents three separate case studies that are unified both by the common theme of using different modelling techniques in a hybrid manner and by the health- and social care context. The first case study, by Geoff McDonnell and Rosemarie Sadsad, combines the use of SD and ABM to better understand and control the spread of a specific drug-resistant pathogen in hospitals; the second case study, by Joe Viana, demonstrates how a hybrid SD and DES model can support clinical and management decision making in the context of sexual health services in Portsmouth, UK; while the third case study, by Shivam Desai, is dedicated to the investigation of a hybrid model that consists of an SD-inspired cell-based population model and a DES model, used to explore the performance of a contact centre for long-term care for people aged 65 and over.

The final chapter is Michael Pidd's personal view of the future – and of the past. Pidd is of course well known as the author of one of the best known and most widely used textbooks on simulation, first published in 1984 and now in its fifth edition (Pidd, 2009). In the 1980s, general OR textbooks tended to have about 10 chapters on mathematical programming, 10 chapters on other analytical techniques, and then a final chapter on the statistical aspects of Monte Carlo simulation, only to be used *when all else fails*. No wonder that Pidd's highly readable, practical and comprehensive book on simulation became a classic. Moreover, unusually for a simulation textbook, even the first edition contained a section on SD. It is hard to think of a more appropriate person to conclude this volume, which we hope will prove useful to academics and practitioners alike.

References

Brailsford, S.C. and Hilton, N.A. (2001) A comparison of discrete event simulation and system dynamics for modelling healthcare systems. Proceedings of ORAHS 2000, Glasgow, Scotland (ed. J. Riley), pp. 18–39.

Brailsford, S.C., Lattimer, V.A., Tarnaras, P. and Turnbull., J.A. (2004) Emergency and on-demand health care: modelling a large complex system. *Journal of the Operational Research Society*, **55**, 34–42.

Lane, D.C. (2000) You just don't understand me: modes of failure and success in the discourse between system dynamics and discrete event simulation. LSE OR Dept Working Paper LSEOR 00-34.

Pidd, M. (2009) *Computer Simulation in Management Science*, 5th edn, John Wiley & Sons, Ltd, Chichester.

2

Discrete-event simulation: A primer

Stewart Robinson
School of Business and Economics, Loughborough University, UK

2.1 Introduction

Discrete-event simulation (DES) grew largely out of a desire to model manufacturing systems. Based upon the foundation of Monte Carlo methods, DES models were developed to improve the design and operation of manufacturing plants. Among the earliest examples is the work of K.D. Tocher who developed the General Simulation Program in the late 1950s at the United Steels Companies in the United Kingdom; see Hollocks (2008) for an excellent summary of these early developments.

Over the years DES has been applied to a much broader set of applications including health, service industries, transportation, warehousing, supply chains, defence, computer systems and business process management. Much of this work has focused on improving the design and operation of the systems under investigation, but there have also been examples of DES for aiding strategic decision making.

DES is seen as a, if not the, mainstream simulation approach in the field of operational research (OR). Indeed, many OR specialists simply refer to it as 'simulation', seemingly ignoring the potential to simulate using other approaches, including system dynamics. DES does imply a very specific approach to simulation from both a technical and a philosophical perspective. In this chapter we will explore both of these perspectives. To set a context, we first present an example of a DES

Discrete-Event Simulation and System Dynamics for Management Decision Making, First Edition.
Edited by Sally Brailsford, Leonid Churilov and Brian Dangerfield.
© 2014 John Wiley & Sons, Ltd. Published 2014 by John Wiley & Sons, Ltd.

model. We then discuss how DES works (the technical perspective) followed by the worldview adopted by DES modellers (the philosophical perspective).

2.2 An example of a discrete-event simulation: Modelling a hospital theatres process

Figure 2.1 shows an example of a DES model, in this case of a hospital theatres process (Burgess *et al.*, 2011). This model can be accessed from www.simlean.org (accessed May 2012) as either a model file (for which the proprietary software is required) or as video files. The model simulates the flow of patients through an outpatients theatre over the period of a day. Patients arrive at a reception area and having registered they are prepared for their operation. Following the operation in one of the four theatres, they are moved to a recovery area. From here they are generally discharged, but in some cases the patient needs to be admitted to a ward.

Figure 2.1 shows the state of the model at around 11:35 in the morning, after an 8 a.m. start. At this time the theatres and preparation area are both fully utilised, resulting in a queue of patients waiting post-reception. There is also a queue for discharge since there is insufficient resource available for discharging patients at this point in the day.

While the model runs it records data on the performance of the system. These performance data typically summarise the experience of the individuals in the system, in this case patients, or the experience of the activities (e.g. preparation, theatres) and

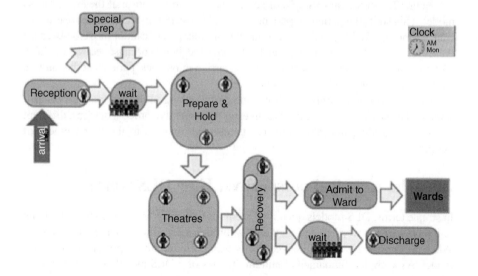

Figure 2.1 A simple DES model of a hospital theatres process (Source: Burgess et al., 2011; Robinson et al., 2012). Reproduced by permission of Elsevier.

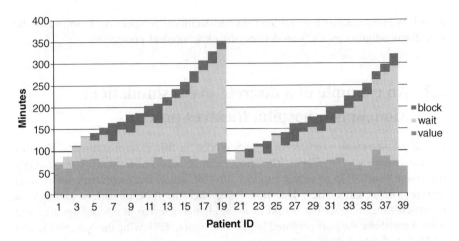

Figure 2.2   Results generated from the DES model of a hospital theatres process (Source: Burgess et al., 2011). Reproduced with permission from SimLean Publishing.

resources (e.g. receptionist, nurses, doctors). We are specifically interested in the amount of time the individuals spend waiting versus being worked on, which must be balanced with the utilisation of the activities and resources. By nature, high utilisation tends to lead to more waiting as the system becomes overloaded, while low waiting times tend to result from underutilised resources. The aim is to find an appropriate balance between these two measures of performance.

Figure 2.2 shows an example of some results from the hospital theatres process model. This splits the patient experience into value (receiving attention or treatment), waiting and blocked (an activity is complete, but the patient cannot move on because there is no space downstream, e.g. unable to leave the theatre because recovery is full). The pattern that emerges is, because patients arrive in batches, patients at the tail of a batch have to wait much longer to be seen.

There is clearly a problem with waiting and blocking in the theatres process as it stands. The simulation can be used to investigate alternative process designs, different ways of scheduling and allocating resources, and changes to the management of the process.

2.3   The technical perspective: How DES works

In simple terms, DES models queuing systems as they progress through time. In doing so it represents the world as *entities* that flow through a network of *queues* and *activities*. Where *resources* are shared between activities, these are also represented in a DES model. As such, the fundamental building blocks of a DES model are as follows:

- *Entities*: individual items that flow through the system, for example widgets in a manufacturing plant, people in a service system or hospital, packets of information in a computer network, vehicles in a transport system, orders in a supply chain.

- *Queues*: areas where entities wait to be worked on, for example buffers and stores, inventory, waiting areas, waiting lists, phone call queuing.

- *Activities*: perform work on entities, for example machines, travelling, moving, serving, printing.

- *Resources*: required to be present to operate activities, for example operators, equipment, doctors.

This structure is very flexible and it can be used to model a wide variety of systems. It is for this reason that DES has been widely used in OR for the types of application listed in the introduction to this chapter.

Figure 2.3 shows an example of a queuing system. Entities arrive at queue 1 and are then directed to either activity 1a or 1b; this could be based on the individual requirements of the entity or at random. Resource 1 is required to be present while activity 1 is taking place, hence it is shared between activities 1a and 1b. Following activity 1, all entities go to queue 2 to await activity 2, after which they leave the system. This system could represent many forms of two-stage queuing process, for instance a doctor's surgery with a reception area (with two distinct checking-in processes) followed by a consultation with the doctor.

An important feature of entities in DES models is that they have *attributes* that describe specific features of an entity. These can describe features such as entity type, dimensions, weight, priority, order number and time in system. The values of these attributes can vary from entity to entity. As such, the attributes can be used to determine the logic of the model, for instance the time an individual entity will spend in an activity, its priority in a queue, or its route through the system. This ability to model the detail of individual entities is a particularly powerful feature of DES models.

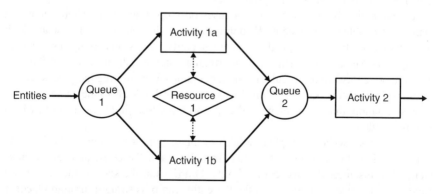

Figure 2.3   An example of a queuing system.

Beyond these fundamental building blocks, there are two other key elements of DES models: the time-handling mechanism and methods for random sampling. We will now explore both of these in turn.

2.3.1   Time handling in DES

A simple way of modelling the progression of time in simulation models is to use a constant time step. This approach works well for modelling systems that change continuously; a fixed time step is used to update the state of a system that is continuously changing. Of course, this only approximates continuous change, but digital computers are not able to truly model continuous time. If the time step is sufficiently small, the approximation enables the system to be modelled with sufficient accuracy. There is a trade-off, however, between run-time and accuracy, both of which increase with smaller time steps.

Such 'continuous' simulation is used widely in, for instance, the physical sciences and engineering for modelling continuous processes such as wave motion or the aerodynamics of a vehicle. It is also used in a business context to model processes that are subject to continuous change, for instance chemical plants or oil pipelines. Inventory systems, including supply chains, can be modelled using this approach if we are satisfied with updating the state of the system at fixed intervals, say daily. System dynamics is itself a form of continuous simulation.

Given that DES focuses on modelling individual entities that flow through a set of queues and activities, the exact timing of a change in the state of the system is difficult to predict. While an entity is being worked on in an activity, the state of the system remains unchanged. It is only when the activity is finished that the system state changes; the entity moves to the next stage of the process and the activity can start work on another entity. To model this accurately might require that time is modelled in very fine-grained units. To achieve this using a continuous simulation approach would require a very small time step. Because the system state only changes relatively rarely in comparison with the scale of the time step, the simulation would be extremely inefficient and slow since the majority of time steps would model points at which there is no change in system state.

To overcome this problem, DES focuses only on time points where there is a change in the state of the system. We define *events* as discrete points in time at which the system state changes. Typical events are an entity arrives, an activity starts and an activity ends. Since these events occur at irregular intervals, the time in a DES model does not move forward in a regular fashion, but jumps forward irregularly between the times at which events occur. It is this focus on events that leads to the name discrete-event simulation; that is, the name of the approach is describing the manner in which time is handled.

Table 2.1 shows an example of a DES of the queuing system in Figure 2.3. The table describes the sequence of events (in italics) and the consequences of those events. For instance, the first entity (Entity 1) arrives at the start of the simulation (time = 0). Because Activity 1a is idle, the entity can pass straight through Queue 1 into the activity which can then start work. For simplicity, Resource 1 is not included

Table 2.1   Example of a DES of the queuing system in Figure 2.3.

Time (min)	Event
0	Entity 1 *arrives*, passes through Queue 1 to Activity 1a, Activity 1a *starts*
2	Entity 2 *arrives*, passes through Queue 1 to Activity 1b, Activity 1b *starts*
7	Entity 3 *arrives* and waits in Queue 1
11	Activity 1a *finishes*, Entity 1 passes through Queue 2 to Activity 2, Activity 2 *starts*, Entity 3 moves to Activity 1a, Activity 1a *starts*
22	Activity 1a *finishes*, Entity 3 waits in Queue 2
23	Entity 4 *arrives*, passes through Queue 1 to Activity 1a, Activity 1a *starts*
27	Activity 1b *finishes*, Entity 2 waits in Queue 2
30	Activity 2 *finishes*, Entity 1 leaves the system, Entity 3 moves to Activity 2, Activity 2 *starts*

in this simulation. The first column of the table shows how time moves forward in an irregular fashion. Entities 1 to 4 arrive at times 0, 2, 7 and 23 respectively. We will show how to model these irregular arrival times in the next subsection. Meanwhile, Activity 1a takes 11 minutes, Activity 1b takes 25 minutes and Activity 2 takes 19 minutes. There is no reason, of course, that the times have to be integer numbers. We would normally model time using real numbers to the level of precision that the data, and computer technology, will allow.

There is a range of ways that are used for managing the progression of time in DES. This requires determining the time at which events are due to occur, maintaining a record of those events and then, at the appropriate time, executing the events along with any consequent changes to the system state. We will not discuss the time-handling mechanisms for DES in detail here. For those that are interested, there is a very good description of these in Pidd (2004). For our purposes, it suffices to understand the key principle of moving time forward based on the instants at which discrete changes in system state (events) occur, which leads to an irregular time step.

2.3.2   Random sampling in DES

A key element of almost all DES models is the need to represent randomness in the system under study, in recognition of the fact that many systems are by nature stochastic. Randomness occurs in the length of time an activity takes, for instance the time to serve a customer, the time to repair a piece of equipment and the time to process an order. It is also common for the arrival of entities to be subject to randomness, for example the arrival time of customers at a service system. Other factors are also random in nature, such as the size of an order, the weight of an item and the route an entity will choose to take through the system. In all these cases we

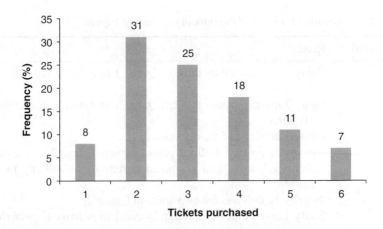

Figure 2.4   Distribution of tickets purchased by customers for sports events.

might know how the data vary, in other words we know the distribution of the data, but we do not know what the exact value of the data will be in the next instance. So, for example, we might know that service time varies between 10 and 20 minutes, we might even know the exact shape of the distribution for service time, but we cannot predict what the exact service time for the next customer will be. To simulate systems that are subject to such randomness we need a mechanism for modelling this variability, for which Monte Carlo methods are employed. We now discuss how this is achieved in DES.

Say that we wish to simulate a ticketing web site and as part of that we want to model the number of tickets a customer will buy for a sports event. We may have looked at historic data for similar events and observed the distribution shown in Figure 2.4.

Given that there are six possible outcomes, an obvious way to model this might be to role a six-sided die. This would give us variability in the right range (1–6), but it would not model the differing probabilities of the six outcomes; it would assume all outcomes had a probability of one-sixth. It is also not very practical to ask a computer to role a die! So we need an approach that allows us to represent the differing probabilities and that can be implemented on a computer for our simulations. To achieve this we need to use random numbers.

2.3.2.1   Random numbers and generating random numbers

Random numbers are a sequence of numbers that appear in a random order. Typically these numbers are on the scale [0,1), the square bracket signifying that 0 is included in the range and the bracket signifying that the highest possible random number is just less than 1. Table 2.2 shows an example of random numbers in this range.

Random numbers have two important properties:

- *Uniform*: there is the same probability of any number occurring at any point in the sequence.

Table 2.2   Example of 100 random numbers on the scale $[0,1)$ to two decimal places.

0.74	0.45	0.21	0.18	0.71	0.12	0.19	0.33	0.63	0.46
0.40	0.54	0.13	0.86	0.29	0.14	0.21	0.98	0.30	0.47
0.84	0.48	0.08	0.04	0.70	0.89	0.07	0.18	0.76	0.36
0.22	0.98	0.61	0.18	0.64	0.85	0.89	0.81	0.21	0.03
0.07	0.80	0.71	0.47	0.82	0.46	0.80	0.34	0.99	0.99
0.12	0.88	0.52	0.02	0.75	0.89	0.60	0.14	0.36	0.58
0.91	0.41	0.18	0.78	0.61	0.05	0.30	0.34	0.32	0.91
0.25	0.55	0.64	0.55	0.87	0.99	0.20	0.40	0.61	0.68
0.44	0.15	0.03	0.80	0.02	0.57	0.41	0.17	0.46	0.28
0.74	0.30	0.33	0.25	0.58	0.07	0.27	0.93	0.70	0.39

- *Independent*: once a number has been chosen, this does not affect the probability of its being chosen again or of another number being chosen.

This can be understood in terms of drawing numbers from a hat. Say we have 100 tickets with the numbers 0 to 99 written on them; we simply divide these numbers by 100 to get the random numbers in Table 2.2. We fold the tickets up so we cannot see the numbers and place them in a hat. We then draw one ticket at random. There is, of course, a 1 in 100 probability of any one of the tickets being selected, hence the probability is said to be *uniform*. In the case of the numbers in Table 2.2, the first ticket chosen is the number 74 (or 0.74). Having chosen this ticket we then fold it up and place it in the hat again. We then pick another ticket at random (number 45 if we write the numbers down across the row in Table 2.2). The importance of placing ticket 74 back into the hat is that the probability of any ticket being chosen second remains as 1 in 100. Otherwise, we would have had a 1 in 99 chance of any ticket being chosen with the exception of 74 which would have a zero probability. So by placing the tickets back in the hat after we have selected them, the probabilities remain unchanged and the selection of the next ticket is in no way affected by the ticket that went before it. In other words, the numbers chosen are *independent* of one another. The procedure then continues and we can, as in Table 2.2, create a sequence of random numbers.

Because of the procedure followed, Table 2.2 will not contain every number in the range $[0,1)$ to two decimal places, despite there being 100 numbers in the table. It is difficult to spot a missing number, but it is quite straightforward to identify numbers that appear more than once: for example, 0.21 appears in the first and second rows. Notice that 0.99 appears twice together at the end of the fifth row. This seems unlikely to occur and indeed it is, with only a 1 in 10 000 (100×100) chance of this happening. That said, there is only a 1 in 10 000 chance of any predefined sequence of two numbers occurring, such as 0.74 and 0.45 at the start of the table. Although not every number appears in Table 2.2, if we continued the procedure and selected thousands or even millions of random numbers, eventually every number would be selected and the percentage of times each number appeared would be close to 1% each.

Random numbers such as these have a third important property and that is that we can generate these numbers using a computer. Strictly speaking, computers cannot generate random numbers, since they are perfectly logical. However, we can fool a computer into generating numbers that to all intents and purposes appear to be random, that is they appear to be uniform and independent. This is achieved through the use of special algorithms. For a description of some of these algorithms see Pidd (2004), Law (2007) and Banks *et al.* (2005). There has been much research into the creation of effective and efficient random number generators. Good examples are L'Ecuyer (1999) and the Mersenne Twister (Matsumoto and Nishimura, 1998).

2.3.2.2   Random sampling with random numbers

Once we have random numbers available to us, it is relatively straightforward to use these to sample from a distribution such as that in Figure 2.4. To achieve this we simply relate the right proportion of random numbers to each outcome in order to ensure that outcome has the correct probability of happening. So, if we wish to have 8% of customers purchasing one ticket, then we associate 8% of the random numbers with that outcome. We can do this by saying that if the random number is less than 0.08 then the customer will purchase one ticket.

Table 2.3 shows the allocation of random numbers required to sample from the distribution in Figure 2.4. Notice how the numbers in the brackets are the cumulative values from the second column. The ranges are interpreted as $0 \leq$ random number $<$ 0.08, 0.08 $\leq$ random number $<$ 0.39, and so on. As such, the random number 0.08 would give a sample value of two tickets. This is correct, since 8% of the random numbers (0.00, 0.01. 0.02, . . . , 0.07) would give a sample value of one ticket.

We can use the random numbers in Table 2.2 to simulate the number of tickets each customer purchases. If we read across the rows in Table 2.2, then we would use 0.74 to determine the number of tickets the first customer purchases and then 0.45 to determine the tickets purchased by the second customer, and so on. The result of this process for the first 10 customers is shown in Table 2.4. Notice how no one purchases one, five or six tickets in this very small sample size. However, if this process was continued for thousands or even millions of random samples, the percentage of

Table 2.3   Allocation of random numbers for distribution of tickets purchased by customers for sports events (Figure 2.4).

Tickets purchased	Frequency (%)	Random numbers
1	8	[0.00,0.08)
2	31	[0.08,0.39)
3	25	[0.39,0.64)
4	18	[0.64,0.82)
5	11	[0.82,0.93)
6	7	[0.93,1.00)

Table 2.4   Sample values for the number of tickets purchased by customers.

Customer	Random number	Tickets purchased
1	0.74	4
2	0.45	3
3	0.21	2
4	0.18	2
5	0.71	4
6	0.12	2
7	0.19	2
8	0.33	2
9	0.63	3
10	0.46	3

customers buying the different numbers of tickets would become increasingly close to the values in the distribution in Figure 2.4.

2.3.2.3   Sampling from continuous distributions

The distribution in Figure 2.4 contains discrete (integer) data. The sampling task is to determine which discrete value to select. What happens if we wish to model continuous data such as those shown in Figure 2.5? This shows a negative exponential distribution with a mean of 1. This distribution is typically used to model the time between entity arrivals when they occur on a random basis, such as telephone enquiries into a call centre. The only parameter required for this distribution is the mean, which in Figure 2.5 is 1. The graph shows the probability density function (PDF) for the sample value x. The probability of a specific sample value increases with the height of the line. So in Figure 2.5, values of x close to 0 (although not exactly 0) are much more likely than higher values of x. When used for simulating the arrival of

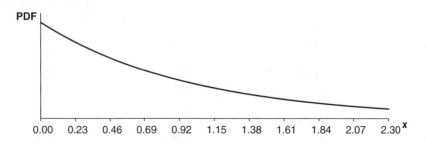

Figure 2.5   An example of a continuous distribution: a negative exponential distribution (mean = 1).

entities, this means that there are generally quite short gaps between entities, but then occasionally there will be quite long periods between entity arrivals.

To sample from a continuous distribution we need to be able to relate a random number to the proportion of the area under the curve. For instance, if our random number is 0.05, then we would find the value of x at which 5% of the area under the curve is to the left of x, giving a small value for x. If the random number is 0.95, then we would find the value of x at which 95% of the area under the curve is to the left of x, giving a large value for x. In the case of the negative exponential distribution this can be achieved by integrating the distribution's PDF to obtain its cumulative distribution function. This gives the formula

$$x = -\mu \log_e(1 - u)$$

where:

x = random sample
μ = mean of the distribution
u = random number, $0 \le u < 1$

Using the first row of random numbers in Table 2.2, 10 samples from the negative exponential distribution are shown in Table 2.5.

The negative exponential distribution is just one example of a standard continuous distribution that is used for sampling in DES. It is also possible to sample from continuous distributions based on empirical data. There is not space here to go into a detailed discussion of the distribution types, their uses and sampling methods. For the interested reader see Robinson (2004), Pidd (2004) and Law (2007) for a more detailed discussion on distributions and random sampling. The key point here is to understand that in DES modelling random numbers are used as the basis of sampling

Table 2.5 Samples from a negative exponential distribution.

Entity	Random number	Inter-arrival time (IAT)	Actual arrival time (cumulative IAT)
1	0.74	1.35	1.35
2	0.45	0.60	1.94
3	0.21	0.24	2.18
4	0.18	0.20	2.38
5	0.71	1.24	3.62
6	0.12	0.13	3.74
7	0.19	0.21	3.96
8	0.33	0.40	4.36
9	0.63	0.99	5.35
10	0.46	0.62	5.97

from discrete and continuous distributions in order to model the random nature of the system under study.

2.4   The philosophical perspective: The DES worldview

Section 2.3 focuses on the technical perspective, in other words how a DES model works. This is important for understanding the method and for being able to compare it with other possible modelling approaches, including system dynamics. At least as important, if not more so, is to gain an understanding of how DES modellers approach a problem situation with their modelling technique. The contention is that the worldview of the modeller will have an impact on how he or she models the world. We might describe this as his or her modelling philosophy.

Although there does not appear to have been a specific debate about the world-view that DES modellers adopt, we can draw some inferences from the nature of the method and what we know about the way problems are modelled in DES. Our own research lends support to the discussion that follows (Tako and Robinson, 2010; Tako, 2009); the former reference is reproduced as a chapter in this book (Chapter 8).

The driving philosophy of a DES modeller is that the world can be understood as a set of interconnected activities and queues that are subject to random (stochastic) variation. As such, when faced with a modelling problem, the DES modeller is immediately aiming to define the system in terms of its queuing structure and the variability that the system is subject to. If the problem does not fit this structure, then DES will not be seen as useful for addressing the problem at hand. Since a wide range of problems do fit this structure, DES is widely applicable, although in some cases its use may not be ideal – there are cases of people fitting the problem to the modelling approach.

The DES worldview is not without a good foundation. Frequently complexity can arise from systems of interconnected randomness that is not evident if the random variability is not taken into account. As an illustration of this let us consider the simplest form of a system that is subject to interconnected randomness, namely a single server queue (Figure 2.6). Here entities arrive into a queue and are then served by a single activity. If the activity is available when the entity arrives then no queuing time is incurred, otherwise the entity must wait in the queue.

We start with the assumption that there is no stochastic variation in the system. If the service time is one time unit and the rate at which entities arrive is 0.9 per time unit (an inter-arrival time of 1/0.9), then we would say the traffic intensity of the system is 0.9 (arrival rate/service rate). The traffic intensity describes the loading of the system.

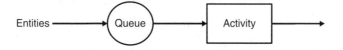

Figure 2.6   A single server queue.

In this example we would expect the activity to be busy 0.9, or 90%, of the time. If we were to ask what the average time is that an entity spends in this system, our expectation would be exactly one time unit. In fact, we would not expect any queuing as the service rate is greater than the arrival rate (traffic intensity < 1) and we would expect every entity to spend exactly one time unit in the system, all of which involves receiving service – perfection!

Now if we add stochastic variation to this system we get a very different result. Under the assumption that the inter-arrival time of entities and the service time of the activity both vary according to a negative exponential distribution, and the mean inter-arrival time is 1/0.9 and mean service time is 1, the average time an entity spends in the system is 10 time units. This is a result of the interaction of random arrivals with random service times, which lead to the build-up of a queue. On average entities spend nine time units in the queue and one time unit receiving service – much less than perfection!

For a simple system such as this, with the assumption that the variability is negative exponentially distributed, we are able to calculate the time in the system for any traffic intensity using the queuing formula (Winston, 1994). Figure 2.7 shows the relationship between mean time in the system and traffic intensity for a system with a mean service time of one time unit. What is immediately obvious is that once the traffic intensity is above about 0.8, the mean time in the system starts to increase to what is likely to be an unacceptable level.

From a DES perspective this is an important result. Typical DES models can include tens, hundreds or even thousands of activities and queues, which are subject to many sources of stochastic variability that can follow any distribution. This implies a multiplication of the complexity in the single server queue problem which cannot be solved analytically, but only through DES. The fact that interconnected randomness

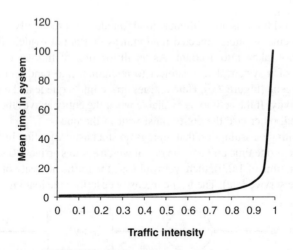

Figure 2.7   Mean time in system for a single server queue with mean service time of 1 time unit.

can lead to very different results than would be expected from simple calculations based on averages is a key reason for using DES. For an example of this see Robinson and Higton (1995), where three designs of a brick factory were shown not to be able to meet the company's production requirements through a DES model, despite the design calculations suggesting all three would suffice.

This need to model the random variation in a system has significant implications for the management of data in DES studies. First, the modeller requires data on the input distributions for all the sources of stochastic variation. Such data can be difficult to obtain and can be difficult to analyse, especially if there are correlations between the sources of variation. Second, the analysis of the simulation output requires careful consideration. A single run of a DES model provides just one sample of the possible range of system performances. This is because it is the result of the random samples taken from the input distributions during a run. By changing the stream of random numbers that are being used during a simulation run, different samples will be taken, leading to a different result for system performance. As such, experimentation with a DES model is a sampling process which aims to estimate the output distribution for the result of interest (e.g. time in system or throughput). The more runs that can be performed with the simulation model, the greater the sample size, and the better the understanding of the output distribution. This adds a layer of complexity to estimating system performance and comparing alternative systems. For an introduction to the issues involved in input modelling and output analysis see Law (2007) and Banks *et al.* (2005).

2.5   Software for DES

We complete this chapter with a brief discussion on the software available for DES. In simple terms there are three options for developing DES models: spreadsheets, specialist simulation software and programming languages. Spreadsheets do have some basic facilities that help with DES modelling (e.g. random number generators and random sampling), but time handling in a spreadsheet is not so simple. For all but the simplest of models, programming constructs (e.g. Visual Basic for Applications) will be required to enhance the capabilities of the spreadsheet.

Programming languages are typically used when specialist simulation software is not available to the modeller or the model's complexity is so great that it is beyond the capabilities of a specialist package. Both cases have become less frequent over the years as a result of the reducing price of simulation software and the increasing capabilities of these packages.

The majority of DES work is carried out using specialist DES packages. Most of these provide facilities for building the model through a menu-based system and an animated visual display of the model as it runs. Programming constructs are normally available for enhancing the logic of the model. Table 2.6 provides a list of some of the packages that are available at the time of writing. This list is not exhaustive and it is recommended that the reader makes a more detailed search of the options before purchasing a package. A useful source of information is the biennial INFORMS

Table 2.6   Examples of specialist DES software packages (listed alphabetically).

Arena	ProModel
Extend	Simio
Flexsim	SIMUL8
MicroSaint	Witness

simulation software survey that is provided in the society's magazine *ORMS Today* (www.orms-today.org, accessed May 2012). The latest version can be found in Swain (2011). DES packages can cost from around $1000 to $20 000.

2.6 Conclusion

This chapter provides a brief introduction to DES. Our focus has been on the technical perspective, how DES works, and the philosophical perspective, the worldview of DES modellers. We have also seen an example of a DES model and discussed the software that is available for DES.

At the centre of DES modelling is the view that it is interconnected randomness that leads to system performance. The term discrete-event simulation describes the time-handling method that the modelling approach employs. Perhaps it would be better to name DES after the worldview that the modelling approach encapsulates – *stochastic dynamics*.

References

Banks, J., Carson, J.S., Nelson, B.L. and Nicol, D.M. (2005) *Discrete-Event System Simulation*, 4th edn, Prentice Hall, Upper Saddle River, NJ.

Burgess, N., Worthington, C., Davis, N. *et al.* (2011) *SimLean Healthcare: Handbook*, SimLean Publishing, www.simlean.org (accessed May 2012).

Hollocks, B.W. (2008) Intelligence, innovation and integrity – KD Tocher and the dawn of simulation. *Journal of Simulation*, **2**(3), 128–137.

Law, A.M. (2007) *Simulation Modeling and Analysis*, 4th edn, McGraw-Hill, New York.

L'Ecuyer, P. (1999) Good parameters and implementations for combined multiple recursive random number generators. *Operations Research*, **47**(1), 159–164.

Matsumoto, M. and Nishimura, T. (1998) Mersenne Twister: a 623-dimensionally equidistributed uniform pseudo-random number generator. *ACM Transactions on Modeling and Computer Simulation*, **8**(1), 3–30.

Pidd, M. (2004) *Computer Simulation in Management Science*, 5th edn, John Wiley & Sons, Ltd, Chichester.

Robinson, S. (2004) *Simulation: The Practice of Model Development and Use*, John Wiley & Sons, Ltd, Chichester.

Robinson, S. and Higton, N. (1995) Computer simulation for quality and reliability engineering. *Quality and Reliability Engineering International*, **11**, 371–377.

Robinson, S., Radnor, Z.J., Burgess, N. and Worthington, C. (2012) SimLean: utilising simulation in the implementation of lean in healthcare. *European Journal of Operational Research*, **219**, 188–197.

Swain, J.J. (2011) Software survey: simulation – back to the future. *ORMS Today*, **38**(5), 56–69.

Tako, A.A. (2009) Development and Use of Simulation Models in Operational Research: A Comparison of Discrete-Event Simulation and System Dynamics. PhD thesis. Warwick Business School, University of Warwick.

Tako, A.A. and Robinson, S. (2010) Model development in discrete-event simulation and system dynamics: an empirical study of expert modellers. *European Journal of Operational Research*, **207**, 784–794.

Winston, W.L. (1994) *Operations Research: Applications and Algorithms*, 3rd edn, Duxbury Press, Belmont, CA.

3

Systems thinking and system dynamics: A primer

Brian Dangerfield
Salford Business School, University of Salford, UK,

3.1 Introduction

This chapter introduces the basics of the system dynamics simulation methodology, together with the adjunct field of systems thinking which emerged subsequently. The field of system dynamics was initially known as *industrial dynamics*, which reflected its origins in the simulation of industrial supply chain problems. The first paper published by the founder of the field, Jay W. Forrester, appeared in 1958 (Forrester, 1958) and it was a precursor to what proved to be a hugely influential book: *Industrial Dynamics* (Forrester, 1961). Forrester sought to apply concepts of control engineering to management-type problems and was very probably influenced by the earlier work of Arnold Tustin (Tustin, 1953). Forrester argued that the field of operations/operational research (OR) at that time was not focused on the sort of problems that sought to inform policy (top-level) issues in an organisation. By its very definition OR was restricted to operational problems. Forrester saw a niche for a methodology which could tackle strategic issues more appropriately addressed to the success or failure of an organisation, as well as prominent national and international policy issues. See Forrester (2007) for his personal recollections of the history of the field.

Discrete-Event Simulation and System Dynamics for Management Decision Making, First Edition.
Edited by Sally Brailsford, Leonid Churilov and Brian Dangerfield.
© 2014 John Wiley & Sons, Ltd. Published 2014 by John Wiley & Sons, Ltd.

The characteristics of system dynamics[1] models can be listed as follows:

- They address issues by considering aggregates (of products, people, etc.) and not individual entities (as in discrete-event simulation – see Chapter 2) or individual agents (as in agent-based modelling – see Chapter 12).

- They primarily reflect the dynamics of a system as having endogenous causes: change over time comes from within the system boundary due to information feedback effects and component interactions, although the initial stimulus for those dynamics may be exogenous. For more on the endogenous perspective see Richardson (2011).

- They carefully distinguish between resource flows and the information flows which cause those resource flows to increase or run down. This is a fundamental (and powerful) feature of the methodology, which means such models can be used to design and evaluate information systems as well as the more usual focus on resource systems.

- The flows are assumed to be continuous and are governed by what are in effect ordinary differential equations. System dynamics (SD) models belong to the broader category of continuous simulation models.

- Although flow rates are included, SD models are primarily concerned with the behaviour of stocks or accumulations in the system. These are described by integral equations. Forrester has famously stated that differentiation does not exist in nature, only integration. Mathematical models characterised by differential equations must be solved in order to determine the stock values; system dynamics puts stock variables to the forefront.[2]

- They do not ignore soft variables (such as morale or reputation) where these are known to have a causative influence in the system.

Before addressing some of these characteristics in greater detail it is sensible to offer an overview of the adjunct field of systems thinking. This is sometimes described as qualitative system dynamics for its provenance is based upon diagramming or mapping techniques, primarily influence diagrams or causal loop diagrams. It was not until the 1970s, nearly 15 years after the publication of Forrester's early industrial models, that such diagrams started to appear. Their origins can be traced back to Maruyama (1963); the text by Goodman (1974) portrays some seminal examples.

---

[1] The name change from industrial dynamics to system dynamics is said to be due to Dana Meadows, a luminary in the field. She was associated with the application of the methodology to other types of systems beyond the purely industrial and argued this progression should be reflected in the name. There are those who mischievously suggest that system (rather than systems) dynamics was chosen so as to highlight those less knowledgeable in the methodology as they would be apt to use the wrong name.

[2] Research has shown that even well-educated people find it difficult to infer the behaviour of a stock variable given known behaviour of the flows affecting that stock.

3.2   Systems thinking

The use of diagramming techniques in the analysis of a system has a long history going back to the block diagrams of control and electrical engineering. However, the qualitative strand associated with SD emphasises the feedback loops present in the system. Feedback is an essential building block of SD whereby information about the current state of the system is used to regulate controls on the resource flows and it underscores the endogenous point of view. For instance, if stocks of manufactured goods are beginning to over-accumulate, it is necessary either to cut back on production throughput or to inaugurate a sales drive, or both.

These mapping techniques are not mandatory in an SD study. On the other hand there are those who argue such methods, of themselves, have the capacity to generate insight and can help form a consensus for policy change in a problem system. See for instance the testimony from Merrill *et al.* (2013) concerning a health application. They state: 'As a tool for strategic thinking on complicated and intense processes, qualitative models can be produced with fewer resources than a full simulation, yet still provide insights that are timely and relevant.' Books have appeared which focus exclusively on such mapping techniques, for instance Ballé (1994) and Sherwood (2002), to the exclusion of formal simulation models which are described in Section 3.3 below. Whether such diagrams alone can be considered advantageous in the overall practice of SD has long been the subject of debate in the field. An exchange between Coyle and Homer and Oliva occupied many pages of the *System Dynamics Review* in 2000–2001. See Coyle (2000, 2001) and Homer and Oliva (2001).

Although the proponents of the need for formal SD models remain implacable, some authors and organisations have prospered in the propagation of systems thinking techniques. Pegasus Communications (now Leverage Networks) has for many years published the magazine *The Systems Thinker* and Peter Senge's reputation as a managerial thought leader was founded on his book *The Fifth Discipline* (Senge, 1990) and its associated *Fieldbook* (Senge *et al.*, 1994). It was these sources that, primarily, introduced 'behaviour over time' sketch graphs together with the notion of system archetypes as additional tools in the armoury of systems thinking.

3.2.1   'Behaviour over time' graphs

Consider the chart in Figure 3.1. It represents some (hypothetical) data for two local authorities showing the percentage of girls who were classed as overweight in 2012.

It is a static graph and, as such, conveys limited and what could be misleading information. At first, an examination of the data would appear to suggest that local authority A has a more serious public health situation on its hands than local authority B. However, reframing the situation using a 'behaviour over time' graph paints an altogether different picture (see Figure 3.2). It is clear that local authority B is more in need of a public health intervention. Consideration of the dynamics in a system is vitally important.

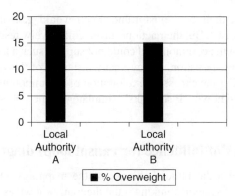

Figure 3.1   Prevalence of overweight girls aged 10–15 years in two local authorities in 2012.

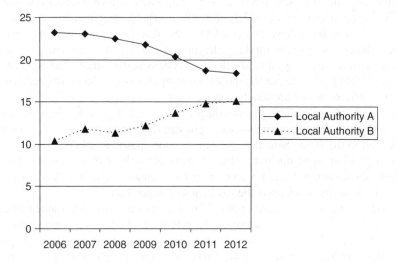

Figure 3.2   'Behaviour over time' graph of the prevalence (%) of overweight girls, 2006–2012.

3.2.2   Archetypes

System archetypes are particular configurations that are capable of being expressed as a mapping in the form of an influence diagram. They commonly recur in various real-world situations and are accorded short descriptors. Besides the reference to Senge above, Wolstenholme (2003) has also contributed to the seminal literature on archetypes.

A classic archetype is the 'tragedy of the commons' which was first described by Hardin (1968). It accounts for how the behaviour of herdsmen in medieval England sowed the seeds of their own misfortune. Each would gain increased utility by purchasing another beast. But their actions, taken collectively, resulted in overgrazing to such an extent that the common land could not support such a large total herd – and many animals died of starvation. Today, many instances of this archetype abound, a notable example being the excessive exploitation of a common fishery. This was the basis for the well-known 'FishBanks' simulator game in SD (Meadows and Sterman, 2012).

3.2.3 Principles of influence (or causal loop) diagrams

This section examines the building blocks of the mappings which have come to constitute the heart of systems thinking – the diagrams known as influence diagrams (IDs) or causal loop diagrams (CLDs). There is no counterpart to an ID or CLD in discrete-event simulation. There one progresses to the development of an activity cycle diagram as the initial framework on which the computer simulation model is constructed. That is to say, the field of discrete-event simulation does not offer an optional diagramming phase which, of itself, is capable of generating insight.

Some practitioners have expressed the view that, in certain instances, an intervention based on systems thinking diagrams is sufficient to unearth the insight necessary to achieve a profound effect on system performance. The argument is bound up with project resources: models as mappings absorb less costs and can still produce insights which are timely and relevant.

A simple example of an influence diagram is given in Figure 3.3. Here we see the basic process underlying a firm's organic growth. As average profits increase they are reinvested to the future benefit of the organisation (positive links).

Other examples of positive links are: sales per unit time of a durable product increase the customer base; revenues received increase the cash balance; students enrolling on a course increase the total student population.

Note that the + sign implies not only that an increase in one variable causes an increase in another, but also, alternatively, that a reduction in one variable causes a reduction in another. In the example in Figure 3.3 a reduction in average profits engenders a reduction in investment. A more obvious example is when rumours of a firm's financial health lead prospective customers to decline to engage with it.

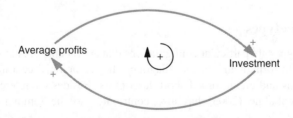

Figure 3.3 Simple positive feedback loop. (Note that the loop descriptor in the middle should flow in the same direction as the loop, in this case clockwise.)

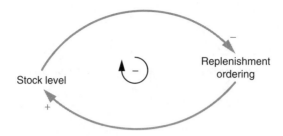

Figure 3.4   Simple negative feedback loop.

Let us now consider a negative loop. The underlying influence created by such a loop is one of a controller. If movement occurs in the dynamics in one direction then a countervailing force pushes against that momentum to establish the original (or a new) equilibrium. The entire discipline of control engineering is concerned with how negative loops can be represented as physical controllers in machinery of all types, for example the autopilot in modern aircraft and the thermostat in a heating system.

Figure 3.4 shows an example of a simple negative loop taken from the domain of stock control. The very word 'control' reflects the nature of what is going on. As stock levels increase then replenishment ordering is cut, or vice versa (negative link). The change in the flow of orders directly affects the stock level and thus completes the loop. Other examples of negative links are: a perceived reduction in the numbers of a particular workforce will lead to an increase in employee recruitment; an increase in spending on wages will lead to a fall in an organisation's cash balances.

In selecting the sign to place on a given arrowhead (establishing link polarity)[3] it is important not to take into account other influences that may be simultaneously operating. The Latin maxim of *ceteris paribus*, so common in elementary economics texts, needs to be adhered to: that is, let other factors remain constant. Therefore the only consideration in assigning link polarity is: what effect will a change in the variable at the tail of the arrow have on the variable at its head?

Two mutually connected negative relationships create a positive loop. Consider Figure 3.5 where an increase in staff turnover (in a close working team) will lead to a fall in morale which in turn will lead to a further increase in staff turnover.

In determining the loop (as opposed to link) polarity there are two methods available. One can enter the loop at any given point, and start with, say, an increase in that variable and trace around the effect. If one returns to that point with a further increase then the loop is positive, but if the initial increase has resulted in a decrease then the loop is negative. An arguably easier approach is to add up the number of negative links in the loop: if the number is zero or is even, then the loop is positive; if it is an odd number, then the loop is negative. The loop polarity is the algebraic product

---

[3] In recent years some authors have replaced the use of + and − by s (same) and o (opposite).

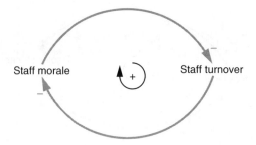

Figure 3.5   Two mutually causative negative relationships create a positive loop.

of the number of negative signs, for example three negatives multiplied together yield a negative result, hence a negative loop.

3.2.4   From diagrams to behaviour

The determination of loop polarity is not merely an exercise for its own benefit but rather serves as a precursor to being able to infer the behaviour mode of the loop if it were to be 'brought to life'. Loop dynamics differ between negative and positive loops so it is essential to determine loop polarity. A positive loop produces dynamics which reinforce an initial change from an equilibrium point and so underpin growth and decay behaviour patterns. A pure positive loop in growth mode will produce exponentially increasing behaviour. A negative loop, on the other hand, will generate equilibrating behaviour such that any shift away from an initial equilibrium point will produce a compensating force driving it back towards that point (or indeed a new equilibrium). Introducing a delay into a negative loop will induce an oscillation in the behaviour. It is this knowledge which can aid in model conceptualisation when time series data is available. After smoothing out any noise which may be present, an oscillatory behaviour pattern is indicative of a system dominated by a negative loop or loops; one which exhibits growth or decay would suggest that a positive loop is at work somewhere. An oscillatory behaviour associated with a trend up or down would suggest the need for a model conceptualisation based around a combination of negative and positive loops.

In order to further develop this idea of behaviour generated by different feedback loops it is necessary to move away from the single loop examples above to a more realistic real-world situation where multiple loops are at play. For instance, the example in Figure 3.6 portrays a simple product diffusion model where initial sales generate further growth through 'word-of-mouth' effects but this growth is ultimately curtailed by market limitations of one form or another. Because this system structure also underpins the dynamics of an epidemic in a closed population (e.g. passengers on a cruise liner), the variables named for the diffusion example have been duplicated by the equivalent epidemic variables: the same system structure can underpin quite widely different situations!

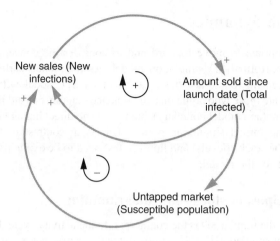

Figure 3.6   Influence diagram showing two loops and two different examples: diffusion dynamics and epidemics underpinned by the same system structure.

Also brought out by Figure 3.6 is the associated concept of loop dominance. As the structure plays out over time, the positive loop is dominant initially – that is to say, it has the control of system behaviour in the early stages while the word-of-mouth effects are at play. Ultimately the market limits begin to take over. There are fewer and fewer people who do not have this product and so the capability of making further new sales is diminishing by each passing week. Now the negative loop assumes dominance in system behaviour and growth slows. Figure 3.7 shows the resultant behaviour: S-shaped growth where the transition from growth to market maturity coincides with the switch in loop dominance. Technically this is at the point of inflection on the cumulative curve, a point where the sales per unit time (not shown) reach a peak and start to fall.

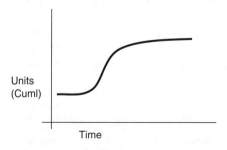

Figure 3.7   S-shaped (or sigmoidal) growth generated by coupled positive and negative loops.

3.3 System dynamics

It is now appropriate to move forward and to consider the conceptualisation and formulation of a formal SD simulation model. As mentioned earlier, there is no essential requirement to preface the creation of an SD model with an influence diagram. There are those who argue that an influence diagram can aid in the definition of system content (and model boundary), but there is no direct linkage between such a diagram and the formal simulation model. This is in contrast to the stock–flow diagram: here the stocks (levels) and flows (rates) need to be explicitly present in the equation listing for the model.

3.3.1 Principles of stock–flow diagramming

The stock–flow diagram in SD is the counterpart to the activity cycle diagram (ACD) in discrete-event simulation (DES). Although the flows may not result in a cycling of resources as such (which is common in DES), each diagram is there to underpin the formal model and the quantitative expressions which define its constituent elements.

SD flow rates are depicted by a tap-like symbol which indicates a device that can control the flow, equivalent to policy controls in the real world. A stock is represented by a rectangle and here there exists an unfortunate misalignment in the DES and SD diagramming conventions. In DES a rectangle is reserved for an activity – an active state. A stock in SD is a 'dead' state, equivalent to a queue in DES. Figure 3.8 is an example of what might be part of the stock–flow diagram underpinning an SD model of a nation's education system.

It is important to note that the boundary of the flow at each edge of the system is represented by a cloud-like symbol. Consideration of the resource beyond these points is outside the scope of the model. Also the stocks and flows must alternate along the sequence. The incoming flow adds to a stock while an outgoing one drains it. Only one resource can be considered along any process flow. So, for example, what starts as a flow of material (or product) cannot suddenly be transformed into a flow of finance. Thus, separate flow lines need to be formulated for the various different resources being considered in the model. 'Resource' can be taken to be a product class, financial flow, human resources, orders, capital equipment, and so on. Clearly the more resource flows being considered, the more complex the model and the more equations it will comprise.

Further possible arrangements are shown in Figures 3.9 and 3.10. One can have an inflow to a stock without an outflow (or vice versa). It is possible to have more than one inflow and/or outflow as in the example of the financial flows into and out of a

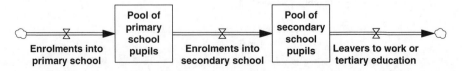

Figure 3.8   Example of a single flow process in a stock–flow diagram.

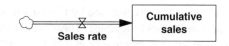

Figure 3.9   An inflow without an outflow.

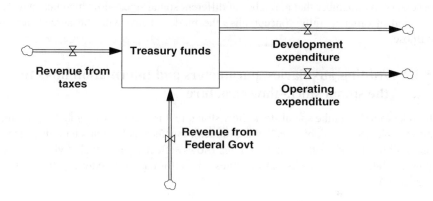

Figure 3.10   Multiple inflows and outflows in a stock–flow diagram.

state treasury department. In certain cases the flow might actually form a cycle. This can happen, for instance, if one is modelling a manufacturing recycling process often described as 'reverse logistics' or a 'closed loop supply chain'. Although such flow arrangements do constitute a loop or cycle they are in no circumstances a feedback loop. As will be described later, a feedback loop is based on *information* feedback.

3.3.2   Model purpose and model conceptualisation

Getting started can be the greatest difficulty in the creation of a useful SD model. One starts with the proverbial blank sheet of paper. Experience over many years has taught the author that two fundamental aspects of SD model conceptualisation are: first, being able to write in one sentence the purpose of the model; and, second, ensuring the stock–flow representation is 'right'. This latter term is deliberately placed within inverted commas because no model can ever be perfectly correct and represent the ultimate truth, but it is meant to suggest that a great deal of thought needs to go into deciding which resource flows to include, and how to structure those flows as bald stocks and flows with no consideration of any other variables or constants at this juncture – these can be usefully termed the spinal stock–flow structures (see examples in Figures 3.8–3.10). Where clients are involved they need to 'buy into' that raw stock–flow diagram and the written definition of model purpose before any further model formulation work is undertaken. Several iterations of this first conceptualisation are typically necessary. The above advice also underlines the point made earlier about influence diagrams – they are not always necessary as a precursor to formal model creation. For this task the stock–flow diagram reigns supreme.

A particularly useful precept, first expressed by Forrester in *Industrial Dynamics* (1961), is to define the level (stock) variables. These would still be visible if the system metaphorically stopped (e.g. employees in a factory, cash in the firm's bank accounts). Next, consider what might be flowing into and/or out of those stocks. These flows would, of course, *not* be visible if the system 'stopped'. All the time it is necessary to remember that a number of different spinal stock–flow modules may be required in order to fully conceptualise the model in line with the agreed model purpose.

3.3.3   Adding auxiliaries, parameters and information links to the spinal stock–flow structure

In order to flesh out the spinal stock–flow structure it is necessary to embellish it with other explanatory variables (called auxiliaries), together with parameters. In general one follows the oft-restated mantra: rates (flows) affect levels (stocks) via resource flows, while levels (stocks) affect rates via information (feedback) links. The sequence is:

Resource flows ≫ System state

System state ≫ Information to management

Information to management ≫ Managerial action

Managerial action ≫ Resource flows.

This is the essential expansion of the concept of the feedback loop which is illustrated in Figure 3.11.

In general, more complexity will be required and other variables, which are neither stocks nor flows, of necessity have to be introduced – these are termed auxiliaries. These reflect variables which, in a business model, lie in the managerial planning and information system. Thus any variable which is intended to represent something planned, desired, a target, or a management goal would be modelled by an auxiliary variable. Consider the augmented stock–flow diagram in Figure 3.12. Here the concept of a desired workforce has been added to explain the recruitment rate on the spinal flow. It would seem intuitive that recruitment policy might be explained by a comparison between the desired workforce and what one currently possessed.

The level of sophistication can increase, however. To jump ahead a little, there is another item which would need to be added, namely the adjustment time for

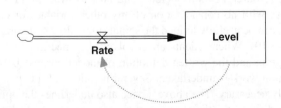

Figure 3.11   A simple feedback loop in stock–flow symbolism.

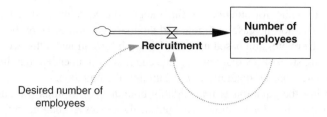

Figure 3.12   Auxiliary variable and information links added to the spinal flow.

eliminating any discrepancy between the desired and actual workforce. *Workforce adjustment time* would be a parameter and would mimic the average time to advertise and recruit new people or to give notice of redundancy and fire them if business conditions dictated it. Moreover, it might be necessary to have two different parameter values if the average time constant were thought to be different for recruitment and firing processes. Additionally, there may be a need to introduce other auxiliary variables in order to better define the *desired number of employees*. In fact chains of auxiliaries are often created in order to effect a proper definition for the flow variable. SD models tend to reflect real-world causes and effects very closely and this is one of the reasons why it is such a powerful methodology and why the total number of variables and parameters can rapidly escalate over and above the original number of variables on the spinal flows.

3.3.4   Equation writing and dimensional checking

Undoubtedly for many the most challenging task in SD model formulation is the composition of the equations for the rates and auxiliary variables. In modern SD software the stock variables are automatically created because the system can 'see' what is flowing into and/or out of a stock. These integration equations take the form

Stock value at current time t = old value of stock at $t - \mathrm{d}t + \mathrm{d}t * (\text{inflows} - \text{outflows})$

SD simulations exhibit a constant time advance (unlike DES) and through this process the equations describing the flow rates (which are, in effect, differential equations) are converted to difference equations and solved to yield the values of the stocks as in the example above. In the earlier SD literature the time increment was termed dt to reflect the 'with respect to' element commonly seen in differential calculus; TIME STEP is often employed nowadays. Its value is normally restricted to a binary fraction ($1/2^n$, for $n = 0,1,2 \ldots$) because of the way computers handle real numbers; by this means the greatest accuracy is achieved in determining the value for the system (reserved) variable *Time*. Clearly at the beginning of the simulation an 'old' value is needed to initialise the stock and this is termed an *initial value*. All stocks must have an associated initial value declared in order for the time advance process of the simulation to get started.

However, while formulation of the integration equations can be left to the software, this is not the case with rate and auxiliary equations. Here the user needs to compose the expression based upon the known informational influences evident in the developing stock–flow diagram. To this end it is recommended that the influence links are entered on the diagram *before* building the equation.

Structuring the equation is unavoidably bound up with units (or dimensional) checking. Most with a background in the physical sciences and engineering will know that any equation describing a real-world process needs to have the units balanced on each side of the '=' sign. Thus, if the units on the left are $/yr then the expression on the right side needs to decompose algebraically to $/yr. Further, if any terms on the right side are added or subtracted then each individual term needs to have the same units as the variable on the left side.

In the integration equation above, the 'dt *' element on the right side is necessary in order for the units to balance since the flows will be in terms of units/time. The dt term is a time interval and so we have time * units/time = units and the entire expression is units = units + units – units.

For the formulation of rate and auxiliary equations the user needs to think in terms of the units involved. If the variable concerned is expressed in terms of units/month then the expression on the right side needs also to be units/month. Thinking along these lines can actually aid in the formulation of the expression. You should know what units the rate or auxiliary is measured 'in; the right side needs duly to conform.

Let's consider some simple examples:

1. The accounts payment rate (APR) is known to be influenced by the value of accounts payable (AP) and a delay in making payment (DMP):

$$APR = AP/DMP \quad and \quad \$/month = \$/month$$

2. The annual out-migration rate (OMR) from a certain region of a country is dependent on the population (POP), the normal fraction of people leaving (FPL) and the departure migration multiplier (DMM).

   The multiplier term could be there to account for periods of time when the normal fraction departing is tweaked as a result of, say, a temporary incentive. Where such constructs are employed in SD models they are inevitably dimensionless, that is to say they have no units. As well as a multiplier, any fraction, proportion, percentage or an index number would be dimensionless and be given units of '1'.

   So we have

$$OMR = POP * FPL * DMM \quad and \quad persons/yr = persons * 1/yr * 1$$

   Why is the FPL term in units of 1/yr? This is because it is the number of persons leaving each year divided by the number there to start with, or (persons/yr)/persons = 1/yr. The same idea applies with an interest rate which is ($/yr)/$ = 1/yr (i.e. a percentage, which is dimensionless, but which can change over time).

Below are listed two possible equations to describe the production rate. It is interesting to note that each is quite different but both are dimensionally balanced.

3. Production rate (PR) is a function of the workforce (WF) and their productivity (PROD). Productivity can crop up in a lot of business models and its dimensions can cause difficulty. It is a compound dimension expressed as (output) units/person/time unit, or (units/(persons * time)).

   So we have

$$PR = WF * PROD \quad and \quad units/time = persons * (units/(persons * time))$$

4. Production rate (PR) is related to the average sales rate (ASR), together with a correction for a stock discrepancy (CSD) and a correction for a backlog discrepancy (CBD). The correction terms will be accounted for separately in the model and they describe the product units produced per time unit that will eliminate any discrepancy between what is desirable and the state of affairs that exists.

   So we have

$$PR = ASR + CSD + CBD \quad and \quad units/time = units/time + units/time + units/time$$

Which formulation for PR is the correct one? *Either* could be and there may indeed be other formulations which occur to the reader. The formulation employed is the one which is most appropriate given the purpose of the model and the circumstances prevalent in the actual system being modelled. A useful categorisation of commonly found formulations for rate and auxiliary equations is set out in the classic SD text by Richardson and Pugh (1981) and also in Sterman (2000). In addition the aspiring modeller should also study the many model listings provided by SD experts in texts and as supplementary material in journal articles.

To conclude, a more complicated equation formulation example is described. It concerns the need to formulate an expression for the extra labour (EL) required to eliminate a greater than normal backlog of orders. Many operations experience this challenge, especially if there are seasonally induced gluts in orders. It is not feasible to employ a large workforce throughout time and it falls to the management to recruit more people when a very high backlog situation arises.

An initial formulation might be

$$EL = (OB - NOB)/PTAB$$

where

OB = Order Backlog
NOB = Normal Order Backlog
PTAB = Planned Time to Adjust the Backlog

EL is obviously dimensioned as 'persons' and the expression is

$$\text{persons} = \text{units}/\text{time} - \text{units}/\text{time}$$

The equation is not balanced dimensionally. It is necessary to introduce another variable (or constant) which will relate units/time to persons. A moment's thought should make one realise that the concept of worker productivity (see example 3 above) is missing and so an additional parameter is required, say normal productivity of labour (NPL). After some further thought it will be established that this parameter needs to be included in the denominator of the expression and as a multiplier. We now have

$$EL = (OB - NOB)/(PTAB * NPL)$$

Extracting just one of the terms in the numerator for the dimensional check yields

$$\text{persons} = \text{units}/(\text{time} * (\text{units}/(\text{persons} * \text{time})))$$

and the two 'time' and 'units' elements cancel, leaving persons = persons and the equation is shown to be dimensionally correct. In the final step above it might be necessary to recall the mathematical dictum often chanted in school: 'Invert the divisor and multiply.'

3.4   Some further important issues in SD modelling

The subsections which follow describe some important additional aspects of SD modelling. Since this is a primer these topics are not covered comprehensively but their flavour is imparted.

3.4.1   Use of soft variables

The utility of SD simulation models is enhanced considerably by a means of representing soft variables not easily capable of being measured. As Forrester once remarked, to ignore these type of variables is equivalent to giving them a value of zero: 'the one value that is certain to be wrong'. So how might these variables be typically represented in an SD model? A common technique is via an X–Y lookup (or table) function.

Consider the influence diagram in Figure 3.5 where a variable, *staff morale*, is shown. The diagram suggests that this 'soft' variable influences *staff turnover*. This

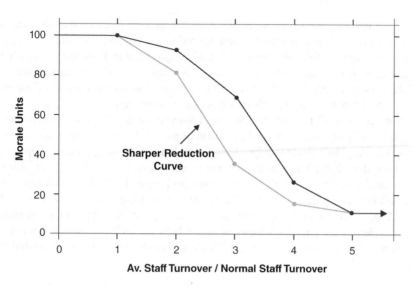

Figure 3.13 Alternative hypothesised X–Y relationships for the effect of staff turnover on team morale.

would seem intuitive since being a member of a team where morale is low is surely going to predispose people to look elsewhere for employment. The loop is closed because the spate of departures further erodes the morale among those who remain. It should be possible to set up an *X–Y* relationship which depicts staff morale as some function of staff turnover. To effect this relationship a preferred approach is to normalise the independent variable, in this case staff turnover. By this is meant creating a ratio of *average staff turnover* to *normal staff turnover*. This avoids the need to determine absolute values for the *x*-axis variable.[4] An average for turnover is employed since the perceptions of team members collectively will not instantaneously change as staff departures increase; it will take some weeks or months for perceptions to become embedded. The larger this ratio becomes, the worse the staff turnover situation is. A plausible relationship might be as depicted in Figure 3.13.

The range on the *Y*-axis for the soft variable is typically between 0 and 1.0 or 0 and 100. The full range is shown on the *X*-axis even though a value of the ratio between 0 and 1.0 will have no effect – it is assumed that an average staff turnover lower than (or equal to) normal staff turnover has no detrimental effect on morale. The range can be extended above 5 if required, but anything beyond two or three times normal staff turnover must be quite serious. The software will linearly interpolate within the line segments as the simulation is on-the-fly and then use the corresponding *Y*-value for staff morale. The *Y*-value for the *X* = 5 point will be used if *X* > 5.0 and a warning message would be given so the user has the option of extending the curve if necessary. It is assumed that at very high rates of turnover the effect on morale cannot get any

---

[4] It may, of course, be represented by a separate variable: *turnover ratio*.

worse, so the negative slope tapers off. Also, the general shape of the curve can be experimented with. For instance, a sharper reduction beyond $X = 1$ or $X = 2$ might be felt more appropriate (see Figure 3.13), although the overall slope must surely be negative. In general any two line segments in these $X–Y$ relationships should not exhibit a slope reversal (i.e. from negative to positive or vice versa). See Chapter 9 where an instance of an exception to this rule is suggested.

The values used for morale units are arbitrary but the diagram reflects that if morale is really high then 100 can be taken to be a typical value, where a value of, say, 10 or less suggests morale is approaching rock bottom. It is important to appreciate that, even though the Y-axis units may not be easily confirmed (although an attitude questionnaire might be possible in this case), so long as the $X–Y$ relationship is not changed between policy runs, its effects are captured but in a neutral way. If the relationship is in some sense considered to be 'wrong' then that applies equally across all runs. Hence, not only does the comparison between different simulations remain intact, but also, most importantly, a difficult-to-measure but influential variable has been captured.

3.4.2   Co-flows

There are occasions when it is necessary to model an associated characteristic of a resource in a stock–flow arrangement. For instance, when modelling the physical flow of production in a manufacturing operation, from material processing through to work-in-progress and on to the finished goods stage, it may also be necessary to incorporate the equivalent book value of this sequence of stocks; in other words, we need a separate financial stock–flow arrangement which mirrors the physical flow. This is known as a co-flow arrangement and is the equivalent in SD simulations of the concept of an attribute for an entity often employed in DES. In this case the attribute for the resource involved is *value*.

The example in Figure 3.14 concerns a totally different application. In a professionally skilled workforce the new employees arrive with a basic level of

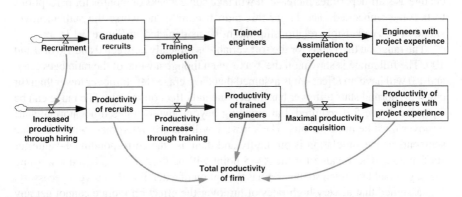

Figure 3.14   Co-flow arrangement for modelling a workforce and their collective productivity.

skill, such as in science-graduate-only recruitment situations, and this is enhanced through, first, in-house training and latterly project experience. One might wish to include not only the numbers at each level of skill, but their collective productivity also. In this example *productivity* is the attribute concerned. The example shown is intended to represent a software engineering firm where an individual's productivity might be measured in terms of lines of error-free code written per person per month. Departures from the firm are ignored here so as not to clutter the diagram.

The co-flow arrangement inevitably involves flouting a normal rule in SD modelling: that is, that direct rate-to-rate connections are not allowed. But it is obvious that as the various physical stocks change, so will the level of the attribute concerned and so a rate-to-rate link is needed. The units for the flows of people in Figure 3.14 are persons/month and the flows representing changes to collective productivity are output/(month * month). This is derived from a multiplication of the person flow (persons/month) by an appropriate individual marginal productivity parameter (output/(persons * month)) which is not shown in the figure.

The co-flow formulation enables an easy assessment of the total productivity (total output capacity) of the entire enterprise. Were it not to be employed it would be necessary to formulate a long auxiliary equation for total productivity which would involve a multiplication and summation of each level of skill's separate contribution to productivity, each based upon its appropriate marginal productivity parameter. However, some may find this more intuitive!

3.4.3   Delays and smoothing functions

All dedicated SD software platforms come replete with a library of functions which the modeller can deploy as he or she thinks fit. Among these functions are those for use to represent delays and smoothing and they merit some separate attention in this subsection because they are very frequently used. Because SD separates resource flows and the information flows required to control the resource flows, delay functions are split between resource delays and information delays, although both yield the same effect – a family of exponential delays.

Let us consider resource delays first. Suppose it is desired to model production completion in a manufacturing model or deaths in a population model. In the former case it could be represented as a delayed function of production starts and in the latter as a delayed function of births (ignoring in- and out-migration). Whether a delay function can properly be used is dependent on a crucial criterion. Any delay function implies the use of an internal stock (level) variable which can be made explicit if required. For instance, it would be work-in-progress in the manufacturing example – see Figure 3.15. Where this internal stock 'drives' the output then using a delay function is permissible. Work-in-progress cannot be allowed to build up indefinitely and so in a sense is driving the completion of production. In the population example the internal stock is, indeed, population which, by dint of the ageing process which affects any living entity, is driving the death rate. Where the delay output is determined by something elsewhere in the system (other than the internal stock) then a delay function is not indicated.

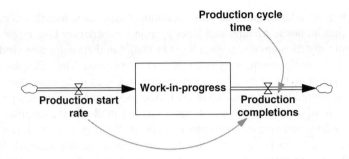

Figure 3.15   Delay construct in the manufacturing example; note rate-to-rate connection and outflow assumed driven by work-in-progress.

Once it has been established that use of a delay function is acceptable, two further determinations need to be made: the value for the *average delay time* and the *order* of the delay to be used. In the case of the manufacturing example, the former is the production cycle (or lead) time (Figure 3.15). Its value would be measured in weeks in respect of, say, consumer durables production and months or even years in respect of an aircraft or a naval warship. The determination of the order of the delay function is bound up with the need to understand the transient behaviour of the function (e.g. following a pulse change in the input) and an ability to relate this to the practical situation under consideration.

Consider Figure 3.16 where the transient output from a pulse change in input is presented for three different delay orders – first, third and infinite order (fixed or pipeline delay).

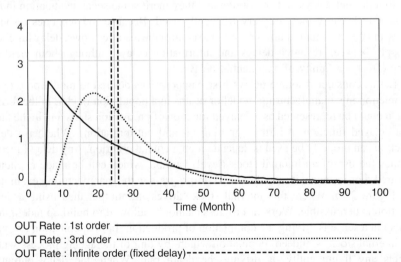

Figure 3.16   Transient output rates from delay functions of various orders (average delay time = 20).

The input PULSE occurs at time $t = 5$. A first-order delayed response occurs immediately and then progressively declines. The third-order output is initially slower to respond but progressively builds up to a peak and then declines. The fixed delay, as its name implies, merely reproduces the input (not shown) as output, in this case 20 months later. Statistically these are a series of exponential functions which belong to the Erlang family of distributions. The negative exponential response (first-order) is equivalent to an Erlang type 1 distribution. As the order increases, the variance of the distribution of output reduces, so higher orders than 3 would yield narrower and more peaked distributions until ultimately reaching the zero-variance distribution of an infinite-order delay. Clearly, whatever enters the delay function must ultimately leave it, even if this involves a long right-hand tail to the output distribution.

The names for these functions can vary with the software platform chosen. Variants of DELAY1, DELAY3 and DELAY FIXED are employed with an upper case 'I' being added at the end of the name where it is required to initialise the delay output to a particular value and this will be included as an additional argument to the function. In the absence of this special case, by default the delay output is initialised to the value of the input. If higher orders than 3 are required then it is permissible to cascade the available functions, but, in general, the three examples offered in Figure 3.16 are the ones most commonly employed.

These functions, which should only ever be used in flow equations, are clearly flexible and can suit different practical circumstances. The first-order function might be used to model deliveries arising from a large order placed for many different items. The third-order version might suit the manufacturing example above, certainly in the case of consumer durables production, whereas the fixed delay might be more appropriate in the case of, say, the construction of a warship.

Turning now to the consideration of information delays, there exists a similar repertoire of functions to those available for resource delays, save for the instance of the infinite delay order. Information delays are used to model the process of decision making. Forrester once famously noted: 'Management is the process of converting information into action.' He was distinctly capturing the real-life processes of information collection and assimilation prior to action being taken. These processes are modelled in SD by chains of auxiliary variables which lead into a flow variable. The auxiliaries effectively define action through the control of the flow rates (the policy variables).

Information collection and assimilation processes take time in real life and hence SD models need to reflect these delays. Flow rates are not changed instantaneously, nor can they be observed instantaneously. Sequences of values arising from a flow (and being reported through the formal or informal information systems) may not predicate a trend. Assimilation of this flow of information is necessary in order to reach agreed actionable conclusions.

Information needs to be processed ahead of any action; this is usually referred to as smoothing. The time delay in processing information is commonly referred to as the 'smoothing time' or 'smoothing constant' and is the second argument in the function following the name of the input variable. First- and third-order smoothing processes are usually available and again have names which can vary with the SD software platform being used: SMOOTH1 and SMOOTH3 are common for first-order and third-order smoothing respectively. The 'I' variant is also available should it be required to initialise the output of the smoothing process. By far the most commonly used is first-order smoothing; in fact this is mathematically equivalent to exponential smoothing of the information input stream. Third-order is much less common and tends to be reserved for the more 'important' decisions and where there is a discrete delay in the reporting sequence, for example quarterly as opposed to daily or weekly information reports.

Finally, it is worth pointing out that information delays and resource delays possess different properties with respect to mass. A resource delay moves mass from the input to the output variable (i.e. it is conserved) whereas an information source can give up the same information to many different destinations in respect of the policy process – it is non-conserved.

3.4.4   Model validation

It is probably true to say that SD models are required to undergo much higher standards of validation than other methodologies and techniques associated with the fields of management science and econometrics. It is a commendation for the SD research community that it has responded and there now exists a significant range of tests which can be used to validate an SD model. It must be emphasised from the outset, however, that no model can be shown to represent the ultimate truth and a primary objective of the process of subjecting a model to various tests is to enhance its suitability for use in policy analysis. Where the work is for clients, it is necessary for them to be sufficiently confident of the internal mechanics and robustness of the model and then to go on and put into action the policy advice which might emerge from the simulation runs. Validation tests can be seen, therefore, as enhancing confidence in the model – nothing more and nothing less.

It is not the intention here to go into great detail of the range of tests now available for SD models. Various references can be consulted. An early, oft-cited contribution is that of Forrester and Senge (1980). Sterman (2000) offers considerable detail on the matter, some material being derived from his earlier article describing appropriate metrics for the historical fit of simulated to reported data (Sterman, 1984). Coyle and Exelby (2000) give attention explicitly to the validation of *commercial* SD models, while Barlas has a number of publications in this domain, a notable one being Barlas (1996). Finally, Schwaninger and Grösser (2009) offer a contemporary overview of SD model validation.

Model verification is a necessary step in DES modelling. It is the task of demonstrating that the coded model is the one intended by the modeller. This is different to model validation, which is more concerned with an external assessment and how well the model corroborates with the real-world system it purports to represent. In respect of SD modelling, a dimensional analysis check (see Section 3.3.4) could be construed as a verification task, but there also exists a test which accords with model verification and which does not always appear in the list of SD model tests published elsewhere: this is the mass-balance check.

It is obvious that, in a model where flows of resources are represented, the amount flowing in must balance with the amount flowing out. In other words, the model must not gain or lose mass. The need for this check is enhanced when the resource flows in the model involve multiple inflows and bifurcations, although there is no harm carrying out the check even with a simple one-way flow system.

It is necessary to cumulate all the inflows and outflows for each resource being modelled. Then, taking account of both initial values and the values of the various stocks at any one time, the balance (or checksum) equation has the form

$$\text{(Sum of all inflows} + \text{initial values of stocks)} - \text{(Sum of all outflows} + \text{current values of stocks)} = 0$$

Given the inevitably large numbers which will be involved either side of the minus sign, it is necessary to deploy a *double precision* version of the software being used. The correct result should be computationally equivalent to zero throughout the run; anything else indicates a flawed model. Where more than one resource flow is involved (e.g. people and finance) then separate checksum equations need to be used – one for each flow.

To conclude this subsection it is worthwhile to align important validation tests against the three main categories of SD model. These categories are: (A) a model representing an existing real-world system where there is knowledge extant for that system and an issue which needs addressing (models built for consultancy assignments, and where a client exists, usually fall into this category); (B) a model of a generic system, addressing a specific issue but not parameterised to any particular real-world system (models developed in academic research are often examples of this type); (C) a model of a system that does not yet exist but where there is a need to anticipate matters which might affect the design of that system (the setting for this type could be either consulting or academic research).

With models of type A there will be an emphasis on behaviour reproduction tests. It is necessary to demonstrate a correspondence between the simulated output and reported data. This correspondence need only be qualitative; point-for-point equivalence is neither sought nor is necessary. Goodness-of-fit metrics can, however, be calculated. The use of R^2 is common but a more appropriate metric is Theil's inequality statistics. This decomposes the mean square error into three components and then determines the fraction of the error due to each. Sterman (2000, p. 876) outlines different possible outcomes and the interpretation of the statistic. Given knowledge of the system being modelled, the model's face validity and its constituent parameter values can also be subject to collective scrutiny by the clients as a further validation check in this case.

In the case of models of type B the output behaviour needs to be plausible. That is to say, important constituent variables should not produce behaviour which is 'obviously wrong'. Generally, a *reference mode* behaviour will be known. This is a sketch graph of the typical dynamic behaviour of this type of system; the model needs to conform qualitatively to this behaviour. In addition it may be possible to assemble a range of experts with an intimate knowledge of such systems and utilise their collective expertise to assess the model's face validity.

Finally, in the case of models of type C, behaviour reproduction tests are clearly impossible and the emphasis will fall solely onto the model's face validity. The formation of an expert group is almost mandatory here and, as far as possible, unanimity among the members as to the model's structure and behaviour is sought.

3.4.5   Optimisation of SD models

The term 'optimisation' when related to SD models has a special significance. It relates to the mechanism used to improve the model vis-à-vis a criterion. This collapses into two fundamentally different intentions. First, one may wish to improve the model in terms of its performance. For instance, it may be desired to minimise the overall costs of inventory while still offering a satisfactory level of service to the downstream customer. So the criterion here is cost and this would be minimised after searching the parameter space related to service level. The direction of need may be reversed and maximisation may be desired as if, for instance, one had a model of a firm and wished to maximise profit subject to an acceptable level of payroll and advertising costs. Here the parameter space being explored would involve both payroll and advertising parameters. This type of optimisation might be described generically as *policy optimisation*.

A separate improvement to the model may be sought where it is required to fit the model to past time series data. Optimisation here involves minimising a statistical function which expresses how well the model fits a time series of data pertaining to an important model variable. In other words, a vector of parameters is explored with a view to determining the particular parameter combination which offers the best fit between the chosen important model variable and a past time series dataset of this variable. This type of optimisation might be generically termed *model calibration*. If *all* the parameters in the SD model are determined in this fashion then the process is equivalent to the technique of econometric modelling. A good comparison between SD and econometric modelling can be found in Meadows and Robinson (1985).

In respect of calibration optimisation, there is also the possibility of fitting the model to projected future behaviour. If a future target is set in terms of some performance metric, then to associate that with a particular trajectory for its attainment is a significant improvement on just determining a target as a single point x time periods ahead. Not least, this approach helps to verify if the chosen target is actually feasible. Model optimisation in this mode can be seen as a synthesis of *policy optimisation* and *calibration optimisation*.

The process by which an SD model is optimised involves pre-selecting the parameters to be included in the search vector, associating each parameter with a range in which the search will be conducted and finally determining the objective function to be optimised. The software will repeatedly simulate, for thousands of runs if required, searching for the values of the particular parameter combination which will conform with the maximum or minimum value of the objective function, as appropriate.

More details on the mechanisms of SD model optimisation can be found in Keloharju and Wolstenholme (1988), Dangerfield and Roberts (1996) and Dangerfield (2009). Examples are reported in Keloharju and Wolstenholme (1989), Dangerfield and Roberts (1999), Graham and Ariza (2003) and also in Dangerfield (2009).

3.4.6   The role of data in SD models

In many cases where the author has suggested an SD approach to a policy issue the rejoinder has been: 'What data will you require to accomplish this?' There is an implicit assumption by lay people that 'modelling' must involve enormous quantities of data and this puts a brake on its possible use. In fact in the case of SD simulation models the answer to the amount of data required in order to produce a working model is: 'Surprisingly little'.

It is a fact, though, that the amount of data required for a modelling exercise varies significantly with the technique being adopted. Confining the discussion to the domain which might be loosely termed 'dynamic systems models', on one extreme lies econometric modelling where every variable requires sufficient historical time series data in order to fit the model. If no historical data exists for an important variable then sometimes a proxy can be formulated from the data which does exist, but, failing this, the variable has to be excluded from the model. Further, there is no provision in econometrics for the incorporation of soft variables as discussed above in Section 3.4.1.

The amount of data required for a DES model lies somewhere on the continuum between econometrics and SD. All activity durations need to be estimated, usually by fitting a theoretical probability distribution to reported data. Further, a maximum number may need to be established for certain entities (e.g. beds on a hospital ward) along with any limits on queue sizes.

In the case of SD the extent of required data can be framed in the knowledge that such models can be created for systems which do not yet exist (i.e. operating in a design mode). Here the data required is simply parameter values, such as productivity of the workforce (output units/person/time unit), delay in adjusting capital stock (time unit), fat quantity in meals outside (grams/meal) or probability of default on debt (dimensionless) together with estimates for the initial values of each of the stocks (levels) in the model. All of these numerical quantities would then be guesstimated by the modeller with the help of the clients if available.

Should the system being studied already exist then statistical estimates or generally accepted and reliable knowledge can replace the guesstimates of parameter values for the SD model. Further, if working on a specific policy issue, past time series data may be available for certain model variables and this data can be embraced as part of the process of model validation (see Section 3.4.4). Therefore, in respect of SD models, the maxim on requisite data is: nothing is required, but if any data exists then by all means use it.

3.5   Further reading

It is impossible in this single chapter to do full justice to what is now a significant methodology in the socio-economic, managerial, health, biological, environmental, energy and military sciences. However, three contemporary books will take the interested reader much further. These are purely the author's choice and they are listed in order of page count.

John Sterman's book (982pp.) has arguably the most comprehensive coverage; see Sterman (2000) in the reference list. John Morecroft's (466pp.) *Strategic Modelling and Business Dynamics: A feedback systems approach* (2007), published by John Wiley & Sons, Ltd, Chichester, offers a very wide coverage of systems thinking and SD and incorporates many practical model examples. Third, Kambiz Maani and Bob Cavana have written a second edition of their offering (288pp.): K.E. Maani and R.Y. Cavana (2007) *Systems Thinking, System Dynamics: Managing change and complexity*, published by Pearson Education NZ (Prentice Hall), Auckland.

All these books come with a CD-ROM and/or a web site which provides specimen models to be run and allows scenario experiments to be conducted. Exercises and instructor's manuals are also available.

References

Ballé, M. (1994) *Managing with Systems Thinking*, McGraw-Hill, London.

Barlas, Y. (1996) Formal aspects of model validity and validation in system dynamics. *System Dynamics Review*, **12**(3), 183–210.

Coyle, G. and Exelby, D. (2000) The validation of commercial system dynamics models. *System Dynamics Review*, **16**(1), 27–41.

Coyle, R.G. (2000) Qualitative and quantitative modelling in system dynamics: some research questions. *System Dynamics Review*, **16**(3), 225–244.

Coyle, R.G. (2001) Rejoinder to Homer and Oliva. *System Dynamics Review*, **17**(4), 357–363.

Dangerfield, B.C. (2009) Optimization of system dynamics models, in *Encyclopedia of Complexity and Systems Science* (ed. R.A. Meyers), Springer, New York, pp. 9034–9043. (Reprinted in *Complex Systems in Finance and Econometrics*, Springer, New York, 2011, 802–811.)

Dangerfield, B.C. and Roberts, C.A. (1996) An overview of strategy and tactics in system dynamics optimisation. *Journal of the Operational Research Society*, **47**(3), 405–423.

Dangerfield, B.C. and Roberts, C.A. (1999) Optimisation as a statistical estimation tool: an example in estimating the AIDS treatment-free incubation period distribution. *System Dynamics Review*, **15**(3), 273–291.

Forrester, J.W. (1958) Industrial dynamics – a major breakthrough for decision makers. *Harvard Business Review*, **36**(4), 37–66.

Forrester, J.W. (1961) *Industrial Dynamics*, MIT Press, Cambridge, MA (now available from the System Dynamics Society, Albany, NY).

Forrester, J.W. (2007) System dynamics – a personal view of the first fifty years. *System Dynamics Review*, **23**(2–3), 345–358.

Forrester, J.W. and Senge, P.M. (1980) Tests for building confidence in system dynamics models, in *System Dynamics* (eds A.A. Legasto Jr, J.W. Forrester and J.M. Lyneis), North-Holland, Amsterdam, pp. 209–228.

Goodman, M.R. (1974) *Study Notes in System Dynamics*, Wright-Allen Press, Cambridge, MA (now available from the System Dynamics Society, Albany, NY).

Graham, A.K. and Ariza, C.A. (2003) Dynamic, hard and strategic questions: using optimisation to answer a marketing resource allocation question. *System Dynamics Review*, **19**(1), 27–46.

Hardin, G. (1968) The tragedy of the commons. *Science*, **162**(3859), 1243–1248. doi: 10.1126/science.162.3859.1243

Homer, J. and Oliva, R. (2001) Maps and models in system dynamics: a response to Coyle. *System Dynamics Review*, **17**(4), 347–355.

Keloharju, R. and Wolstenholme, E.F. (1988) The basic concepts of system dynamics optimisation. *Systems Practice*, **1**, 65–86.

Keloharju, R. and Wolstenholme, E.F. (1989) A case study in system dynamics optimisation. *Journal of the Operational Research Society*, **40**(3), 221–230.

Maruyama, M. (1963) The second cybernetics: deviation-amplifying mutual causal processes. *American Scientist*, **51**(2), 164–179.

Meadows, D.L. and Sterman, J. (2012) *Fishbanks: A Renewable Resource Management Simulation*, https://mitsloan.mit.edu/LearningEdge/simulations/fishbanks/Pages/fish-banks.aspx (accessed 30 May 2013).

Meadows, D.M. and Robinson, J.M. (1985) *The Electronic Oracle*, John Wiley & Sons, Ltd, Chichester (now available from the System Dynamics Society, Albany, NY).

Merrill, J.A., Deegan, M., Wilson, R.V. *et al.* (2013) A system dynamics evaluation model: implementation of health information exchange for public health reporting. *Journal of the American Medical Informatics Association*, **20**, e131–e138. doi: 10.1136/amiajnl-2012-001289

Richardson, G.P. (2011) Reflections on the foundations of system dynamics. *System Dynamics Review*, **27**(3), 219–243.

Richardson, G.P. and Pugh, A.L. (1981) *Introduction to System Dynamics Modelling with DYNAMO*, MIT Press, Cambridge, MA (now available from the System Dynamics Society, Albany, NY).

Schwaninger, M. and Grösser, S. (2009) System dynamics modelling: validation for quality assurance, in *Encyclopedia of Complexity and Systems Science* (ed. R.A. Meyers), Springer, New York, pp. 9000–9014. (Reprinted in *Complex Systems in Finance and Econometrics*, Springer, New York, 2011, 767–781.)

Senge, P. (1990) *The Fifth Discipline*, Currency Doubleday, New York.

Senge, P., Kleiner, A., Roberts, C. *et al.* (1994) *The Fifth Discipline Fieldbook*, Currency Doubleday, New York.

Sherwood, D. (2002) *Seeing the Forest for the Trees: A Manager's Guide to Applying Systems Thinking*, Nicholas Brealey, London.

Sterman, J.D. (1984) Appropriate summary statistics for evaluating the historical fit of system dynamics models. *Dynamica*, **10**(2), 51–66.

Sterman, J.D. (2000) *Business Dynamics: Systems Thinking and Modeling for a Complex World*, Irwin McGraw-Hill, New York.

Tustin, A. (1953) *The Mechanism of Economic Systems*, Harvard University Press, Cambridge, MA.

Wolstenholme, E.F. (2003) Towards the definition and use of a core set of archetypal structures in system dynamics. *System Dynamics Review*, **19**(1), 7–26.

4

Combining problem structuring methods with simulation: The philosophical and practical challenges

Kathy Kotiadis and John Mingers
Kent Business School, University of Kent, Canterbury, UK

4.1 Introduction

Combinations of problem structuring methods (PSMs) such as soft systems methodology (SSM) with simulation methods such as discrete-event simulation (DES) and system dynamics (SD) report benefits. These benefits include: the PSM enabling the participation of staff in the modelling process (Lehaney and Paul, 1996), enabling a better understanding of the situation of interest (Paucar-Caceres and Rodriguez-Ulloa, 2007; Kotiadis, 2007) and enabling the situation to be expressed and structured (Paucar-Caceres and Rodriguez-Ulloa, 2007; Kotiadis, 2007). A 10-year review (1998–2008) of mixing operational research (OR) methods in practice (Howick and Ackermann, 2011) only found around 15 papers to involve a simulation method (DES and/or SD). From this we can extrapolate that, although hundreds of simulation papers have been published in that time period, only a handful of these involve a

Discrete-Event Simulation and System Dynamics for Management Decision Making, First Edition.
Edited by Sally Brailsford, Leonid Churilov and Brian Dangerfield.
© 2014 John Wiley & Sons, Ltd. Published 2014 by John Wiley & Sons, Ltd.

multimethodology. In addition, the review (Howick and Ackermann, 2011) also revealed that most of these simulation combinations involved a PSM.

In this chapter we reflect on the barriers to such combinations, which can be seen at various levels such as the philosophical level – paradigm incommensurability – and the cognitive level – type of personality and cognitive difficulty of switching paradigm. We explore the literature and argue that the philosophical problems are not insurmountable. In addition we examine the practical aspect to such 'multiparadigm multimethodology' combinations. More specifically we explore the potential benefits and the possible barriers through a case study where SSM was combined with DES modelling within a healthcare context – community-based intermediate care. Reflecting on the practical application at the philosophical level reveals a multiparadigm multimethodology, with an interplay strategy for mixing the soft and hard paradigms. Although many of our reflections focus on DES, some are also applicable to SD.

PSMs are introduced next. Following that, the chapter focuses on the theoretical and practical concerns when combining OR methodologies from different paradigms. These concerns apply to combinations of simulation with PSMs. The next section reviews some relevant empirical examples of such work. Then our case study is described, and in the final section we reflect on its relevance to the concerns described earlier.

4.2   What are problem structuring methods?

PSMs have evolved within OR over the last 30 years in order to better deal with messy, wicked and complex problems that are not amenable to the traditional, largely quantitative, OR techniques. The term 'PSM' was popularised by Rosenhead (1989) in his book *Rational Analysis for a Problematic World* as an alternative to the terms 'soft OR' or 'soft systems' which were felt to have unfortunate connotations.

PSMs are defined by a range of characteristics that are in contrast to those of traditional 'hard' techniques (Mingers and Rosenhead, 2004). Very briefly these are as follows:

- They deal with unstructured problems characterised by multiple actors, multiple perspectives, conflicts of interest, major uncertainties and significant unquantifiable factors.

- They must enable the modelling of alternative perspectives.

- They, and their associated models, must be accessible to the actors involved to facilitate their genuine participation.

- They must be flexible and iterative.

- They must allow local rather than global improvements.

The most well-known PSMs (Rosenhead and Mingers, 2001) include soft systems methodology (SSM), strategic options development and analysis (SODA – now developed as 'Journeymaking'), strategic choice analysis (SCA) and drama theory. Others that could be included are system dynamics, viable systems model (VSM), interactive planning and critical systems heuristics.

While the development and use of PSMs in themselves have been extremely successful (Mingers, 2000b), it has been argued by many people (Jackson, 1999; Midgley, 1997b; Mingers, 2000b; White and Taket, 1997) that significant benefits can accrue by combining different methodologies together – what has been termed 'multimethodology'. This can consist of simply combining several PSMs together, which is relatively unproblematic, or it can involve combining PSMs with more traditional, hard techniques. Empirical evidence shows that this has occurred less often (Munro and Mingers, 2002; Howick and Ackermann, 2011) for reasons to be discussed below, but considerable benefits could be gained by encouraging such combinations.

4.3   Multiparadigm multimethodology in management science

Several other authors have proposed the idea of combining together methodologies or parts of them using different terms: 'coherent pluralism' (Jackson, 1999), 'creative design of methods' (Midgley, 1997a), 'pragmatic pluralism' (White and Taket, 1997) and 'complementarity' (Pidd, 2004b). All of these accept the general arguments for the combination of different methods from different paradigms and most have gone to great lengths to explain or suggest how to achieve a combination at the methodology level, though they differ in the underlying rationale and the particular approach taken. Their views fall into two camps: those that believe that genuine multiparadigm multimethodology is possible, that is that methodologies, and parts of methodologies, from different paradigms, can be combined together coherently; and those who believe that methodologies can only be combined together having respect for their underpinning paradigms. Mingers and Brocklesby (1997) make four general arguments for multiparadigm multimethodology. First, that the world is complex and multidimensional and using different paradigms enables one to focus attention on different aspects of the situation. Second, a problem goes through different phases and more than one methodology might be required to tackle all phases. Third, multiparadigm multimethodology is common practice even if there are limited reports in the literature. Finally, triangulation of the situation using different methodologies can generate new insights while enhancing confidence in the results through a reciprocal validation (Mingers, 2002).

This leads us to explore the problems associated with the feasibility of multiparadigm multimethodology work, particularly at the philosophical level, which generally has not received as much attention even though it was discussed by Mingers and Brocklesby in 1997. They reported different levels of problems: (a) philosophical (particularly the issue of paradigm incommensurability); (b) cultural (the extent to which organisational and academic cultures militated against multiparadigm work); (c) cognitive (the difficulties that the individual experiences when moving from one paradigm to another); and (d) practical (takes more time, may be a lack of experience of several methods, innate conservatism – especially of funding bodies and journals, pressure to do something not 'risky' by organisation/client) (Mingers, 2001). Each of these will be explored briefly in the following subsections.

4.3.1   Paradigm incommensurability

The notion that paradigms are incommensurable is said to have originated at the beginning of the last century (Schultz and Hatch, 1996) but the debate is more extensive in organisational theory compared with management science. At the centre of the debate, the seminal book by Burrell and Morgan (1979) (*Sociological Paradigms and Organisational Analysis*) challenged the dominant functionalist (positivist) orthodoxy by putting forward four paradigms which were claimed to be incommensurable, that is mutually exclusive, unable to be combined or linked because their underlying assumptions are thought to be incompatible. However, Willmott (1993) argues that 'Paradigms (Burrell and Morgan's book) assumes, and strongly endorses, a restriction of analysis within the confines of four, mutually exclusive "ways of seeing"' (Willmott, 1993, p. 682) and objects to the polarisation of either a subjective or objective approach to social science. He also makes the point that a close reading of the thesis by Kuhn (1970) reveals that Burrell and Morgan's conceptualisation of paradigms is constrictive and ignores Kuhn's account of scientific activity as

> a process of movement in which 'new' paradigms emerge, phoenix-fashion, from the ashes of those they replace. It is important to recognise that Kuhn (1970) stresses the substantial continuity and overlap between the paradigms in the mediation of 'normal' and 'revolutionary' moment of scientific practice.
>
> *(Willmott, 1993, p. 686)*

In the remaining part of this section we will be concerned particularly with the way in which others view the approach for combining the hard and soft paradigms in interventions. The views that will be presented will not include those that generally accept multiparadigm multimethodology but are not particularly interested or engaged in the debate about how it is done at the paradigm level.

Pidd (2004b), who also believes Kuhn to have suggested that paradigms are incommensurable, acknowledges that this notion is a contributing factor to the difficulty of combining methodologies. More importantly, he uses the point made by Ormerod (1997) about people managing to work successfully in both soft and hard approaches to propose that either Kuhn was wrong or soft and hard OR do not belong to different paradigms. In his book he portrays the relationship between soft and hard OR/MS as shown in Figure 4.1.

Describing from left to right: (a) represents incommensurability; (b) describes hard and soft feeding on each other 'in an eclectic and pragmatic way' (Pidd, 2004a, p. 19) in (c) the soft methods are seen to contain the hard methods, which means that 'understanding of meanings gained in soft OR/MS enables a sensible attempt at hard

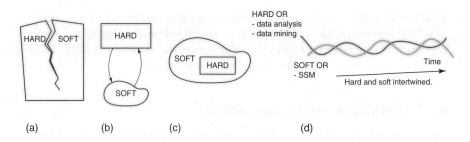

Figure 4.1   The soft and hard paradigm. From Pidd (2004a) and Brown, Cooper and Pidd (2005).

OR/MS' (p. 19); and (d) illustrates soft and hard OR being intertwined and is similar to the view of Checkland and Scholes (1999) presented in *Soft Systems Methodology in Action* (p. 282). The latter illustration (d) in Figure 4.1 specifically applies to a case study involving data mining and SSM in the public sector (Brown, Cooper and Pidd, 2005) and visually seems closer to illustrating multiparadigm multimethodology than the other representations.

Similarly Schultz and Hatch (1996) distinguish three metatheoretical positions for doing multiparadigm research: (a) paradigm incommensurability; (b) paradigm integration; and (c) paradigm crossing. We will now focus on (b) and (c). Paradigm integration, according to Schultz and Hatch, enables the synthesis of 'a variety of contributions, thus ignoring the differences between competing approaches and their paradigmatic assumptions' (Schultz and Hatch, 1996, p. 532). They criticise this position on two accounts. First, in a few cases it 'represents simple resistance to multiparadigm thinking' (p. 533), for example to abandon all but one paradigm as suggested by Pfeffer (1993). Second, in many cases it 'provides an overall framework that mixes and combines terms and implications of arguments grounded in different paradigmatic assumptions without considering the relationship between the assumptions themselves' (Schultz and Hatch, 1996, p. 533). We will focus on paradigm crossing as it provides a theoretical basis and explanation of multiparadigm work in practice.

Paradigm crossing is about how multiple paradigms are dealt with by an individual researcher without ignoring them, as the integrationist position, or refusing to confront them, as the incommensurability position. Schultz and Hatch (1996) focus on strategies for paradigm crossing and they add a new one to the existing ones in the literature. The existing strategies are the sequential, the parallel, the bridging, and interplay (Schultz and Hatch's strategy). In the sequential strategy the relationship between paradigms is linear and movement from one paradigm to another is unidirectional. In the parallel strategy different paradigms are all applied on an equal basis instead of sequentially. The sequential and parallel strategies leave the boundaries of each paradigm deployed intact but in the bridging strategy the boundaries are more permeable. In fact, Gioia and Pitre (1990), who first articulated the bridging strategy, argued that paradigm boundaries are ill defined and blurred, and that it is 'difficult, if not impossible, to establish exactly where one paradigm leaves off and another begins' (p. 592). Therefore the boundaries are better conceived as transition zones.

The interplay strategy 'refers to the simultaneous recognition of both contrasts and connections between paradigms rather than differences' (Schultz and Hatch, 1996, p. 534). Researchers using this strategy will transpose the findings from one paradigm in such a way that they inform the research conducted in a different paradigm. Therefore the researcher can move back and forth between paradigms allowing cross-fertilisation between the paradigms while maintaining diversity. The main difference of the interplay with the other strategies is 'in the nature of the relationship it constructs between the researcher and the multiple paradigms it specifies' (Schultz and Hatch, 1996, p. 535). It is this approach that is closest to that advocated by Mingers (1997b) and appears to fit the last illustration of Pidd (2004a) (Figure 4.1) which is essentially a representation of a paradigm strategy.

Tashakkori and Teddlie (1998) also explore the paradigm strategies for combining qualitative and quantitative methodologies and categorise the sequential and parallel under *mixed method studies*, but they do not refer to the bridging or interplay strategies although they acknowledge a *mixed model studies* category that could incorporate them.

This debate about strategies is an attempt to understand and communicate the multimethodology work undertaken that could be classified as multiparadigm. This debate is in its infancy and it is interesting to see the authors that are currently engaged attempting to communicate with the audience not only by words, but also with pictures, as if words cannot go deep enough under the surface into the subconscious where most of the work arguably took place. A picture is worth a thousand words and perhaps in such situations a more useful communication tool.

4.3.2   Cultural difficulties

Peoples' assumptions about the world and how to deal with its problems are to some extent a cultural issue that has resulted through socialisation and education. In management science there are communities, particularly educational ones, that are perceived as more hard OR focused or to have a more balanced number of hard and soft OR specialists than others. However, the latter are thought to be fewer, particularly as many working in the field of management science emerge from a variety of positivist disciplines, for example mathematics, computer science, engineering, and so on. Undoubtedly this must affect both the type of research or projects undertaken in these departments as well as the student's experience and attitude to problem solving immediately after university. It is logical to assume that this culture also feeds into industry. PSMs are probably not even considered by the majority of management scientists in the first instance when reviewing a problematic situation and many will simply turn to the old familiar approaches. In the event that PSMs are considered for use either alone or with a traditional hard OR method, there is probably a degree of fear about being competent in their use.

On the other hand, Pidd (2004a) makes some interesting points on the clients' view about someone being able to shift from hard to soft work. The first one is that the client might be unwilling to believe that someone might be competent enough to carry out both technical work and high-quality soft work. He points out that 'competence in using soft approaches cannot be picked up by reading a chapter in a book important though that is. As Eden and Ackermann point out, there must be skills to be practised and practicalities that must be attended to' (Pidd, 2004a, p. 205). If clients do feel this way it is understandable because many MS/OR degree programmes weigh in favour of quantitative approaches and their questioning the analyst's competence is not unreasonable as it would be very difficult to prove success in problem structuring skills. We cannot resist reminding the reader of the famous anecdote of Checkland (1999a) that if someone told him that they had used his methodology (SSM) and it worked he would have to reply:

> How do you know that better results might not have been obtained by an *ad hoc* approach? If the assertion is: The methodology does not work, the author may reply, ungraciously but with logic, How do you know the poor results were not due to simply your incompetence in using the methodology?
>
> *(p. A12)*

4.3.3   Cognitive difficulties

This category of problems will be divided into difficulties in shifting paradigms and the personality of the management scientist.

4.3.3.1   Difficulties in shifting paradigms

Ormerod (1997) believes that mixing methods is possible and that switching between paradigms was not an issue for him, and that the limitations lie in the competence of the consultant and the participants rather than the methods themselves. However, Brocklesby (1995) acknowledges that shifts in paradigms can be a 'painful experience for the individual concerned' (p. 1290) but that it is possible for a person to become multimethodology literate given sufficient determination. In a later work he says that 'the process of transforming an agent who works within a single paradigm into someone who is multimethodology literate is perhaps unlikely, although by no means impossible' (Brocklesby, 1997, p. 212). However, for this to happen a number of obstacles must be overcome.

First, the agent must become paradigm conscious. Second, the agent must believe that the new paradigm offers something worth having and fits with the agent's personality and beliefs. Third, effective performance in a paradigm necessitates learning its propositional and common-sense knowledge. Brocklesby uses the work by Varela, Thompson and Rosch (1991) that identifies two types of knowledge: propositional and common-sense, needed for someone to act effectively in a 'new paradigm'. He explains that in soft OR (otherwise known as problem structuring methods) the propositional knowledge needed to create rich pictures and produce root definitions can be acquired from textbooks, but in order to be effective in soft OR one must work directly with people, and respond to developing situations, which cannot be captured in a propositional format. However, to become proficient in a new paradigm the newcomer has to acquire relevant propositional knowledge, but this is only the first step and that 'really' knowing the paradigm and acting effectively in it means active bodily involvement, experience and practice. He also says that acting effectively within a new paradigm requires both learning and unlearning.

In addition, Pidd (2004a), similarly to Brocklesby (1995), also believes that it is harder than some people think to move from 'one intellectual universe to another' (p. 205) and suggests that perhaps the answer is to have different people carry out the hard elements of the work and others to undertake the soft (PSM) parts. Although he does also point out that the two must respect one another's insights, this could be construed as an indirect warning that these two groups will find it difficult to collaborate.

4.3.3.2 Personality

The empirical evidence cited earlier shows that the combination of a hard and soft methodology is less common than that of two hard or two soft methodologies. This finding may be an indication that operational researchers find it difficult to work across two paradigms and there is evidence that this may be related to particular personality types (Mingers and Brocklesby, 1997). The first personality described is the 'analytical scientist' personality type who prefers quantitative, aggregate data and shows distaste for qualitative data because he or she values precision, accuracy and reliability. The second personality is described as the 'particular humanist' and prefers to conduct research via personal involvement with other people, and prefers qualitative data. Furthermore, he or she is consultative and zealously promotes consensus and acceptance. While it is likely that most management scientists overlap these categories, there will be some that will fit into one category more than the other:

> for such people it may be surmised that they will experience some difficulties in moving from one paradigm to another, and/or experience a certain internal tension or discomfort if they are compelled to work in a paradigm that calls for actions and behaviours that do not 'fit' their cognitive processing preferences.
>
> *(Mingers and Brocklesby, 1997, p. 499)*

An example of these difficulties is mentioned by Doyle (1990) who found that users experienced considerable difficulty when trying to cope with both SSM and Jackson's system development (JSD).

Consideration of the importance of the personal characteristics of the practitioner in both choosing and using multiple methods and, in particular, of how difficult it is for individuals to work across paradigms combining technical, quantitative analysis with soft facilitation skills, will be further explored in the case study discussion.

4.3.4   Practical problems

Finally, there are practical difficulties that constrain multimethod work (Mingers, 2001). It takes time to undertake such work, which means that many, particularly academics, might choose the clean-cut single method work, which is easier to explain and sell to clients, funding bodies and journals. Furthermore, multimethod work requires the knowledge and experience of several methods, which can be a problem, especially if there is one analyst.

In this section we have discussed some of the problems involved in multiparadigm work. One of the main arguments against it – paradigm incommensurability – has been shown to be flawed as Kuhn's work was misinterpreted and therefore the issue is still open for discussion. The issue of cultural feasibility draws attention to obstacles that are socially constructed. We have also covered problems of personality type in working within different paradigms and the difficulties of switching from one paradigm to another. The main purpose of the case study is to enable us to contribute our experience on these points. In the next section we will describe some practical applications relevant to our case study.

4.4   Relevant projects and case studies

There are examples of PSMs and hard OR used together (Mingers, 2000b), with the most popular PSM being SSM (Bennett and Kerr, 1996; Coyle and Alexander, 1997; Pauley and Ormerod, 1998). A more recent survey (Howick and Ackermann, 2011) also found that SSM was in recent years the most popular method in multimethodologies (12 out of 30 papers reviewed from 1997 to 2008). The review reports only four instances of SSM involving simulation (two SD and two DES), which is an underestimate of the actual number as only papers in four general OR journals were used. However, we will now review a relevant subset of the papers describing a multimethodology involving a PSM and simulation. Of course, in light of the case study to follow we are particularly interested in those using SSM and applied to healthcare and those that express views about their paradigm strategy.

The starting point in the simulation methodology is about *understanding* the problem, which could also be described as the problem structuring phase. For this stage there are no guidelines or a structured technique to help the analyst in this task. It is surprising therefore that it is not common practice for problem structuring methodologies and techniques to be used in simulation methodology. However, Lehaney and Hlupic (1995) review the use of DES for resource planning in the health sector and suggest the use of SSM as an approach for improving the process and project or research outcomes. Lehaney and Paul (1996) examine the use of SSM in the development of a simulation of outpatient services at Watford General Hospital. The paper is concerned with the hypothesis that simulation can be developed through the use of SSM and that the acceptability of the final model may be increased through the participate nature of SSM. In this case the aim was to see if SSM could be used in the model building process even though simulation had been selected as the tool of analysis. The authors argue that this multimethodology allows the modelling of the actual patient experience and the participation of staff in the modelling process. In addition, the participation paved the way for acceptance of the conceptual model and gave rise to the final simulation being credible.

Lehaney, Clarke and Paul (1999) report on an intervention that utilised simulation within a soft systems framework and they call it soft-simulation. The project was for an outpatient dermatology clinic and the SSM approach was to use primary task root definitions. The paper includes a lengthy discussion on critiques, challenges and responses to issues about SSM or DES or their combination. The only challenge relating to their paradigm approach relates to the issue of paradigm incommensurability. They explain that SSM and simulation modelling are not completely different approaches as 'they are both useful in facilitating debate and decisions, and are therefore both useful in similar areas, with each strengthening the other' (p. 889) and, in their case, the notion of different paradigms is inappropriate. Because of this last statement one might think that their paradigm position is that of paradigm integration.

Lane and Oliva (1998) have explored the methodological synthesis of SSM and SD. They identify with the position of paradigm integration. They assert that SD does not belong to one paradigm and can therefore be recrafted and applied to others. Mingers takes a similar view (Mingers, 2000a; Mingers and White, 2010), arguing that SD exemplifies critical realism. Critical realism is a new paradigm, which aims to go beyond the current paradigms while recognising both their strengths and weaknesses (Archer et al., 1998; Bhaskar, 1978; Fleetwood and Ackroyd, 2004; Mingers, 2004). Methodologically, critical realism encourages a plurality of research methods and is therefore entirely compatible with multimethodology.

A study by Paucar-Caceres and Rodriguez-Ulloa (2007) put forward a framework for combining SSM with SD which they call soft systems dynamics methodology. Unlike Lane and Oliva (1998), they explain that their approach is multiparadigm multimethodology and test the framework on a small enterprise involved in commercialising steel products. Although they describe paradigm crossing, the specific strategy adopted is difficult to pinpoint with any certainty. In our opinion both the interplay and bridging strategy are likely candidates. This study proposes that a key benefit of using SSM is that it enables the problem situation to be expressed and structured.

On the other hand simulation is also commonly used to develop our understanding of a problem, which is a function often attributed to PSMs. Robinson (2001) describes using simulation modelling in a soft intervention to aid the improvement of a user support helpline service at Warwick Business School. He describes his approach as qualitative simulation and a multimethodology because the first part of his study follows the hard OR methodology, though he departs at the validation stage from the hard methodology due to the lack of accurate data making it impossible to validate the simulation model. The intervention from then onwards focuses on a facilitated discussion centred on the model. The most striking difference to most traditional DES users is that the model was used as a focus for debate, a means for learning about the problem situation and for reaching an agreement to act. Robinson argues that the power of the simulation methodology and the DES technique means that lack of accurate data does not need to be a hindrance to DES. Although he does not use a specific PSM, Robinson compares his steps and outcomes with SD and SSM. The purpose of his paper as he describes it is to demonstrate that DES can be used to support soft OR interventions. He places his approach closest to pragmatic pluralism because the mixing of paradigms was not intended. It could be described as an example of paradigm crossing but it is more difficult to identify the precise strategy, although the author describes features from both the sequential strategy and the interplay. In terms of matching his approach to one of the diagrams featured in Figure 4.1, one only needs to look at the title of his paper 'Soft with a hard centre'.

The final study we review does not involve simulation but expresses views that are important to our study because of the paradigm strategy discussion. The study combined SSM with data mining and was applied to the public sector (Brown, Cooper and Pidd, 2005; Pidd, 2004a) with an the aim to modernise the UK's personal tax system. Data mining was used to obtain representative models of the UK taxpayer groups and their interaction with the current taxpayer system, and SSM was used to extract models including the features that are considered desirable and necessary in the future tax system. The methods were not deployed sequentially but in parallel; they declare their strategy as two interwoven lines which represent hard and soft methods (see Figure 4.1). Brown, Cooper and Pidd (2005) explain that 'though both the hard and the soft OR made valuable contributions to the study it was significantly enhanced by their combination' (p. 4) and those findings from each approach fed into the other.

The studies that link hard and soft approaches report clear benefits in dealing with complex situations. However, we argue that not all previous authors have adequately declared or tackled their paradigm position and it is therefore not always clear if they are describing multiparadigm multimethodology – critical pluralism (Mingers, 1997a; Mingers, 1997b; Mingers, 1999; Mingers and Brocklesby, 1996) or pragmatic pluralism (White and Taket, 1997) – (crossing or integration) or coherent pluralism (Jackson, 1999) (incommensurability). This is understandable as their work was groundbreaking and there were other, more practical and theoretical issues to tackle which have paved the way for philosophical ones. We will describe a case study that uses DES and SSM to evaluate a healthcare system for older people called Intermediate Care. We will then argue that it is a multiparadigm multimethodology and the position adopted is that of paradigm crossing with an interplay strategy.

4.5   The case study: Evaluating intermediate care

The case study that follows describes the use of SSM and DES modelling to evaluate the operational function of an intermediate care (IC) system, which is a health and social care system for older people in the UK. In the description of the following case the first author will describe her own feelings and experiences as this is clearly seen as important in applying a multimethodology.

4.5.1   The problem situation

The Department of Health (DH) had allocated a significant amount of money and resources to this area. Specifically, it pledged to invest £900 million a year by 2003–2004 in IC services (Department of Health, 2001a). The creation of IC services for older people was in order to relieve the over-utilised acute hospitals and long-term institutional care settings from those who do not benefit from these services. These would include: 'Hospital day units and community based services aimed at maintaining people in their home communities in good health, preventing avoidable admissions, facilitating early discharge and active rehabilitation post-discharge and supporting a return to normal community-based living wherever possible' (Department of Health, 2000, p. 4). The DH published guidance on service models for intermediate care (Department of Health, 2001a), which included services with specific goals of preventing hospital admissions, enabling earlier discharge from hospital by providing rehabilitation closer to home and preventing admission to long-term care.

An Elderly Strategic Planning Group and its Joint Planning Board commissioned the University of Kent in 2000 to evaluate one of their IC systems. The project team based at the Centre for Health Service Studies had four main team members including the first author who was the only person with OR/MS knowledge. The others were a health service researcher and two geriatricians. Each of the members of the team had a different role/task and the first author's task was to evaluate the operational function of the IC system, which is broadly described in this case study.

At the start of the project, the first author had recently graduated with a management science degree with little soft OR content; her practical experience in OR was a final-year project applying DES modelling in a traditional 'hard' way in a manufacturing environment, but she was eager to apply simulation modelling again in research to obtain a PhD. In fact, she joined the team because the project leader, a geriatrician, thought that simulation modelling could be useful in healthcare – especially as exploring resource requirements, a typical use of simulation modelling, is also important to health and social care managers.

Initially, there was a considerable amount of vagueness about what the operational function was and what they (stakeholders) wanted to evaluate. This vagueness is not unusual in a complex societal problem, particularly if it is a large and important real-life problem. DeTombe (2001) argues that

> complex societal problems are often ill defined or multi-defined, hard to analyse and to handle. Knowledge and data are missing or contradictory, the causes of the problem vague and it is often not clear in which direction the problem is going. Many phenomena, many parties, private and governmental are involved. The problem often has or will have a large impact on (parts of) society.
>
> *(p. 231)*

This description fits this problem accurately and can perfectly describe the starting point to this research. DeTombe mentions that these complex societal problems involve interdisciplinary study and that the methodology for handling complex societal problems is multidisciplinary. This lack of understanding and vagueness could have described IC in 2000, at the start of the project, as there was hardly any IC literature available and many of the health and social care employees did not even recognise the term 'intermediate care'.

The main problem with structuring the problem was that both the project team and the stakeholders did not understand the DH's conceptual vision of an IC system and its functions. The IC stakeholders had translated the DH's vision into practice by simply setting up the individual services but had not been able to examine the system as a whole, which is why the DH had commissioned the project. In fact one might say that they had set up services but not a system to 'hold' them together.

Part of the evaluation could be achieved through the simulation methodology but the vagueness encountered prevented an understanding of what needed modelling in that system. It was then realised that SSM did not provide any tools for extremely confusing and complex situations such as the one encountered. After some consideration of the methodologies and techniques taught (e.g. SSM, SODA/cognitive mapping and robustness analysis) the first author felt that SSM was more appropriate as it provided a much more structured approach to understanding than the other PSMs. Deciding to use SSM was not a comfortable decision as she was not experienced in applying it and, furthermore, she would have little supervisory support in using it due to the lack of expertise at that time in the academic department. It was clear, though, that a useful simulation model could not be built unless the system and its operations were understood. However, it was almost immediately evident that engaging in SSM meant that her attitude changed and the line of questioning moved from 'how does this work?' to 'how could this work?', not just because of the SSM requirements, but also in terms of the simulation modelling. In simulation modelling this line of questioning is usually left to the end of the model development phase. This new line of questioning benefited employees that previously were not able to explain the situation, as it was confusing even to them, but they nevertheless had ideas and an understanding of what could work and was desirable. The next subsection will examine the SSM approach in this investigation.

4.5.2   Soft systems methodology

Initially, the IC system was explored through interviews and stakeholder group meetings and, following Checkland (1999a), CATWOE root definitions were produced, which helped construct an activity model. The aim was to find out about the stakeholder group's operational and strategic activities to manage, coordinate and improve the systems operations. Otherwise, the only link between the services would be the target population of older people. These activities could be considered as the primary tasks (Checkland, 1999b) of the IC function and if these were missing from the real system but were considered desirable and culturally feasible, then action should be taken by the system owners to put these tasks in place. Adopting a primary task approach to SSM makes the process even more focused. One might even argue that there is a hint of the hard paradigm when applying this focused SSM, and Checkland and Scholes (1999) hint at this by saying that it 'has the advantage of providing a highly structured entry, which reassures the nervous' (p. 66). The nervous can really be none other than those that want to abolish future uncertainty and are more comfortable with hard OR (see Table 4.1).

Table 4.1   The six key characteristics of the hard and soft paradigms (Rosenhead, 1999).

Characteristics of the hard paradigm	Characteristics of the soft paradigm
Problem formulation with a single objective and optimisation. Multiple objectives, if recognised, are subject to trade-off on to a common scale	Non-optimising; seeks alternative solutions which are acceptable on separate dimensions, without trade-offs
There are overwhelming data demands, with consequent problems of distortion, data availability and data credibility	Reduced data demands, achieved by greater integration of hard and soft data with social judgements
Scientisation and depoliticisation; there is an assumption of consensus between stakeholders	Simplicity and transparency aimed at clarifying the terms of conflict
People are treated as passive objects	Conceptualises people as active subjects
There is a single decision maker with abstract objectives from which concrete actions can be deduced from implementation through a hierarchical chain of command	Facilitates planning from the bottom up
There is an attempt to abolish future uncertainty, pre-taking future decisions	Accepts uncertainty, and aims to keep options open for later resolution

Interviews and observation revealed that one of these operational activities should be about the decision-making mechanism for patient referral to each of the services in the system as each service served different rehabilitation needs. However, it was found that there was no formal decision-making process in place to allocate patients to services. All IC employees and stakeholders were keen to change this 'ad hoc' referral because patients entering the 'wrong' service for their needs is a waste of resources and it can also put the patients' rehabilitation and even life at risk. Therefore, in addition to the tasks of determining the efficiency of the IC resources, it was decided to explore the patient referral mechanism to the IC services.

This process of understanding the IC operational function through the SSM activities model also provided the evaluation tasks for the project. Of course some activities like the evaluation of the resources were evident from the start, but others like evaluation of the decision-making mechanism for referral resulted from the SSM process. Furthermore, some of the evaluation tasks could be explored in the SSM phase. For example, the DH IC literature (Department of Health, 2002) suggested that a single standardised tool should be used for patient data collection and assessment across all services and, although desirable and culturally feasible by IC employees, this was not in place in that IC system (or any other at that time). Therefore in the SSM comparison phase the current patient assessment approach was evaluated by comparing it with a different approach that was more desirable and culturally feasible than the existing one. The outcome was to recommend the different approach, which was to adopt a comprehensive and standardised assessment instrument throughout the IC system. During the SSM comparison phase many differences were found between the real world and the systems world which in itself formed part of the evaluation.

Although SSM was only initially thought to be useful to understand the system and enable the simulation modelling, it helped structure the problem into evaluation tasks and it also enabled part of the evaluation. The SSM phase also confirmed the need to explore the efficiency of the resources and that the evaluation would require more than just a comparison. This confirmed the team's original thought that simulation modelling could be useful for this evaluation. It was decided to build a simulation model for each of the services to evaluate their resources (beds/places and staff) and a 'whole system' model to explore how the decision-making mechanism of patient referral affects the services. The next subsection will discuss the use of DES in the evaluation process.

4.5.3   Discrete-event simulation modelling

The modeller in the initial stage of building a simulation model will dedicate a length of time to understanding the system of interest. The process of understanding the IC system during the SSM phase also doubled as the first step of building a simulation model. Therefore, the soft approach helped satisfy the needs for the hard paradigm method.

The simulation modelling phase for this case study can be divided into the phase of building the models of the IC services in order to assess their resources and the second phase of building a model of the system to examine the patient referral decision-making process. The first phase was straightforward because there were employees that had a good understanding of how each of the services worked and what the future might hold for these services that could be explored through 'what if?' scenarios. This meant that the simulation models of the services were built adopting the traditional hard paradigm approach.

However, the second phase had no IC employee(s) or stakeholder(s) with an in-depth understanding of the IC system, so this second phase was explored more analytically as simulation modelling was not approached in the traditional manner. As mentioned earlier, the SSM phase revealed that there was no formal decision-making mechanism in place for patients to be sent to the most appropriate service, but patients were referred in an ad hoc manner, which did not benefit the services or the patient and was not in line with DH thinking (Department of Health, 2001b). Therefore, if a model of the actual situation were built it would not actually benefit anyone, because everyone was aware of the resulting problems, particularly of inappropriate patients entering the services. The team decided to build a simulation model that did not depict the actual situation but gave the most agreeable (desirable and culturally feasible) conceptual view of the patient referral decision-making process. This would allow further exploration and evaluation of this view in a virtual environment. The remaining methodology steps after obtaining the conceptual model followed the hard paradigm. This meant that the team had moved from initially approaching simulation modelling with the hard paradigm (of the services) to applying simulation modelling under the soft and hard OR paradigms for the system model. Therefore, the simulation methodology was used with a mixture of the soft and hard paradigms.

4.5.4   Multimethodology

The SSM and DES methodology were not applied consecutively but were both done concurrently, although some steps exceeded others. They both fed into each other at different times and each methodology enriched the findings of the other. For example, one of the SSM findings revealed that a vital operation in the system was to determine Service Entry Eligibility Criteria (Kotiadis, Carpenter and Mackenzie, 2004), which are essentially tailor-made rules for each service to ensure that only suitable older people at a physical, mental and social level are admitted. However, it was not possible to evaluate these with SSM, though they could be evaluated using the simulation models. By adding this finding to the simulation models the team were able to build models that answered questions that were more important to the stakeholders. It is undoubtedly useful to determine the utilisation of a service but even more useful to determine it while also knowing the proportion of the people admitted that are inappropriate to that service. The simulation model revealed that many of the services in that system admitted patients that did not meet all the Service Entry Eligibility Criteria, which was fed back into the SSM process in order to decide on what desirable and culturally feasible action to take. To sum up, within the simulation methodology the soft paradigm fed into the hard paradigm and vice versa. Without doubt, the project would not have been possible with just one of the two methodologies or with just one of the two paradigms.

Despite the positive impact that each methodology had on the other, the multi-methodology was not without its problems. The procedure was 'painful' for the first author and at first the outcome of the multimethodology was largely uncertain in case the theoretical underpinnings of the simulation model in particular were flawed with this non-standard approach to the methodology. Furthermore, even if the primary task version of SSM was applied for the 'nervous' it was still difficult to change an attitude developed over a lifetime of positivist training in such a short period of time. Many factors enhanced the level of pain, such as providing the stakeholders with value for their money, the enormity of the impact that false results could have on this real system that provided a lifeline to many older people, and so on. This experience fits with the description by Brocklesby (1995) of paradigm shifts as a painful experience. The validation and verification of the simulation model and the acceptance of the findings by the stakeholders from both the SSM and simulation model brought some relief that at the very least the project has helped them gain a better understanding of their IC system. In fact the project achieved much more than that because it underpinned the region's IC strategy and the findings were used to introduce new services to the region. Furthermore, a one-day workshop was organised by the stakeholders to present the system simulation model and its findings not only to the participants in the study, but also to all service providers of other localities within the stakeholders' jurisdiction. It was generally thought that the new conceptual vision of how the system could work was best communicated visually and in the spirit of SSM it brought about acceptance of the region's IC strategy.

4.6   Discussion

We will now contribute our learning about combining SSM with DES with respect to the various issues that we explored earlier in the chapter. These are the paradigm position and strategy, the cognitive and cultural difficulties affecting multiparadigm work. We have not dedicated a specific section to practical difficulties as they emerge in the others that we explore.

4.6.1   The multiparadigm multimethodology position and strategy

Lehaney, Clarke and Paul (1999) explored paradigm incommensurability by discussing whether SSM and simulation as they applied them could be considered as different approaches and concluded that in their case the notion of different paradigms is inappropriate. Similarly, Lane and Oliva (1998) also expressed paradigm integration beliefs about the synthesis of SSM with SD. In our case study a multimethodology was used and both paradigms were adopted at different stages of the project. The paradigm approach to this multimethodology can be explained with the paradigm interplay strategy. The process of obtaining our conceptual model for the simulation methodology required the soft paradigm, which is an example of cross-fertilisation of the paradigms while accepting their diversity. Without doubt this is an example of the interplay strategy, although this was not something recognised at the time or even something that the researcher strived for.

Interestingly, we can deduce that the simulation method (DES or SD) combined with SSM does not necessarily lead to similar views on the paradigm positioning. Our study is more aligned to the view of Paucar-Caceres and Rodriguez-Ulloa (2007) who also put forward paradigm crossing.

In looking retrospectively on our experience and trying to make sense of it we have found it helpful to communicate our multiparadigm approach using the 'yin yang' symbol which represents two opposing elements of the Universe and is used to explain 'how things work' (Figure 4.2). In fact Zhu (2001) also uses the 'yin yang' metaphor to discuss information systems design at the philosophical level. The outer circle represents 'everything', while the black and white shapes within the circle represent the interaction of two energies, called 'yin' (black) and 'yang' (white), which cause everything to happen. They are not completely black or white, just as things in life are not completely black or white, and they cannot exist without each other. The soft and hard paradigms in OR can be considered as the two opposite philosophical principles (Table 4.1) 'yin' (hard) and 'yang' (soft), and the outer circle can be considered as the problematic situation (Figure 4.3). It should be noted that the colour of the outer circle is irrelevant and does not mean that the problematic situation is hard.

Figure 4.2   The yin and yang symbol.

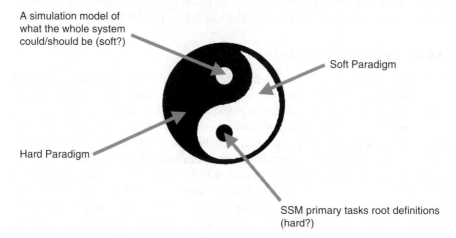

Figure 4.3   The soft *and* hard *paradigms in the intermediate care case study.*

In our study there was an element of the hard paradigm (yin) when SSM was applied (primary task approach to SSM) and an element of the soft paradigm (yang) during the simulation modelling (building a simulation model of the most agreeable patient referral mechanism). Therefore, unlike Lehaney, Clarke and Paul (1999) the notion of different paradigms in this case is appropriate and furthermore the two methods in question could not be applied just with one paradigm and satisfy the objective of the study. Although both of us used SSM and DES in healthcare, there are considerable differences in our application areas and objectives that might have warranted such a different approach and opinion about our paradigm positioning. The differences in terms of the application area include the scale of our system (we modelled the interaction of several independent healthcare services versus one outpatient department) and the larger number of stakeholders and participants in our study, which also influenced our objectives. It is also possible that we might have come to the same conclusions had we tackled each other's problem. We propose that in fact the requirements of a problematic situation might dictate the paradigm strategy rather than the methodologies used, which might explain why we share more similarities with the multiparadigm approach of Brown, Cooper and Pidd (2005). However, we believe that the 'yin yang' symbol is a richer representation of the muliparadigm multimethodology process.

This study shows that multiparadigm multimethodology is possible and was largely motivated by the need to tackle the problematic situation. We also believe that learning about the philosophical approach is important and that the 'journey' (the paradigm strategy) is just as important as the 'destination' (the outcome). Therefore, although the approach to the project shares characteristics of the pragmatic pluralism (White and Taket, 1997) approach we are not ignoring the philosophical issue, which therefore makes it closest to critical pluralism (Mingers, 1997b).

4.6.2 The cultural difficulties

The study was 'advertised' as a simulation study and not a multimethodology of SSM and DES for two main reasons. First, we knew from the start that given the questions about resource requirements simulation would be used but the need to use SSM came a few months later when the project was underway. Therefore, in the initial meetings and presentations organised, only simulation modelling was explained to the participants. The second reason, though, can be attributed to the culture in healthcare. The simulation modelling approach generated a lot of interest and was perceived by stakeholders as a new, exciting scientific approach to explore their whole system, which they knew would not be possible using the traditional community healthcare approaches, qualitative studies or randomised control trials. Explaining simulation to more than 40 healthcare professionals and stakeholders was a challenge and a priority in order to get as many 'on board', particularly as the data collection was a huge expense and undertaking on their part. SSM was initially considered secondary to the simulation modelling and was not explained or presented to the participants, with the exception of the key stakeholders, mainly because there was not enough time, even though over the course of the study they inadvertently provided valuable information for it. In addition, conducting studies in healthcare means that communication requires the investigator to understand a new language and attitude in healthcare systems, and the feeling is very much the same on their (healthcare stakeholders') part when explaining OR to them. The very valuable time that they offered was better spent understanding and extracting information about the system rather than explaining the merits of SSM.

4.6.3 The cognitive difficulties

There are a number of issues mentioned in the first section of this chapter which we will reflect on: Mingers (1997b), Brocklesby (1997) and Ormerod (2001) consider the actor's personality, feelings and competences when working across two paradigms and we will discuss whether these affected this study. We will initially consider the impact of the agent's personality. The first author, the agent for the intervention, might be considered as an 'analytical scientist' type if her previous hard OR experience was taken into account, but it is also possible that she is a 'particular humanist' type that has learnt to apply hard OR successfully but would be better at soft OR – particularly when one considers that as a female she is predisposed to communication skills (Sojka and Tansuhaj, 1997) which are essential to the SSM process. In addition, the majority of this healthcare system's employees and stakeholders are female, which might have made communication even easier than if it were a male-dominated environment. The reasons for making such a suggestion are the following two points. First, men and women are thought to communicate differently because they listen for different information; for example, a woman might concentrate more on the feelings projected during a conversation and a man more on the facts (Shakeshaft, Nowell and Perry, 1991). Second, research has shown that there might be some discomfort when communicating with a member of the opposite sex (Shakeshaft, 1987). Differences in communication between males and females such as the conversation style, the non-verbal communication, perceptions (Hartley, 1996) and the fact that women are better at one of the most important elements of communication, which is listening, because they do not interrupt as much as men (Ellis and Beattie, 1986), could have a drastic affect on such studies (Bergvall, 1999). Although the analyst's gender may have had an effect on the SSM we are not able determine objectively if the analyst is really an analytical scientist and consequently whether that affected (positively or negatively) her work across the two paradigms.

The second consideration is the agent's feelings during the intervention about working across two paradigms. Brocklesby (1997) suggests that we must question the level of discomfort that the agent felt during the paradigm shifts. The analyst in our case study felt extreme and long-lasting stress and anxiety about applying the methods in a different way to the one taught during her training and written about in books. However, the analyst had very little experience of real-life projects and of course that must have contributed to the level of discomfort, which brings us to the third issue, namely the agent's competence. As experience brings competence, Ormerod (1997) may be correct in suggesting that this is an issue when working across methods and paradigms. However, despite the analyst's feelings and lack of experience, the project was completed on time and produced results that have been used by the stakeholders in their strategy and decision making. Although we might link feelings to experience and experience to competence, competence is not just about experience and other competences should be explored.

Psychology and organisational behaviour might be useful in providing a direction for selecting the competences to explore. For example, organisational behaviourists believe that personality is a dynamic interaction of a set of traits which include gender, abilities, physique, motivation, attitudes and perception with adult experiences at work (achievements, roles, working experiences) and early development experiences (social, family, culture) (Mullins, 1993). In our intervention a personal competence and early development experience that may have positively affected the process is that the agent is bilingual and therefore accustomed to moving from one language and culture to another, which is a similar concept to moving between the hard and soft paradigm. It may be possible that such skill is transferable and that if one can master the ability to learn and unlearn new rules, just as Brocklesby (1997) suggests, one might be able to learn to move between paradigms.

4.7   Conclusions

This chapter has considered some of the barriers to multiparadigm multimethodology such as paradigm incommensurability, cognitive style and difficulties of switching paradigms. It has done this in the context of an actual application – a combination of SSM and DES within intermediate care. The main conclusions from the success of this application are: (i) That paradigm commensurability is not a barrier to such combinations. Indeed, the application was an example of a particular strategy for working across paradigms – what has been characterised as the 'interplay' strategy. (ii) That the notion of yin and yang is an appropriate metaphor for this approach. Overall, the multimethodology combined a hard method with a soft method but, as well as this, within the hard there was some soft and vice versa. (iii) That one's past experience and training as well as personality and perhaps gender all affect one's approach to interventions in general and multimethodology in particular. They do not, however, prevent movement towards being able to work successfully across paradigms.

A note of caution is that although some of the above conclusions have resulted from the study of a particular combination of SSM with DES, the insights on paradigm positioning may not be relevant to all simulation studies (DES and SD). Indeed we have found SD and DES combinations with a PSM, SSM in particular, that align themselves with the position of paradigm integration. More research in this area would benefit both the DES and SD communities.

Acknowledgements

Chapter text, figures and tables are reproduced by permission of Palgrave Macmillan from the article: Kotiadis, K. and Mingers, J. (2006) Combining PSMs with hard or methods: the philosophical and practical challenges. *Journal of the Operational Research Society*, **57**(7), 856–867.

References

Archer, M., Bhaskar, R., Collier, A. *et al.* (1998) *Critical Realism: Essential Readings*, Routledge, London.

Bennett, L.M. and Kerr, M.A. (1996) A systems approach to the implementation of total quality management. *Total Quality Management*, **7**(6), 631–665.

Bergvall, V. (1999) Towards a comprehensive theory of language and gender. *Language in Society*, **28**(2), 273–293.

Bhaskar, R. (1978) *A Realist Theory of Science*, Harvester Press, Hemel Hempstead.

Brocklesby, J. (1995) Intervening in the cultural constitution of systems – methodological complementarism and other visions for systems science. *Journal of the Operational Research Society*, **46**(11), 1285–1298.

Brocklesby, J. (1997) Becoming multimethodology literature: an assessment of the cognitive difficulties of working across paradigms, in *Multimethodology: The Theory and Practice of Combining Management Science Methodologies* (eds J. Mingers and A. Gill), John Wiley & Sons, Ltd, Chichester, pp. 189–216.

Brown, J., Cooper, C. and Pidd, M. (2005) A taxing problem: the complementary use of hard and soft OR in the public sector. *European Journal of Operational Research*, **172**(2), 666–679.

Burrell, G. and Morgan, G. (1979) *Sociological Paradigms and Organisational Analysis*, Heinemann, London.

Checkland, P. (1999a) *Soft Systems Methodology: A 30-year Retrospective. Systems Thinking, Systems Practice*, 2nd edn (ed. P. Checkland), John Wiley & Sons, Ltd, Chichester, pp. A1–A66.

Checkland, P. (1999b) *Systems Thinking, Systems Practice: Includes a 30-Year Retrospective*, John Wiley & Sons, Ltd, Chichester.

Checkland, P. and Scholes, J. (1999) *Soft Systems Methodology in Action: Includes a 30 Year Retrospective*, John Wiley and Sons, Ltd, Chichester.

Coyle, R. and Alexander, M. (1997) Two approaches to qualitative modelling of a nation's drug trade. *System Dynamics Review*, **13**, 205–222.

Department of Health (2000) *Shaping the Future NHS: Long Term Planning for Hospitals and Related Services*, Department of Health, London.

Department of Health (2001a) *Intermediate Care*, HSC 2001/01:LAC(2001)1, Department of Health, London.

Department of Health (2001b) *National Service Framework for Older People*, LAC (2001)12, Department of Health, London.

Department of Health (2002) *Guidance on the Single Assessment Process*, HSC 2002/001, Department of Health, London.

DeTombe, D.J. (2001) Introduction to the field of methodology for handling complex societal problems. *European Journal of Operational Research*, **128**, 231–232.

Doyle, K.G. (1990) Modelling integrated information systems for institutions of higher education. MPhil thesis. Bristol Polytechnic.

Ellis, A. and Beattie, G. (1986) *The Psychology of Language and Communication*, Weidenfeld & Nicolson, London.

Fleetwood, S. and Ackroyd, S. (2004) *Critical Realism in Action in Organizations and Management Studies*, Routledge, London.

Gioia, D. and Pitre, E. (1990) Multiparadigm perspectives on theory building. *Academy of Management Review*, **15**(4), 584–602.

Hartley, P. (1996) *Interpersonal Communication*, Routledge, London.

Howick, S. and Ackermann, F. (2011) Mixing OR methods in practice: past, present and future directions. *European Journal of Operational Research*, **215**(3), 503–511.

Jackson, M. (1999) Towards coherent pluralism in management science. *Journal of the Operational Research Society*, **50**(1), 12–22.

Kotiadis, K. (2007) Using soft systems methodology to determine the simulation objectives. *Journal of Simulation*, **1**, 65–66.

Kotiadis, K., Carpenter, G.I. and Mackenzie, M. (2004) Examining the effectiveness and suitability of referral and assessment in Intermediate Care services. *Journal of Integrated Care*, **12**(4), 42–48.

Kuhn, T. (1970) *The Structure of Scientific Revolutions*, Chicage University Press, Chicago.

Lane, D. and Oliva, R. (1998) The greater whole: towards a synthesis of system dynamics and soft systems methodology. *European Journal of Operational Research*, **107**, 214–235.

Lehaney, B., Clarke, S.A. and Paul, R.J. (1999) A case of an intervention in an outpatients department. *Journal of the Operational Research Society*, **50**, 877–891.

Lehaney, B. and Hlupic, V. (1995) Simulation modelling for resource allocation and planning in the health sector. *Journal of the Royal Society of Health*, **115**, 382–385.

Lehaney, B. and Paul, R.J. (1996) The use of soft systems methodology in the development of a simulation of out-patients services at Watford General Hospital. *Journal of the Operational Research Society*, **47**, 864–870.

Midgley, G. (1997a) Developing the methodology of TSI: from oblique use of methods to creative design. *Systems Practice*, **10**(3), 305–319.

Midgley, G. (1997b) Mixing methods: developing systemic intervention, in *Multimethodology: The Theory and Practice of Combining Management Science Methodologies* (eds J. Mingers and A. Gill), John Wiley & Sons, Ltd, Chichester, pp. 250–290.

Mingers, J. (1997a) Critical pluralism and multimethodology, post postmodernism, in *Systems for Sustainability: People, Organizations, and Environments* (eds F. Stowell, R. Ison, R. Armson*et al.*), Plenum Press, New York, pp. 345–352.

Mingers, J. (1997b) Towards critical pluralism, in *Multimethodology: Theory and Practice of Combining Management Science Methodologies* (ed. A. Gill), John Wiley & Sons, Chichester, pp. 407–440.

Mingers, J. (1999) Synthesising constructivism and critical realism: towards critical pluralism, in *World Views and the Problem of Synthesis* (eds E. Mathijs, J. Van der Veken and H. Van Belle), Kluwer Academic, Amsterdam, pp. 187–204.

Mingers, J. (2000a) The contribution of critical realism as an underpinning philosophy for OR/MS and systems. *Journal of the Operational Research Society*, **51**(11), 1256–1270.

Mingers, J. (2000b) Variety is the spice of life: combining soft and hard OR/MS methods. *International Transactions in Operational Research*, **7**, 673–691.

Mingers, J. (2001) Combining IS research methods: towards a pluralist methodology. *Information Systems Research*, **12**(3), 240–259.

Mingers, J. (2002) Multimethodology – mixing and matching methods, in *Rational Analysis for a Problematic World Revisited* (eds J. Rosenhead and J. Mingers), John Wiley and Sons, Ltd, Chichester, pp. 289–309.

Mingers, J. (2004) Real-izing information systems: critical realism as an underpinning philosophy for information systems. *Information and Organization*, **14**(2), 87–103.

Mingers, J. and Brocklesby, J. (1996) Multimethodology: towards a framework for critical pluralism. *Systemist*, **18**(3), 101–132.

Mingers, J. and Brocklesby, J. (1997) Multimethodology: towards a framework for mixing methodologies. *Omega*, **25**(5), 489–509.

Mingers, J. and Rosenhead, J. (2004) Problem structuring methods in action. *European Journal of Operational Research*, **152**(3), 530–554.

Mingers, J. and White, L. (2010) A review of the recent contribution of systems thinking to operational research and management science. *European Journal of Operational Research*, **207**(3), 1147–1161.

Mullins, L.J. (1993) *Management and Organisational Behaviour*, Pitman, London.

Munro, I. and Mingers, J. (2002) The use of multimethodology in practice – results of a survey of practitioners. *Journal of the Operational Research Society*, **59**(4), 369–378.

Ormerod, R. (1997) Mixing methods in practice: a transformation-competence perspective, in *Multimethodology: Theory and Practice of Combining Management Science Methodologies* (eds J. Mingers and A. Gill), John Wiley & Sons, Ltd, Chichester, pp. 29–58.

Ormerod, R. (2001) Mixing methods in practice, *Rational Analysis for a Problematic World Revisited* (eds J. Rosenhead and J. Mingers), John Wiley & Sons, Ltd, Chichester, pp. 289–310.

Paucar-Caceres, A. and Rodriguez-Ulloa, R. (2007) An application of soft systems dynamics methodology (SSDM). *Journal of the Operational Research Society*, **58**(6), 701–713.

Pauley, G. and Ormerod, R. (1998) The evolution of a performance measurement project at RTZ. *Interfaces*, **28**, 94–118.

Pfeffer, J. (1993) Barriers to the advance of organizational science: paradigm development as a dependent variable. *Academy of Management Review*, **18**(4), 599–620.

Pidd, M. (2004a) Bringing it all together, in *Systems Modelling: Theory and Practice* (ed. M. Pidd), John Wiley & Sons, Ltd, Chichester, pp. 197–207.

Pidd, M. (2004b) Complementarity in Systems Modelling, in *Systems Modelling: Theory and Practice* (ed. M. Pidd), John Wiley & Sons, Ltd, Chichester, pp. 1–19.

Robinson, S. (2001) Soft with a hard centre: a discrete event simulation in facilitation. *Journal of the Operational Research Society*, **52**, 905–915.

Rosenhead, J. (ed.) (1989) *Rational Analysis for a Problematic World*, John Wiley & Sons, Ltd, Chichester.

Rosenhead, J. and Mingers, J. (eds) (2001) *Rational Analysis for a Problematic World Revisited*, John Wiley & Sons, Ltd, Chichester.

Schultz, M. and Hatch, M.J. (1996) Living with multiple paradigms: the case of paradigm interplay in organisational culture studies. *Academy of Management Review*, **21**(2), 529–557.

Shakeshaft, C. (1987) *Women in Educational Administration*, Sage, Newbury Park, CA.

Shakeshaft, C., Nowell, I. and Perry, A. (1991) Gender and supervision. *Theory into Practice*, **30**(2), 134–139.

Sojka, J.Z. and Tansuhaj, P. (1997) Exploring communication differences between women and men sales representatives in a relationship selling context. *Journal of Marketing Communications*, **3**(4), 197–216.

Tashakkori, A. and Teddlie, C. (1998) *Mixed Methodology: Combining Qualitative and Quantitative Approaches*, Sage, Thousand Oaks, CA.

Varela, F., Thompson, E. and Rosch, E. (1991) *The Embodied Mind*, MIT Press, Cambridge, MA.

White, L. and Taket, A. (1997) Critiquing multimethodology as metamethodology: working towards pragmatic pluralism, in *Multimethodology: The Theory and Practice of Combining Management Science Methodologies* (eds J. Mingers and A. Gill), John Wiley & Sons, Ltd, Chichester, pp. 379–407.

Willmott, H. (1993) Breaking the paradigm mentality. *Organizational Studies*, **15**(5), 681–719.

Zhu, Z. (2001) Towards an integrating programme for information systems design: an oriental case. *International Journal of Information Management*, **21**, 69–90.

5

Philosophical positioning of discrete-event simulation and system dynamics as management science tools for process systems: A critical realist perspective

Kristian Rotaru,[1] **Leonid Churilov**[2] **and Andrew Flitman**[3]

[1]*Department of Accounting, Monash University, Melbourne, Victoria, Australia*
[2]*Florey Institute of Neuroscience and Mental Health, Melbourne, Australia; RMIT University, Melbourne, Victoria, Australia*
[3]*Florey Institute of Neuroscience and Mental Health, Melbourne, Victoria, Australia*

5.1 Introduction

Management science (MS) has historically developed as a scientific approach to analysing management problems and making management decisions, and is distinct from other disciplines by applying scientific principles in the context of practical management decision making. In line with Kuhn (1970) and Meadows (1980), a

Discrete-Event Simulation and System Dynamics for Management Decision Making, First Edition.
Edited by Sally Brailsford, Leonid Churilov and Brian Dangerfield.
© 2014 John Wiley & Sons, Ltd. Published 2014 by John Wiley & Sons, Ltd.

scientific discipline is traditionally supported by a set of explicitly formulated assumptions underlying its approach to the main phenomena under investigation. These assumptions, in particular, encapsulate ontological and epistemological bases of a given scientific discipline, that is postulate 'the way human beings comprehend knowledge about what is perceived to exist' (Becker and Niehaves, 2007, p. 201; Burrell and Morgan, 1979). The main phenomenon of investigation for MS is management decision making. MS achieves its aims through the use of scientific methodologies. Explicit articulation of the fundamental philosophical assumptions underlying MS methodologies is therefore an important requirement for further development of MS as a scientific management decision-making discipline. Buchanan, Henig and Henig (1998) concisely express this requirement as: 'Our convictions about the nature of the world . . . should indeed be made explicit as a necessary prerequisite for any proposed decision making methodology' (p. 343).

The importance of adequate philosophical positioning for the science of management has been traditionally recognised, in particular, on the pages of *Management Science* journal. This tradition goes back to the first Editor-in-Chief of *Management Science*, W. Churchman, who believed that: 'Philosophy should be used to study serious problems like war, security, and human living' (Churchman, 1994). Mitroff (1994) underlines direct dependency of the quality of management theories upon the quality of underlying philosophical notions. The importance of philosophical debates in MS was emphasised at different times by Churchman (1955), Mitroff (1972) and Hopp (2008).

Over the years simulation modelling has become clearly recognised as an effective and robust part of the MS toolkit widely used in management practice (Altinel and Ulas, 1996; Dittus *et al.*, 1996; Lehaney, Malindzak and Khan, 2008; Brailsford *et al.*, 2004; Peña-Mora *et al.*, 2008). One of the most successful areas of application of simulation modelling is providing scientific tools for management thinking and decision support for process systems (Van Horn, 1971; Doomun and Jungum, 2008; Gorunescu, McClean and Millard, 2002; Haraden and Resar, 2004; Pidd, 2003a). Simulation methods have been applied in virtually every process industry – from manufacturing production lines and logistics networks to call centres, air traffic control and patient flows in hospitals (Lehaney, Malindzak and Khan, 2008; Doomun and Jungum, 2008; Brailsford *et al.*, 2004; Greasley, 2005). Simulation of a process system involves building a valid model of such a system and, subsequently, using this model in order to gain insight into the system's functioning under alternative conditions and courses of action, thus providing scientific support for management decision-making activities.

Depending on the management decision support needs, different simulation approaches are used, potentially resulting in several alternative models for a given decision-making situation (Rohleder, Bischak and Baskin, 2007; Morecroft and Robinson, 2005; Karpov, Ivanovsky and Sotnikov, 2007; Popkov and Garifullin, 2006). In addition to the published research on the application of individual simulation approaches, there is a growing body of literature that explores the possibilities and implications for providing management decision support based on more than one simulation approach (Venkateswaran and Son, 2005; Lorenz and Jost, 2006). In the context of process systems, the existing MS literature mainly explores the possibilities of the individual, combined (Tako and Robinson, 2008, 2009; Brailsford and Hilton, 2001; Morecroft and Robinson, 2005; Lane, 2000) or even integrated use (Peña-Mora et al., 2008; Coyle, 1985; Barton and Tobias, 2000; Borshchev, Karpov and Kharitonov, 2002; Brailsford, Churilov and Liew, 2003) of discrete-event simulation (DES) and system dynamics (SD) modelling approaches.

In line with the inherent subjectivity of the very notion of a model (e.g. defined by Pidd (2003b, p. 12) as 'an external explicit representation of part of reality *as seen by people* who wish to use that model to understand, to change, to manage and to control that part of reality [emphasis added]'), any given simulation approach assumes 'a particular worldview that prescribes how a model should be developed and expressed' (Pidd, 2003a, p. 77). These 'worldviews' are also referred to in the literature as modelling philosophies (Morecroft and Robinson, 2005) or modelling paradigms (Meadows, 1980; Lorenz and Jost, 2006). The idea of simulation worldviews emerged from intelligent thinking about practical management experiences (Pidd, 2004a, p. 2). This echoes a more general observation by Buchanan, Henig and Henig (1998) who emphasise that 'the worldview that we hold determines the process we will advocate for solving a decision problem' (p. 343).

The MS literature includes quite a number of contributions illustrating how a combination of a particular simulation engine and a corresponding worldview can be applied to facilitate management decision support in process systems (e.g. Borshchev, Karpov and Kharitonov, 2002; Popkov and Garifullin, 2006). Unfortunately, much less attention has been paid historically to the scientific philosophical assumptions underlying the use of respective simulation worldviews. As early as in 1980, Meadows (1980) noted that these assumptions are often implicit and made by modellers on the subconscious level. Almost 30 years later, despite the recognised need to make explicit the fundamental assumptions underlying the use of simulation worldviews by management scientists (Pidd, 2004a) in order to effectively apply different simulation approaches to support managers in recognising and solving specific decision-making problems (Morecroft and Robinson, 2005), there is still a clear gap in the MS literature symptomatic of a very limited understanding of fundamental assumptions underlying various simulation worldviews.

In particular, this leads to the situation where the systematic philosophical position-ing of DES and SD remains largely unexplored. The relevant literature is scarce. Specifically, Lane (1999) maps a range of diverse SD research streams into a four-quadrant framework proposed by Burrell and Morgan (1979) based on the interaction between social science and society, thus depicting four paradigms for the analysis of social theory and philosophical schools that correspond to such social theories, but not explicitly addressing epistemological foundations of SD. Mingers (2003) provides a high-level epistemological framework for characterising the philosophical assumptions underlying OR/MS methodologies and techniques, including, but not specifically targeting, DES and SD. When it comes to DES, Morecroft and Robinson (2005) compare the advances of the SD 'philosophy of practice' (based on Lane, 1999) with the current advances in DES, concluding that while such a philosophical position does not yet exist in DES, it might be beneficial for extending the scope of DES.

Overall, lack of research into philosophical assumptions underlying the use of respective simulation worldviews precludes management scientists from using simulation methodology in a truly scientific manner, thus going to the core of MS as a scientific discipline that is aimed at supporting managers in analysing management problems and making management decisions.

This chapter sets out to contribute to MS knowledge by explicitly examining the main philosophical assumptions underlying the DES and SD simulation worldviews while utilising practical simulation experience in process systems as described in the relevant body of the published literature. More specifically,

> the objective of this chapter is to investigate how critical realist philosophy of science (Bhaskar, 1978, 1979) facilitates explicit articulation of the fundamental philosophical assumptions underlying system dynamics and discrete-event simulation worldviews, thus contributing to more effective use of simulation for intelligent thinking about, and management decision support in, process systems.

The novelty and original contribution of this research is in using the elegance and power of critical realism as a philosophy of science, and, in particular, critical realist stratified ontology and abductive mode of knowledge generation, to examine explicitly the philosophical bases of DES and SD simulation worldviews. The outcomes of this research are targeted at both the manager, who is the contributor to, as well as the end user of, the simulation model of a real-world process system and, as such, could benefit from clear understanding of how management knowledge is generated through the modelling process, and the management scientist who chooses to use simulation modelling to support management decision making in real-world process systems and requires an in-depth understanding of the scientific bases of respective modelling methodologies to apply it in a truly scientific manner.

Three observations regarding the *scope* of this study should be made explicit at the outset. First, the differences between the study by Mingers (2003) and our study are that the former is a broad review covering the entire spectrum of operational research methodologies, while the latter is specifically targeted towards simulation methodologies and is using critical realism to address their foundations in the context of process systems. Second, this chapter supports a management scientist by focusing on simulation worldviews and treating all the issues of simulation engines, however important they might be, beyond its scope. Third, while 'any system that can be modelled by one approach can also be expressed in one of the others' (Pidd, 2003a, p. 77), resolution of the popular argument regarding the *potential* use of DES in place of SD (or vice versa) is well beyond the scope of this chapter. When analysing both simulation modelling methodologies, our analysis is based on the findings reported in the literature on *what simulation methodology was used for a specific purpose*, rather than *what it could have been used for*. In this sense, this study is driven by the empirical evidence provided in the literature that explores the use of DES and/or SD modelling methodologies as MS tools for intelligent thinking and management decision support in process systems.

The remaining part of this chapter is organised as follows. In Section 5.2 stratified ontology and abductive modes of knowledge generation are introduced as the bases of critical realism (CR). Section 5.3 presents process system modelling with SD and DES through the prism of CR scientific positioning by superimposing the outcomes of each phase of a relevant simulation modelling lifecycle upon the stratified ontological representation of CR. Section 5.4 uses patient care process simulation with SD and DES as an example of a complex process system modelling activity and specifically explores the relevant MS literature through the ontological/epistemological lens of CR; this section also provides the analysis of apparent process system simulation modelling trends and formulates subsequent implications for MS from the adopted CR ontological/epistemological perspective. Finally, summary and conclusions are presented in Section 5.5.

5.2   Ontological and epistemological assumptions of CR

CR is a philosophy of science that is most commonly associated with the work of Bhaskar (1978, 1979, 1986, 1989). Epistemology, or theory of knowledge, is concerned with the nature and scope of knowledge, in particular addressing the important question of *how knowledge is acquired* (Meredith, 2001; Becker and Niehaves, 2007; Burrell and Morgan, 1979). Epistemological foundations of a modelling method cannot be considered in isolation from the *ontological assumptions about the nature of existence* (Mingers and Brocklesby, 1997; Meredith, 2001). Thus, for the purposes of this discussion, two fundamental elements of CR scientific positioning are of major relevance: *the stratified CR ontology* and *CR epistemology*, primarily presented by the *abductive mode of reasoning*.

5.2.1   The stratified CR ontology

The stratified CR ontology divides reality into three distinct 'domains of being' (Bhaskar, 1979) that are commonly referred to as the domains of *real, actual* and *empirical* (Danermark *et al.*, 2002; Mingers, 2000, 2006a, 2006b).

The domain of *real* consists of underlying generative mechanisms and causal structures activated by these mechanisms. In the context of process system simulation, the causality underlying the behaviour of a process is not necessarily a reflection of the succession of distinct causes and effects, but rather an emergent property of the complex interaction between the agency and technical components of an open socio-technical process system. Taken from this perspective, the 'cause thus has an ontological depth' (Downward and Mearman, 2007, p. 88). For SD simulation, this perspective on causality results in the idea of systems archetypes (Mingers, 2000), generic structures (Lane, 1998), or recurring structures (Morecroft and Robinson, 2005).

The activation of the causal powers of the generative mechanisms triggers patterns of events and behaviours that reside in the ontological domain of *actual* and represent the *universum of all* events and behaviours that could be possibly produced by causal structures and generative mechanisms. In the context of process simulation, the ontological domain of *actual* is represented by the set of *all* possible simulation scenarios of a given process.

The domain of *empirical* is made up of those events and behaviours that are actually experienced or observed by humans. In particular, it is the domain where the empirical data is available to validate the simulation models.

Contrary to the realist philosophy of science grounded in the positivist/empiricist tradition (Meredith, 2001), the CR stratified ontology argues that data-free conceptualisation of the problem behaviour of a system is ontologically separated from the experienced/perceived behaviour of the system. This ontological gap explains the difficulty of building realistic conceptual models based on the limited and personalised view of the real world. The abductive mode of reasoning which is at the core of the CR epistemological position provides the link between the knowledge acquired at the empirical and real levels of the CR ontology.

5.2.2   The abductive mode of reasoning

The general direction of scientific generalisation adopted by CR is from the available empirical data to the postulation of 'actual events' and 'real causes' (Downward and Mearman, 2007, p. 93). CR regards empirical knowledge as antecedent for generating the new knowledge, that is the knowledge about generative mechanisms and the *universum* of all possible events and behaviours generated by generative mechanisms' actualised causal powers. This position has been well articulated by Lawson (1998, p. 156):

> The aim is not to cover a phenomenon under a generalization . . . but to identify a factor responsible for it, that helped produce, or at least facilitated, it. The goal is to posit a mechanism (typically at a different level to the phenomenon being explained) which, if existed and acted in the postulated manner, could account for the phenomenon singled out for explanation.

In CR this type of scientific generalisation is supported by a mode of inferential logic referred to as *abduction*, also known as *retroduction* (Collier, 1994; Lawson, 1998; Mingers, 2000; Danermark *et al.*, 2002; Downward and Mearman, 2007). Abduction aims to obtain knowledge about 'what properties are required for a phenomenon to exist' (Danermark *et al.*, 2002, p. 206). Mingers (2000, 2006b) provides a more detailed account of abduction (referring to it as retroduction) as the *realist method of science* in which cognition about the phenomenon moves from the often limited empirical data about the perceived behaviour of the phenomenon to the postulation and testing of the generative mechanism(s) that, if it (they) existed, would cause this behaviour to be expressed empirically.

Bhaskar (1994) suggests a method of elicitation of generative mechanisms based on the abductive mode of reasoning mostly suitable for scientific generalisation in the context of applied science including, in particular, MS. This approach (*RRREI*) is based on the following steps: *Resolution, Redescription, Retroduction, Elimination* and *Identification*. In this study, the *RRREI* approach is used to direct the abductive mode of logical inference that allows the acquisition of knowledge about the generative mechanisms that underlie the behaviour of the process systems being modelled. According to *RRREI*, the abductive process starts by classifying the empirical data available about the process or phenomenon under study. The general understanding about how the process/phenomenon is shaped is required at this stage (*Resolution*) (Mingers, 2000, 2006b). This is followed by the representation of the empirical data through the prism of a particular theory (*Redescription*) (Mingers, 2000, 2006b). Based on these empirically viable results, a 'creative model' of the possible generative mechanism leads to a new supposition about the phenomenon under generalisation, that is the process system's behaviour. Herein the reasoning moves from the acquired empirical event data to the postulation of the underlying generative mechanisms, which, if they existed, would causally generate the given empirical data (*Retroduction*) (Collier, 1994; Mingers, 2000; Downward and Mearman, 2007; Danermark *et al.*, 2002). Next, the existence of generative mechanism(s) is demonstrated by: first, isolating the hypothetical generative mechanism; and, second, eliminating alternative hypotheses by comparing it with the available empirical data (*Elimination*). At this stage, the inductive mode of logical inference plays a role as part of the abductive process. The elimination process is iterative to the point when the eliminated generative mechanism soundly corresponds to the available empirical data. This is followed by identifying a correct causal structure in the model under study (*Identification*).

In the next section philosophical assumptions underlying DES and SD world-views are considered from a CR standpoint through superimposing relevant elements of SD and DES modelling constructs and methods upon the relevant domains of the stratified CR ontology.

5.3 Process system modelling with SD and DES through the prism of CR scientific positioning

In order to better understand the nature of knowledge acquired as a result of each phase of the simulation lifecycle, the outcomes of each simulation phase are to be imposed upon the three-layered ontological representation of CR (see Figure 5.1). The stratified CR ontology clearly separates the steps of the knowledge building process according to the three ontological levels where this knowledge originates. For instance, the empirical nature of the *Data Inputs* modelling outcome pertains to the CR 'empirical' ontological domain. The *Conceptual Mode/Dynamic Hypothesis* output, being non-empirical by definition and reflecting the underlying mechanism(s) and structure(s) that determine(s) the origins of the system behaviour, is related to the CR domain of 'real'. Yet, the process of building the *Computerised Simulation Model* is inherently empirical and therefore pertains to the 'empirical' domain of CR

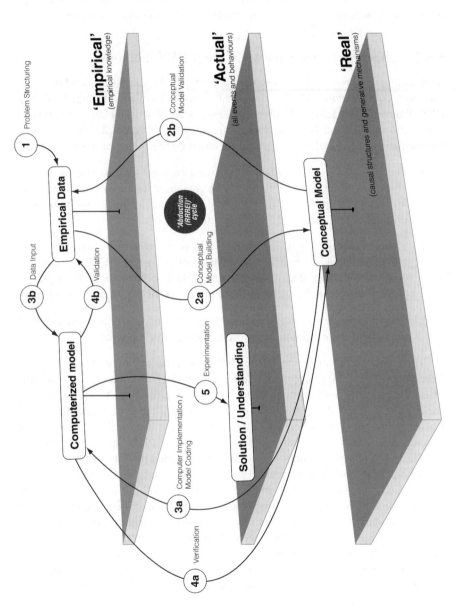

Figure 5.1 DES and SD model building and implementation phases through the prism of a CR ontological/epistemological continuum.

ontology. And finally, the aim of the *Solutions/Understanding* output is to cover as many as possible alternative scenarios generated by the modelled system. This places this latter phase in the ontological domain of 'actual' where all the events and behaviours produced by underlying generative mechanism(s) are actualised.

5.3.1   Lifecycle perspective on SD and DES methods

In line with previous studies that discussed the lifecycle-based perspectives for the SD (Sterman, 2000; Randers, 1980; Lane, 2000; Größler, Thun and Milling, 2008) and DES (Wainer, 2009; Tako and Robinson, 2008; Pidd, 2004b) simulation methods, in Figure 5.1 the SD and DES modelling lifecycles are represented. The respective representation of SD and DES lifecycles is based on the following assumptions:

(a) The granularity of the lifecycle perspective may vary depending on the objectives of the simulation project (the granularity of the reported simulation lifecycles varies from two (Größler, Thun and Milling, 2008) to eight (Wainer, 2009; Tako and Robinson, 2008) phases.

(b) The phases of the simulation lifecycle are not executed in a strictly sequential but rather iterative manner and may involve feedback transitions to the opposite directions of the lifecycle view (Tako and Robinson, 2008; Sterman, 2000; Wainer, 2009). For an in-depth discussion of this point, see Chapter 8.

(c) The principles and assumptions underlying SD and DES methods are unfolded throughout the execution of the simulation lifecycle phases.

The implementation of the SD modelling method begins with the *Problem Articulation (Boundary Selection)* phase (Figure 5.1) (Sterman, 2000). At this phase the rationale and objective of the simulation project are specified, the key variables that reflect the behavioural characteristics of the system and that are of particular interest to modellers are determined, and the historical behaviour of the key variables – the system's reference modes (Randers, 1980; Sterman, 2000) – are specified. The reference modes may well be represented by both quantitative data describing past system performance and qualitative data that represents expert knowledge on the system's performance (Sweetser, 1999). This phase is followed by the *Formulation of Dynamic Hypothesis* phase that aims to generate a hypothesis about the underlying causal (feedback) structure(s) and mechanism(s) that generate (s) the behaviour of the modelled system (Randers, 1980; Sterman, 2000). Herein, Randers (1980) stresses the importance of the accumulated empirical data on the system's reference modes for the hypothesis building process. The dynamic hypothesis may be built using both causal loop and stock and flow SD grammatical constructs (Sterman, 2000).

The DES lifecycle starts in the *Problem Formulation/Structuring* phase. Similar to the SD lifecycle initial phase, it requires the general nature of the simulation project to be understood ('a precursor to conceptual modelling' (Robinson, 2008, p. 283)) and the objectives of the project to be identified (Wainer, 2009; Tako and Robinson, 2008).

This phase is followed by the *Conceptual Model Building* phase (Pidd, 2003a; Wainer, 2009; Tako and Robinson, 2008). Pidd (2003a, p. 35) refers to it as 'an activity in which the analyst tries to capture the essential features of the system that is being modelled'. Robinson (2008, p. 283) defines the conceptual model building as developing 'a non-software specific description of the computer simulation model (that will be, is or has been developed), describing the objectives, inputs, outputs, content, assumptions and simplifications of the model'.

Combining the perspectives on the conceptual model provided by the definitions of both Pidd (2003a) and Robinson (2008), it becomes clear that within the DES method the conceptual model is a non-empirical representation of 'the essential features of the system that is being modelled'.

Contrary to the modelling logic inherent to the SD method, the DES method implementation prescribes that the *Conceptual Model Building is not preceded but followed by* the collection and analysis of *Data Inputs* (Wainer, 2009; Tako and Robinson, 2008). To complete this 'super-phase', Wainer (2009, p. 27) suggests to 'observe and collect the attributes chosen in the previous phase'.

The logic of the *Implementation and Experimentation* 'super-phase' of the SD and DES lifecycles is quite similar. The lifecycle perspective on both SD and DES modelling methods refers to the *Formulation/Implementation* of the simulation model based on computer simulation packages as the following simulation phase.

The next phase is referred to in the DES literature as *Validation and Verification* (e.g. Wainer, 2009) and in the SD literature as *Testing of the Dynamic Hypothesis* (e.g. Randers, 1980). In this phase: (a) the correctness of transferring the non-software-specific conceptual model into a computerised simulation model is verified; and (b) in order to check whether the executable computer model adequately reproduces the modelled system's behaviour, the executable computer model is validated against the empirical data (the so-called 'comparison to reference modes' (Sterman, 2000, p. 86)) collected in the previous phase of the simulation lifecycle (*Problem Formulation* phase for SD and *Data Input* phase for DES).

The final phase of both the SD and DES simulation lifecycles is the *Experimentation and Output Analysis (DES)/Policy Formulation and Experimentation (SD)*. In this phase possible behavioural patterns exhibited by the system under consideration are simulated. The ultimate aim of this phase is to attain an improved *understanding* of the behaviour of the system as a consequence of a given management decision. This is achieved through the analysis of the 'alternative histories' of the simulated system's behaviour generated by the executable computer system in response to alternative courses of action chosen by a manager in the process of decision making.

Achieving better understanding of real-world management systems is a key goal for MS as a scientific discipline (e.g. Hall, 1985; Commission on the Future Practice of Operational Research, 1986; Meredith, 2001; Walsham, 1992). Based on the discussed lifecycle perspective of SD and DES modelling methods, simulation models can be regarded as tools for systematically attaining a better understanding of the *real* world and thereby for providing support for addressing the *real*-world management decision-making problems.

The ontological and epistemological assumptions of SD and DES worldviews informed by CR are discussed below in a stepwise manner in accordance with the phases depicted in Figure 5.1.

Phase 1: Problem structuring (DES)/problem articulation (boundary selection) (SD)

Within phase 1 of the simulation lifecycle, irrespective of the chosen simulation approach, each simulation project starts by determining the goal and scope of the project. If the simulation model is built to analyse the behaviour of an existing system, the data that reflects the historical behaviour of this system is accumulated in this phase of the simulation lifecycle. This accumulation of the relevant empirical data on the past system's behaviour is paramount, in particular in SD modelling. To illustrate, Forrester (1961, 1992) refers to accumulated empirical knowledge as the 'mental database' of the simulation project. Randers (1980) argues that the conceptual modelling in SD should be preceded by solid empirical data expressed in the form of 'reference mode'. Hence, both SD and DES approaches determine the goal and boundaries of the simulation study based on the acquired *empirical knowledge* of the past behaviour of the process system under study (Sweetser, 1999). This evidences that, in the context of both SD and DES simulation, the knowledge acquired in this phase of the simulation lifecycle corresponds to the 'empirical' domain of CR ontology (step 1 in Figure 5.1).

There is a strong similarity between the *phase 1* of the SD/DES simulation lifecycle and the *Resolution* and *Redescription* steps of the *RRREI* abductive process discussed in Section 5.2. Indeed the focus of the first phase in both SD and DES modelling lifecycles is on specification of the objective of the simulation study and on defining the boundaries of the real-world system described by the simulation model. The common objective of the simulation study includes the understanding and improvement of the performance of the real-world system. The realised necessity to model and improve the behaviour of the real-world system is always supported by some empirical evidence of its past performance upon which a call for improvement is made. This directly matches the scope of the *Resolution* step of the *RRREI abductive process*. Moreover, the *phase 1* requirement to specify the boundaries of the real-world system that is to be reflected in the simulation model is in line with the requirement to acquire a general understanding about the behaviour of the phenomenon under study as part of the *Resolution* step of the *RRREI abductive process*. In the following step of the abductive process, the empirical data about the behaviour and structural characteristics of the phenomenon under study is represented in the context of a particular theory. Hence, the process systems, which are the focus of this study, are grounded in one of the two theoretical approaches: SD as a structural theory (Größler, Thun and Milling, 2008) or DES as another theoretical approach to modelling and understanding complex dynamic systems (Pidd, 2003a, 2004b). The choice of the modelling techniques for building the conceptual model of the process system therefore reflects the parallel between the *RRREI abductive process* and the lifecycle view on simulation modelling methods.

Phase 2: Conceptual model building (DES)/formulation of dynamic hypothesis (SD)

The *Conceptual Model Building* phase (Figure 5.1) aims to capture the origins of the behaviour of the system being modelled (e.g. Pidd, 2004b). In *SD* the formulation of the dynamic hypothesis is regarded as the product of the conceptualisation phase and implies 'creating an initial model that is capable of reproducing the major dynamics of the reference mode' (Randers, 1980, p. 122) (step 2a in Figure 5.1). The principle underlying the SD method, and especially the conceptual model building stage, corresponds to the core epistemological premise of CR: that is, *structure (including the relationships between structural components of the system) determines the system's behaviour* (Richardson and Pugh, 1981). This is in line with the CR method informed by its ontological/epistemological position that 'proceeds by trying to discover underlying structures that generate particular patterns of events (or non-events)' (Mingers, 2000, p. 1266). In order to identify the dominant feedback structures in causal loop diagrams as well as the relationships between the stock and flow variables, the logic of the model building process progresses from the available empirical data to the postulation of the dynamic hypothesis (see step 2a).

In *DES* the conceptual model building phase has been reported as 'a vital' but at the same time 'the most difficult and least understood' phase of the simulation modelling (Robinson, 2008, p. 278). This phase 'aim[s] to identify the main entities of the system and *to understand the logical ways in which they interact* [emphasis added]' (Pidd, 2004b, p. 36). This is in line with the reported characteristic of 'DES methodology [as] . . . a *disciplined means of capturing the structure* of an existing or proposed system [emphasis added]' (Sweetser, 1999, p. 3). According to Brailsford and Hilton (2001), for *process systems*, it is *the structure of the underlying queuing system (network)* that *represents the underlying control flow principle and determines all possible pathways of the simulated system behaviour.* Therefore, while DES adopts different tools of conceptual model representation than SD, it is evident that the primacy of structure over behaviour as a core principle of conceptual model building still holds.

As the universe of all possible system's pathways is primarily generated by the *structure of the underlying queuing system*, the CR epistemological lens refutes the reported argument that postulates the primacy of the inherently empirical concept of *randomness* in determining the origins of the system's behaviour. Indeed in the context of DES simulation, randomness determines the *probability of choosing one or another pathway* whose existence is enabled by the *underlying generative mechanism* (e.g. a network of queues for a process system). As such, CR clarifies the role of *randomness* as an empirical factor that allows one to operationalise generative mechanisms in the later phases of the simulation project lifecycle.

The process of developing the SD dynamic hypothesis from the available empirical data has not been explicitly formalised in the MS literature. However, Mingers (2000) demonstrated how the abductive mode of reasoning (*RRREI*) supported by CR corresponds to the logic of building and testing the SD dynamic hypothesis. In the same manner, the structure of the underlying conceptual model in

DES may be based on the *Resolution* of the empirical data on the past behaviour of the modelled system. In light of the adopted CR epistemological position, the *queuing systems (networks)* which are at the core of the 'non-software-specific' conceptual model representation of process systems, can be viewed as *generative mechanisms*. Hence, the *Retroduction* step would involve an understanding of the structural mechanisms (i.e. the structure and interconnectedness of the underlying queuing system) that, if they existed, would generate the problem behaviour of the modelled process system. In terms of CR this kind of relationship is referred to as *generative causality*. Then, in the subsequent phase of the simulation process, the most feasible generative mechanism(s) (i.e. adopted conceptual model) is (are) eliminated and identified when the conceptual model is validated against the available *Data Inputs*.

Summarising the discussion of this simulation phase, CR epistemology informs the process of conceptual model building for both SD and DES by stressing the commonality of two knowledge building concepts:

(a) The primacy of structural factors that generate the behaviour of the modelled systems.

(b) The requirement of the empirical data (accumulated in the preceding modelling phase) necessary to hypothesise about the underlying structural factors that could have generated the available data about the past system's behaviour.

From the point of view of the CR *RRREI abductive process*, these two factors form the necessary basis for the *Retroduction* and the following steps of the *RRREI* abductive process. Thus, the empirical data that reflects the behaviour of the real-world process system (or rather the part of this system that is of interest to the modeller) is used as a basis for hypothesising about the nature of the underlying generative mechanisms, which, if they existed, would causally generate this empirical data at hand. This is in direct correspondence with step 2a of the simulation modelling lifecycle (see Figure 5.1) referred to in the SD literature as *'Initial hypothesis generation'* (Sterman, 2000, p. 86). Next, in line with the *Elimination* step, the hypothesised generative mechanism(s) is represented in isolation from any other components of the process system (which are not part of the generative mechanism(s)) and then the suggested hypothesised generative mechanism is validated against the initial set of the empirical data. In SD (e.g. Randers, 1980) and DES (e.g. Sargent, 2004) simulation modelling (step 2b in Figure 5.1) this refers to the necessary validation of the hypothesised conceptual model (initial dynamic hypothesis(es)). And finally, based on the results of the *Elimination* phase the correct generative mechanism that is tested against the actual behaviour of the modelled system is identified.

Overall, there are strong similarities between the abductive mode of logical inference that was adopted by CR as a tool for knowledge building and the phases of the SD and DES lifecycles contained within the *Data Collection and Conceptualisation* 'super-phase' of the combined simulation modelling lifecycle depicted in Figure 5.1.

Phase 3: Model coding/computer implementation and data input

In this stage of the simulation project the computerised simulation model is implemented/coded (step 3a) and the relevant historical (empirical) data is populated in the computerised model (step 3b). As shown in Figure 5.1, from the point of view of CR, both of these steps are performed at the *empirical* level of CR ontology. While DES modellers traditionally spend most of their project time in this simulation modelling phase sampling probability functions into the computerised model (e.g. Tako and Robinson, 2008; Sweetser, 1999), the epistemological premises underlying the knowledge building process with DES and SD do not considerably differ. In both DES and SD the incorporation of the relevant *empirical* data into the computerised simulation model allows one to operationalise the non-empirical generative mechanisms identified in the previous phase of the simulation lifecycle.

Phase 4: Verification and validation (DES)/testing (SD)

In the *Verification (DES)/'Testing for consistency' (SD)* step (step 4a in Figure 5.1) the correctness of the computer programming and implementation of the conceptual model/dynamic hypothesis is checked (Sargent, 2004; Tako and Robinson, 2008; Sterman, 2000; Randers, 1980). From the point of view of CR epistemology, the goal of this step is to verify whether the built-in parameters of the computerised model reflect the nature of the underlying generative mechanisms identified in step 2a of the simulation process.

Validation (DES)/'Testing through comparison to reference modes' (SD) (step 4b) implies checking that the 'model's output behaviour has sufficient accuracy for the model's intended purpose over the domain of the model's intended applicability' (Sargent, 2004, p. 132; Randers, 1980). As the computerised model and relevant data utilised for model validation are empirically driven, the knowledge generated in this step of the simulation project lifecycle refers to the CR ontological domain of 'empirical'.

Phase 5: Experimentation

The final step in the model building process implies running the simulation and the exploration of the 'what-ifs' based on the simulation results.

From the CR viewpoint this step aims at unveiling all possible events and behaviours that may be produced by *generative mechanisms* identified during the *Conceptual Model Building*. The available simulation knowledge and the computational power of simulation packages thus allow experimental identification of how a hypothetical generative mechanism may be actualised. Varying the impact of different experimental factors (simulation model parameters) aims to reveal *all possible 'what-if' scenarios* produced by a certain generative mechanism (or interplay of multiple generative mechanisms). Therefore the *Experimentation* phase in both SD and DES allows the representation of the patterns of events that have been outside of the manager's empirical scope prior to the simulation project. In other words, this step aims at getting access to the knowledge on the behaviour of the process system that corresponds to the 'actual' domain of CR ontology (Figure 5.1).

This section demonstrated in detail how the CR scientific position facilitates explicit articulation of the fundamental epistemological assumptions underlying SD and DES worldviews. In the following section the implications of the CR philosophical position for more effective use of DES and SD simulation to support managers' understanding of decision making in process systems are discussed.

5.4   Process system modelling with SD and DES: Trends in and implications for MS

In this section the CR-enabled phases of the SD and DES simulation lifecycle as described in Figure 5.1 are illustrated using a specific segment of the process system modelling literature: *patient care process modelling*. Patient care process modelling is a challenging area of research where simulation has traditionally played a role as a tool to support management decision making (Davies and Davies, 1994; Jun, Jacobson and Swisher, 1999). For each phase of the DES and SD modelling and implementation lifecycle, we first review the selected research literature that explored the operational perspective on patient care process simulation with SD and DES through the ontological/epistemological lens of CR and then discuss how the suggested CR ontological/epistemological perspective contributes to more effective use of simulation for intelligent thinking about, and management decision support in, process systems.

For simplicity reasons, the modelling context discussed in this section is limited by the operational perspective on patient care processes (Cote, 2000). For instance, the patient flow can be represented from the operational point of view as the movement of patients (entities) through a set of locations in a health care facility (Cote, 2000). A more complete operational definition of a patient care process was suggested by Little (1986, p. 6) who categorised it as a type of *designed flow process* that 'include[s] a wide variety of situations where tangible items . . . [i.e. patients or materials] or intangible entities (such as information) flow in some purposive way along paths that may be individually charted or collectively established but proceed with the intention of achieving some outcome'. The perspective chosen for illustrative purposes leaves the clinical perspective on patient care processes (ibid.) outside the focus of this discussion.

Phase 1: Problem structuring (DES)/problem articulation (boundary selection (SD))

SD: Lattimer *et al.* (2004) start with problem articulation and boundary selection that is followed by the accumulation of the empirical data on how the system operates as well as on the nature of interdependency between the components of the system. Initial interviews and hospital site visits are used in order to accumulate necessary empirical evidence on how the acute patient flows could be mapped. As described in

Chapter 6, Brailsford *et al.* (2004) ran 30 interviews with key decision makers from across health and social care prior to developing a 'conceptual map' of the emergency health care system in Nottingham. Lane, Monefeldt and Rosenhead (2000) reveal that in order to build and refine the initial conceptual model of the process system, the 'mental databases' that reflected the knowledge of the process owners as well as the 'formal sources' of empirical data were accessed. Prior to developing a conceptual model (influence diagram) of 'the flow of elderly people from the community into UK National Health Service (NHS) and out into community care', Wolstenholme (1993, p. 927) refers to the source of the empirical data that allowed understanding the behaviour of the modelled system as a result of the collaboration with a senior manager who conversed with both the NHS and Personal Social Services.

On the other hand, not all of the reviewed SD studies clearly specify the sources of the empirical data. For example, Wolstenholme (1999) provides a conceptual map of the patient flow model but remains silent on the nature of the empirical data used to build the conceptual model. Moreover, one of the reviewed studies did not follow in a consecutive manner the phases of the SD lifecycle. Worthington (1991) introduces the basic conceptual hospital waiting list management model before the description of the data collection and analysis stage. This results in a schematic and over-simplistic representation by Worthington (1991) of the basic waiting list model as part of the description of the 'overall approach' (p. 835) adopted in this study.

DES: The review of the DES studies in patient care process simulation reveals that, as is the case with SD modelling, there is no uniformity in whether the empirical data sources are described at this stage or not. For example, Moreno *et al.* (2000) and Ferreira *et al.* (2008) report on the methods of data collection and sources of the empirical data in order to better understand how the process system under study operates. The review of more than 350 papers on simulation modelling (mainly using DES) by Fletcher and Worthington (2009) reveals that data collection in patient flow modelling was generally performed through computer records, but also occasionally through work studies and consultation with the experts (e.g. clinicians).

On the other hand, a considerable number of the reviewed DES studies report on the results of the problem formulation and even conceptualisation *before* the sources of the empirical data are described (e.g. Dittus *et al.*, 1996; Coelli *et al.*, 2007; Werker *et al.*, 2009; Cardoen and Demeulemeester, 2008; Levin *et al.*, 2008). This poses questions on the availability of the necessary empirical support at the following conceptual model building phase of the DES simulation modelling.

Trends and implications for management science: As discussed in Section 5.3, this phase of SD and DES simulation modelling lifecycle aims at specifying the objective of the simulation study as well as to determine the boundary of the simulated system. The epistemological position of CR calls for the acquisition of empirical evidence on the problem behaviour of the modelled system as a prerequisite for further exploration of the mechanisms that trigger this behaviour. In most cases the objectives of the reviewed simulation projects are triggered by calls for understanding the reasons for the poor performance of the patient care processes and by the attempts to increase this performance. Hence, collecting empirical data at a very early stage of the simulation project allows one effectively to identify the objective of the project and to understand which part of the real world is to be covered by the project.

Based on the review of the SD and DES simulation literature, not all of the reviewed SD- and DES-based patient flow simulation projects specify data sources in this phase of the simulation lifecycle. This does not necessarily imply that such data does not exist, as in most cases it is specified in the later phases of the simulation modelling lifecycle. However, as follows from the CR philosophical position, the logic of scientific enquiry calls for specification of the available empirical evidence on the modelled system upfront, as such data may well determine the research design and the selection of the specific simulation method (cf. Brailsford *et al.*, 2004), as described in Chapter 6.

The nature of the empirical data sources in SD and DES projects often varies. Generally, the data sources used in SD simulation are more value based than those used in DES simulation: while SD modellers refer to the 'mental databases' (Forrester, 1961) collected through the interviews with the process owners and/or clinicians, DES modellers tend to utilise the computer records that contain historical data on the performance of the system of interest.

Phase 2: Conceptual model building (DES)/formulation of dynamic hypothesis (SD)

SD: Lane and Husemann (2008) demonstrate how the conceptual models of the patient flow have been refined throughout a series of interviews, site visits and workshops. Lane, Monefeldt and Rosenhead (2000) also demonstrate how the initial conceptualisation of the modelled Accident and Emergency system elements, processes and pathways revealed the need to include in the conceptual model other factors that were unnoticed in the initial conceptualisation phase. While building the conceptual model, Wolstenholme (1993) concentrates on the development of the initial or 'first type' conceptual model of the flow of elderly people from the community into NHS and out into community care and its transformation into an archetype-driven conceptual model. The focus is on refining the conceptual model that provides a very high-level description of the interaction of the components of the modelling process into a more detailed feedback loop diagram that demonstrates key SD archetypes that influence the behaviour of the process system under study. From the point of view of the abductive mode of reasoning supported by CR, the initial conceptual model of the process system suggested by Wolstenholme (1993) reflects the *Resolution* step (where the boundaries of the modelled process systems were shaped and the key variables defined) and *Redescription* (where the SD modelling has been selected as an appropriate theory to reflect the behaviour of the modelled system). *Redescription* also results in the transformation from the initial 'first-order' conceptual model into the model that reveals the nature of the underlying system archetypes.

DES: In order to determine how bed demand affects Emergency Department (ED) patients' access to patient cardiac care, Levin *et al.* (2008) built a conceptual model of the patient flow between the cardiology macro-system units. The model is built using queuing principles. Bailey (1952), Jackson, Welch and Fyr (1964), Vissers and

Wijnaard (1979), Brahimi and Worthington (1991) and Walter (1973) formulate the problem of outpatient and general practice/hospital department appointment scheduling as a queuing system. Lowery and Martin (1992), El-Darzi *et al.* (1998), Altinel and Ulas (1996), Cardoen and Demeulemeester (2008), Coelli *et al.* (2007), Ferreira *et al.* (2008), Werker *et al.* (2009), Levin *et al.* (2008) and de Bruin *et al.* (2007) regard the operational model of patient flow as a simulated queuing system. Jiang and Giachetti (2008) suggest an alternative structure of the queuing network model which handles the fork/join structure of patient flows and provides rapid analysis of alternative patient flows. Edwards *et al.* (1994) compare the performance of patient processing based on two different queuing system structures and found that patient waiting times can be reduced up to 30% when using quasi-parallel processing as opposed to serial processing.

These reviewed DES studies of patient care processes strongly emphasise the *structure vs randomness* debate when building and/or simulating the conceptual model of the process system. They acknowledge that the structural characteristics of the queuing system, that is the structure of the entities' flow through their active and passive states, play a decisive role in behaviour generation of process systems. For example, de Bruin *et al.* (2007), based on expert meetings with cardiologists, develop a structural model of the emergency cardiac in-patient flow. According to the authors, the critical role of the provided conceptual model is not in a complete representation of all cardiac in-patient flows in the university medical centre, but rather in understanding the major underlying queuing mechanisms that affect the dynamics of the in-patient flow system under study. Hence, while not a completely accurate representation of a real process system, the suggested structural model allows one to 'reduce complexity without losing integrity by focusing on the most critical patient flows' (p. 127). Kolker (2008, 2009), by using in a combined manner process model simulation and queuing theory, identifies the core queuing mechanisms that determine the structure and behaviour of the intensive care unit and Emergency Department patient flows. Haraden and Resar (2004, p. 4) argue that 'it is a common but an incorrect assumption that the healthcare flow is a result of what appears to be the randomness and complexity of the disease presentation'. Instead they argue that the health care flow is a result of more structured factors that can be consistently represented and managed based on historical data and queuing methods. In the same manner, according to Ryan and Heavey (2006) the application of queuing methods is 'vital to the modelling of a discrete-event system when gathering requirements or building a conceptual model for the purposes of a simulation project' (p. 441). This said, the randomness still plays a role in determining the process system dynamics; however, its primacy as a behaviour generating mechanism is questionable in the process system context.

The review of the DES studies also demonstrates that there is a discrepancy between the representation of the overly simplistic conceptual model on the one hand and the computerised model on the other. According to Pidd (2003a) and Robinson (2008), the representation of the conceptual model reflects the essence of the behaviour of the dynamic system under study. We demonstrated above the role of the queuing mechanism for understanding the behaviour of the process system.

However, in a number of studies (e.g. Dittus *et al.*, 1996; Cardoen and Demeule-meester, 2008), while the *queuing mechanism* is used within the computerised model, it is omitted in the conceptual model. When reviewing the conceptual models suggested by these authors, a question arises whether the reported relationships between the model components indeed reflect the essence of the modelled system.

Trends and implications for MS: The review of the SD and DES studies reveals a number of epistemological differences when approaching this phase of the simulation modelling. Generally, SD simulation studies are more concerned with the link between the empirical data (both 'hard' and 'soft') and the conceptual model. Thus, in SD studies the development of the conceptual model undergoes concep-tualisation (Wolstenholme, 1993) or the initial conceptual model is validated against the reference modes of the simulated process system (Lane, Monefeldt and Rosenhead, 2000; Lane and Husemann, 2008; Wolstenholme, 1993), that is against the sources of empirical data that describe the behaviour of the key concepts and variables of the system. The specification of reference modes can be run on a continuous basis, thus allowing a comparison of the hypothesised conceptual model against novel reference modes of the system under study (e.g. Lane, Monefeldt and Rosenhead, 2000; Lane and Husemann, 2008) (see Figure 5.1, step 2b).

On the other hand, most of the reviewed DES projects suggest a conceptual model for the patient care processes without mentioning the sources of the empirical data used in order to reflect the key features of the simulated system (e.g. Werker *et al.*, 2009; Cardoen and Demeulemeester, 2008; Dittus *et al.*, 1996; Levin *et al.*, 2008; Coelli *et al.*, 2007). Neither of the reviewed DES studies demonstrates whether and how the conceptual model is refined as a result of comparing the suggested conceptual model with the collected historical data that describes the modelled behaviour of the process system. This questions the role of the conceptual model as the one that aims 'to capture the essential features of the system that is being modelled' (Pidd, 2003a, p. 35). This also explains why SD modellers dedicate a considerably larger part of their project time to the conceptual model building phase than DES modellers do (Tako and Robinson, 2008), as described in Chapter 8.

From the point of view of CR epistemology, while the conceptualisation phase in reviewed SD studies corresponds to the complete abductive cycle (both *steps 2a and 2b* in Figure 5.1), the DES studies cover only *step 2a*. Hence, the reviewed DES studies formally ignore the *Elicitation* and *Identification* steps of the abduction process in the conceptual model building phase. Instead, these knowledge building activities are undertaken in the two subsequent phases of the DES simulation lifecycle.

Phase 3: Formulation of a simulation model/computer implementation

SD: Lane, Monefeldt and Rosenhead (2000) in their study of an Accident and Emergency department refer to this phase as 'Formulation of the structure and equations'. This does not necessarily reflect the true nature of the described activity, as the structure has been already formulated in the previous phase of the simulation modelling when the initial dynamic hypothesis was suggested and subsequently refined. In this phase the qualitative conceptual model expressed as a causal loop diagram is transformed into a quantitative stock and flow diagram (Lane, Monefeldt and Rosenhead, 2000). The latter is based on 194 equations and designed using the iThink software simulation package. When describing this phase of simulation model building, Wolstenholme (1993) notes that the conceptual model supported by systems thinking falls short when comparing and contrasting the alternative interventions. Hence, the model parameters and variables are to be specified as a precondition for testing the modelled system's logic upon which the simulation model was built. Brailsford *et al.* (2004) refer to this phase as a 'Quantitative phase' defining its aim as 'to facilitate experimentation with various potential changes in service configurations and demand rates' (p. 37). They used the iThink simulation package to develop the computerised simulation model. The model was populated with data obtained from health care providers in Nottingham, UK, as described in Chapter 6.

DES: In DES modelling this phase also presupposes translation of the conceptual model into the formal computerised model while specifying the necessary process tasks and preserving the underlying queuing mechanism (Werker *et al.*, 2009; Lowery and Martin, 1992; El-Darzi *et al.*, 1998; de Bruin *et al.*, 2007; Altinel and Ulas, 1996; Kolker, 2008, 2009). These and all additional aspects of the model (such as the availability of resources, the arrival patients' data) are often represented in DES simulation studies by adopting one of the existing simulation packages, for example Arena (Werker *et al.*, 2009; Cardoen and Demeulemeester, 2008), SIMUL8 (Katsaliaki *et al.*, 2009) as well as MATLAB (Levin *et al.*, 2008) and MedModel (Levin *et al.*, 2008; Coelli *et al.*, 2007). Some more complex modelling situations require additional model coding. For example, while investigating the opportunities to shorten the average time patients spend in the urgent care facility, Tan, Gubaras and Phojanomongkolkij (2002) built a simulation model using the SIMAN programming language. The model is then simulated on the Arena software package.

Trends and implications for MS: The major task of the computerised model is to reflect the essence of the modelled system that was represented within the conceptual model. Built on the empirical data inputs, the computerised model is formulated within the 'empirical' domain of CR ontology. At the same time the computerised model should reflect the premises inferred by the conceptual model formulated within the ontological domain of 'real'. As defined when discussing the previous phase of the DES and SD simulation lifecycle, the SD *Dynamic Problem Definition* step (step 2b in Figure 5.1) is often omitted in DES simulation projects. The DES simulation literature provides evidence that, while they are formulated, DES conceptual models are seldom compared with the historical behaviour of the key concepts and variables reflected within these models. Therefore, other factors being equal, the dynamic hypotheses suggested within SD simulation studies are more empirically grounded than most of the DES conceptual models that are not empirically validated (see step 2a in Figure 5.1). This may well explain why in the majority of the reviewed DES studies the discrepancy between the complexity of the computerised (empirically based) model and the very high-level representation of the conceptual model was greater than in SD studies (to the extent that not all the key features responsible for the dynamics of the process system were reflected within the DES conceptual model). Hence, in terms of the decision making, the process of gradual development of the SD conceptual model (including the *Dynamic Hypothesis Definition* phase – see step 2b in Figure 5.1) potentially infers a tighter relationship between the representation of the computerised and conceptual model of the process system, that is between the computerised system and the part of the real-world system that is being modelled.

Phase 4: Verification and validation (DES)/testing (SD)

SD: Brailsford *et al.* (2004, p. 38) refer to the validation of the SD models as a 'thorny topic'. The fact is that the validation of the qualitative and quantitative SD models requires a different degree of understanding of the simulation model by managers. For example, building and validation of the qualitative SD simulation model require a detailed understanding of the model's structure and outputs by managers as their 'mental databases' are regarded as a paramount tool not only for model building, but also for model validation (e.g. Wolstenholme, 1993; Lane and Husemann, 2008). In this regard, the SD study performed by Lane and Husemann (2008) is particularly illustrative. It discusses the activities performed before and during a series of workshops that allowed one to build, refine and validate a complex qualitative model of pathways for acute patients referring to expert knowledge as the major source of model building and validation.

At the same time, the quantitative SD models can be validated 'by a "black box" process' (Brailsford *et al.*, 2004, p. 38), that is by comparing the simulated behaviour of the modelled system with the historical behaviour of the corresponding real-world system (e.g. Lane, Monefeldt and Rosenhead, 2000). This can also be the case when an SD study combines both approaches to model building and validation (e.g. Brailsford *et al.*, 2004), as described in Chapter 6.

DES: While Pidd (2003a) refers to verification in DES modelling as an almost outdated phase of simulation modelling due to the introduction of visual interactive modelling systems (VIMSs), the DES studies in the area of patient care process simulation still use model verification as the core step. This step assures the correct transition from the conceptual model to the computerised representation. The discrepancy between the conceptual model and computerised model revealed in most of the reviewed studies, therefore, calls for verification of the transformation process from the conceptual to the computerised model.

The validation of the simulation model in DES studies is predominantly quantitative and, as is the case with quantitative SD, includes a comparison of the behaviour exhibited by the simulation model. Examples of such an approach to validation of the DES simulation models can be found in Dittus *et al.* (1996), Altinel and Ulas (1996), Cardoen and Demeulemeester (2008), Coelli *et al.* (2007), Ferreira *et al.* (2008), Kolker (2008, 2009), Werker *et al.* (2009) and Levin *et al.* (2008). As validation of DES is predominantly quantitative, a considerable amount of the simulation time in DES studies is spent on data accumulation and validation (Tako and Robinson, 2008), as discussed in Chapter 8.

Trends and implications for MS: Evidently, the validation performed in SD and DES simulation projects is *empirical* by nature as it questions whether the suggested computerised model can reproduce the problem behaviour of the process system (see step 4b in Figure 5.1). The empirical data describing the problem behaviour of the process system may be in the form of 'mental databases' (in the case of qualitative SD) or in the form of quantitative data about the past system's behaviour (in the case of quantitative SD and DES). Due to the discrepancy in the level of detail between the DES conceptual and computerised models, as well as the lack of empirical grounding in a large part of DES conceptual models, verification in DES simulation is more problematic than in SD. Indeed, in the case when such properties of a process system as the nature of the built-in queuing mechanism are not covered within the conceptual model, but are reflected in the computerised model (e.g. Cardoen and Demeulemeester, 2008), the value of the computerised model verification may be highly questionable.

Phase 5: Experimentation

The following step within both quantitative *SD* studies (Worthington, 1991; Lane, Monefeldt and Rosenhead, 2000; Brailsford *et al.*, 2004) and *DES* (Werker *et al.*, 2009; de Bruin *et al.*, 2007; Cardoen and Demeulemeester, 2008; Kolker, 2008, 2009) studies includes the analysis of alternative scenarios by using model simulations. In this phase a set of simulation experiments is run in order to determine the possible scenarios of the system's behaviour under changing conditions. Therefore both SD and DES simulation support the management decision-making process when evaluating different operational alternatives in order to improve existing patient care processes or to assist in designing and planning new ones.

Trends and implications for MS: Both SD and DES reportedly help the managers to analyse the design of the patient care process and to come up with a more effective design in terms of use of the limited resources in order to achieve the objectives that drive the operational patient care processes. DES and SD modelling help to get an insight into the functioning of the system under the alternative conditions and courses of action. The ultimate aim of adopting SD and DES as tools for support for management decision making in the context of patient care processes is to realise the alternative scenarios of the behaviour of the process system under changing conditions. In order to understand *how* the alternative courses of action of the process system can be enabled and *why* they are enabled in a specific way, the understanding, modelling and operationalisation of the generative mechanisms that enact the 'what-if' scenarios are required.

5.5 Summary and conclusions

This research set out to investigate how the CR philosophy of science facilitates explicit articulation of the fundamental philosophical assumptions underlying SD and DES worldviews. In doing so, we shared the view of many management scientists that 'the worldview that we hold determines the process we will advocate for solving a decision problem' (Buchanan, Henig and Henig, 1998).

This research postulated the primacy of ontological positioning (i.e. 'our convictions about the nature of the world') that, in turn, strongly influences the nature of the produced knowledge (epistemology) about a specific decision-making problem as well as the choice of the methodology used to tackle this problem. In the context of this research, simulation models were regarded as tools for systematically attaining a better understanding of the *real* world and thereby for providing support for *real*-world management decision-making activities, in particular contributing to more effective use of simulation for intelligent thinking about, and management decision support in, process systems.

We have demonstrated how CR provides a unique response to a call (made in Buchanan, Henig and Henig, 1998) for a scientific position that would allow the combined use of the objectivist worldview that is based on the assumption of existence of 'an objective reality which can be measured and described' and a more subjectivist position that allows 'the mind of the decision maker' to interpret the reality based on the available, including value-laden, data. CR addresses this research call by explicitly articulating the fundamental philosophical assumptions underlying SD and DES simulation worldviews.

The adopted CR scientific position allowed an interpretation of the logic underlying the SD and DES simulation methods through the prism of its epistemological position dictated by its stratified ontology. Mapping distinct phases of SD and DES lifecycles onto the ontological domains of CR allowed a comparison of SD and DES knowledge-generating/decision-making steps and the respective outcomes. Particularly, this relates to the use of the available empirical data (generally more value laden in the case of SD than DES) in order to identify the generative mechanisms that trigger the puzzling behaviour of the complex process systems. While using different sets of grammatical constructs, both simulation methodologies aim at understanding the underlying structural pattern that triggers the behaviour of the complex process systems.

Without denying the criticality of understanding the impact of randomness (the factor that operates at the ontological level of empirical) on the behaviour of the system under study (Morecroft and Robinson, 2005), our perspective informed by CR epistemology demonstrated the need for a more structured enquiry into the logic that guides conceptual model building as part of the DES-induced decision-making process. The abductive mode of logical inference was therefore suggested as an established method for identification of the underlying entities of the modelled system and the logic of their interaction.

Based on the literature that reports the use of DES in the context of health care process systems, it was concluded that DES MS models are often less explicit in relating the existing data with the suggested conceptual model of the process system as well as in relating the conceptual model and the computerised model. Hence, one of the suggestions for improving the results of using DES for decision support in process systems would be to focus on providing a better link between the data accumulation and conceptual model building phases (steps 2a and 2b in Figure 5.1) and between conceptual model building and computerised (simulation) model building (steps 3a and 4a in Figure 5.1). A tighter relationship between the available empirical data and conceptual and computerised models would arguably assure a better understating of the existing preferences and overall better decision-making results when adopting DES in the context of process systems.

In the case of SD simulation, the adoption of the abductive mode of logical inference as part of the CR position demonstrated the importance of acquiring the most possibly compete dataset (whether it is value based or not) to assure reliability of dynamic hypothesis building (step 2a in Figure 5.1) and testing (step 2b in Figure 5.1). Herein, the importance of random factors is not denied and the possibility to reflect them as part of the empirical dataset would mean the acquisition of better empirical bases for enquiry about the hypothetic causal mechanisms that triggered the given behaviour of the process system under study. The adoption of the CR epistemology also indicates that building the computerised model and running the experiment are strongly required in order to expand the set of decision-making alternatives concerning the behaviour of the process system under study.

This chapter is the first one to examine explicitly the philosophical bases of DES and SD simulation worldviews by adopting a particular philosophical position and to report its findings through the prism of CR stratified ontology and the abductive mode of knowledge generation. By doing so it provides a better understanding of how management decision-making knowledge is generated through the modelling process. This contribution is important for managers who, as 'intelligent thinkers' about real process systems, both contribute to and use simulation models for decision making. It is also important for management scientists who require an in-depth understanding of the scientific bases of modelling methodologies in order to choose an appropriate simulation worldview to support management decision making in real-world process systems in a truly scientific manner, thus living the original promise of MS as a scientific discipline.

References

Altinel, I.K. and Ulas, E. (1996) Simulation modeling for emergency bed requirement planning. *Annals of Operations Research*, **67**, 183–210.

Bailey, N.T.J. (1952) A study of queues and appointment systems in hospital outpatient departments, with special reference to waiting times. *Journal of the Royal Statistical Society, Series B*, **14**, 185–199.

Barton, P.M. and Tobias, A.M. (2000) Discrete quantity approach to continuous simulation modeling. *Journal of the Operational Research Society*, **51**(4), 485–489.

Becker, J. and Niehaves, B. (2007) Epistemological perspectives on IS research: a framework for analysing and systematizing epistemological assumptions. *Information Systems Journal*, **17**, 197–214.

Bhaskar, R. (1978) *A Realist Theory of Science*, Harvester Press, Brighton.

Bhaskar, R. (1979) *The Possibility of Naturalism: A philosophical critique of the contemporary human sciences*, Harvester Press, Brighton.

Bhaskar, R. (1986) *Scientific Realism and Human Emancipation*, Verso, London.

Bhaskar, R. (1989) *Reclaiming Reality: A Critical Introduction to Contemporary Philosophy*, Verso, London.

Bhaskar, R. (1994) *Plato Etc: The Problems of Philosophy and their Resolution*, Routledge, London.

Borshchev, A., Karpov, Y. and Kharitonov, V. (2002) Distributed simulation of hybrid systems with AnyLogic and HLA. *Future Generation Computer Systems*, **18**, 829–839.

Brahimi, M. and Worthington, D.J. (1991) Queuing models for out-patient appointment systems: a case study. *Journal of the Operational Research Society*, **42**(9), 733–746.

Brailsford, S.C., Churilov, L. and Liew, S.-K. (2003) Treating ailing Emergency Departments with simulation: an integrated perspective, in *Health Sciences Simulation 2003* (eds J. Anderson and M. Katz), Society for Modeling and Computer Simulation, San Diego, CA, pp. 25–30.

Brailsford, S.C. and Hilton, N.A. (2001) A comparison of discrete event simulation and system dynamics for modelling health care systems, in *Planning for the Future: Health Service Quality and Emergency Accessibility* (ed. J. Riley), Glasgow Caledonian University, Glasgow.

Brailsford, S.C., Lattimer, V.A., Tarnaras, P. and Turnbull, J.C. (2004) Emergency and on-demand health care: modelling a large complex system. *Journal of the Operational Research Society*, **55**, 34–42.

Buchanan, J.T., Henig, E.J. and Henig, M.I. (1998) Objectivity and subjectivity in the decision making process. *Annals of Operations Research*, **80**, 333–345.

Burrell, G. and Morgan, G. (1979) *Sociological Paradigms and Organisational Analysis: Elements of the Sociology of Corporate Life*, Heinemann, London.

Cardoen, B. and Demeulemeester, E. (2008) Capacity of clinical pathways – a strategic multi-level evaluation tool. *Journal of Medical Systems*, **32**, 443–452.

Churchman, C.W. (1955) Management Science, the journal. *Management Science*, **1**(2), 187–188.

Churchman, C.W. (1994) Management science: science of managing and managing of science. *Interfaces*, **24**(4), 99–110.

Coelli, F.C., Ferreira, R.B., Almeida, R.M.V.R. and Pereira, W.C.A. (2007) Computer simulation and discrete-event models in the analysis of a mammography clinic patient flow. *Computer Methods and Programs in Biomedicine*, **87**, 201–207.

Collier, A. (1994) *Critical Realism: An Introduction to Roy Bhaskar's Philosophy*, Verso, London.

Commission on the Future Practice of Operational Research (1986) Report of the 1986 Commission on the Future Practice of Operational Research. *Journal of the Operational Research Society*, **37**, 829–886.

Cote, M.J. (2000) Understanding patient flow. *Decision Line*, **31**, 8–10.

Coyle, R.G. (1985) Representing discrete events in system dynamics models: a theoretical application to modeling coal production. *Journal of the Operational Research Society*, **36**(4),307–318.

Danermark, B., Ekstrom, M., Jakobsen, L. and Karlssn, J.C. (2002) *Explaining Society: Critical Realism in the Social Sciences*, Routledge, London.

Davies, R. and Davies, H. (1994) Modeling patient flows and resources in health systems. *Omega*, **22**, 123–131.

de Bruin, A., van Rossum, A., Visser, M. and Koole, G. (2007) Modeling the emergency cardiac in-patient flow: an application of queuing theory. *Health Care Management Science*, **10**, 125–137.

Dittus, R.S., Klein, R.W., DeBrota, D.J. and Fitzgerald, M.A. (1996) Medical resident work schedules: design and evaluation by simulation. *Management Science*, **42**(6), 891–906.

Doomun, R. and Jungum, N.V. (2008) Business process modeling, simulation and reengineering: call centres. *Business Process Management Journal*, **14**(6), 838–848.

Downward, P. and Mearman, A. (2007) Retroduction as mixed-methods triangulation in economic research. *Cambridge Journal of Economics*, **31**, 77–99.

Edwards, R.H., Clague, J.E., Barlow, J. *et al.* (1994) Operations research survey and computer simulation of waiting times in two medical outpatient clinic structures. *Health Care Analysis*, **2**, 164–169.

El-Darzi, E., Vasilakis, C., Chaussalet, T. and Millard, PH. (1998) A simulation modeling approach to evaluating length of stay, occupancy, emptiness and bed blocking in a hospital geriatric department. *Health Care Management Science*, **1**, 143–149.

Ferreira, R.B., Coelli, F.C., Pereira, W.C.A. and Almeida, R.M.V.R. (2008) Optimizing patient flow in a large hospital surgical centre by means of discrete-event computer simulation models. *Journal of Evaluation in Clinical Practice*, **14**, 1031–1037.

Fletcher, A. and Worthington, D. (2009) What is a 'generic' hospital model? A comparison of 'generic' and 'specific' hospital models of emergency patient flows. *Health Care Management Science*, **12**(4), 374–391.

Forrester, J.W. (1961) *Industrial Dynamics*, MIT Press, Cambridge, MA.

Forrester, J.W. (1992) Policies, decisions and information sources for modeling. *European Journal of Operations Research*, **5**, 42–63.

Gorunescu, F., McClean, S.I. and Millard, P.H. (2002) A queuing model for bed-occupancy management and planning of hospitals. *Journal of the Operational Research Society*, **53**(1), 19–24.

Greasley, A. (2005) Using system dynamics in a discrete-event simulation study of a manufacturing plant. *International Journal of Operations and Production Management*, **25**(6), 534–548.

Größler, A., Thun, J.-H. and Milling, P.M. (2008) System dynamics as a structural theory in operations management. *Production and Operations Management*, **17**(3), 373–384.

Hall, R.W. (1985) What's so scientific about MS/OR? *Interfaces*, **15**, 40–45.

Haraden, C. and Resar, R. (2004) Patient flow in hospitals: understanding and controlling it better. *Frontiers of Health Services Management*, **20**, 3–15.

Hopp, W.J. (2008) Management science and the science of management. *Management Science*, **54**(12), 1961–1962.

Jackson, R.R.P., Welch, J.D. and Fyr, J. (1964) Appointment systems in hospitals and general practice. *Operational Research Quarterly*, **15**, 219–237.

Jiang, L. and Giachetti, R. (2008) A queuing network model to analyze the impact of parallelization of care on patient cycle time. *Health Care Management Science*, **11**(3), 248–261.

Jun, J.B., Jacobson, S.H. and Swisher, J.R. (1999) Application of discrete-event simulation in health care clinics: a survey. *Journal of the Operational Research Society*, **50**(2), 109–123.

Karpov, Y.G., Ivanovsky, R. and Sotnikov, K.A. (2007) Application of simulation approaches to creation of decision support system for IT service management, in *Parallel Computing Technologies*, Lecture Notes in Computer Science, vol. 4671, Springer, Berlin, pp. 553–558.

Katsaliaki, K., Mustafee, N., Taylor, S.J.E. and Brailsford, S. (2009) Comparing conventional and distributed approaches to simulation in a complex supply-chain health system. *Journal of the Operation Research Society*, **60**, 43–51.

Kolker, A. (2008) Process modeling of emergency department patient flow: effect of patient length of stay on ED diversion. *Journal of Medical Systems*, **32**, 389–401.

Kolker, A. (2009) Process modeling of ICU patient flow: effect of daily load leveling of effective surgeries on ICU diversion. *Journal of Medical Systems*, **33**, 27–40.

Kuhn, T.S. (1970) *The Structure of Scientific Revolutions*, 2nd edn, Chicago University Press, Chicago.

Lane, D.C. (1998) Can we have confidence in generic structures? *Journal of the Operational Research Society*, **49**, 936–947.

Lane, D.C. (1999) Social theory and system dynamics practice. *European Journal of Operations Research*, **113**, 501–527.

Lane, D.C. (2000) *You just don't understand me: Models of failure and success in the discourse between system dynamics and discrete event simulation*, LSE OR Department, Working Paper LSEOR 00-34.

Lane, D.C. and Husemann, E. (2008) System dynamics mapping of acute patient flows. *Journal of the Operational Research Society*, **59**(2), 213–224.

Lane, D.C., Monefeldt, C. and Rosenhead, J.V. (2000) Looking in the wrong place for healthcare improvements: a system dynamics study of an accident and emergency department. *Journal of the Operational Research Society*, **51**, 518–531.

Lattimer, V.A., Brailsford, S.C., Turnbull, J.A. *et al.* (2004) Reviewing emergency care systems: insights from system dynamics modelling. *Emergency Medicine Journal*, **21**, 685–691.

Lawson, T. (1998) Economic science without experimentation, in *Critical Realism: Essential Readings* (eds M. Archer *et al.*), Routledge, London, pp. 144–169.

Lehaney, B., Malindzak, D. and Khan, Z. (2008) Simulation modeling for problem understanding: a case study in the East Slovakia coal industry. *Journal of the Operational Research Society*, **59**, 1332–1339.

Levin, S.R., Dittus, R., Aronsky, D. *et al.* (2008) Optimizing cardiology capacity to reduce emergency department boarding: a systems engineering approach. *American Heart Journal*, **156**, 1202–1209.

Little, J.D.C. (1986) Research opportunities in the decision and management sciences. *Management Science*, **32**(1), 1–13.

Lorenz, T. and Jost, A. (2006) Towards an orientation framework in multi-paradigm modelling. Proceedings of the 24th International Conference of the System Dynamics Society, Nijmegen, The Netherlands.

Lowery, J.C. and Martin, J.B. (1992) Design and validation of a critical care simulation model. *Journal of the Society for Health Systems*, **3**, 15–36.

Meadows, D.H. (1980) The unavoidable a priori, in *Elements of the System Dynamics Method* (ed. J. Randers), Productivity Press, Cambridge.

Meredith, J.R. (2001) Reconsidering the philosophical basis of OR/MS operations research. *Operations Research*, **49**(3), 325–333.

Mingers, J. (2000) The contribution of critical realism as an underpinning philosophy for OR/MS and systems. *Journal of the Operational Research Society*, **51**, 1256–1270.

Mingers, J. (2003) A classification of the philosophical assumptions of management science methods. *Journal of the Operational Research Society*, **54**, 559–570.

Mingers, J. (2006a) A critique of statistical modeling in management science from a critical realist perspective: its role within multimethodology. *Journal of the Operational Research Society*, **57**, 202–219.

Mingers, J. (2006b) *Realising Systems Thinking: Knowledge and Action in Management Science*, Springer, New York.

Mingers, J. and Brocklesby, J. (1997) Multimethodology: towards a framework for mixing methodologies. *Omega*, **25**(5), 489–509.

Mitroff, I.I. (1972) The myth of objectivity or why science needs a new psychology of science. *Management Science*, **18**(10), B613–B618.

Mitroff, I.I. (1994) The cruel science of world mismanagement: an essay in honor of C. West Churchman. *Interfaces*, **24**(4), 94–98.

Morecroft, J.D.W. and Robinson, S. (2005) Explaining puzzling dynamics: comparing the use of system dynamics and discrete event simulation, in *Proceedings of the 23rd International Conference of the System Dynamics Society* (eds J.D. Sterman, M.P. Repenning, R.S. Langer*et al.*), System Dynamics Society, Boston, MA.

Moreno, L., Aguilar, R.M., Martín, C.A. *et al.* (2000) Patient-centered simulation to aid decision-making in hospital management. *Simulation*, **74**, 290–304.

Peña-Mora, F., Sangwon, H., Lee, S.H. and Park, M. (2008) Strategic-operational construction management: hybrid system dynamics and discrete event approach. *Journal of Construction Engineering and Management*, **134**(9), 701–710.

Pidd, M. (2003a) *Computer Simulation in Management Science*, 4th edn, John Wiley & Sons, Ltd, Chichester.

Pidd, M. (2003b) *Tools for Thinking: Modeling in Management Science*, 2nd edn, John Wiley & Sons, Ltd, Chichester.

Pidd, M. (2004a) Simulation worldviews – so what? Proceedings of the 2004 Winter Simulation Conference.

Pidd, M. (2004b) *Computer Simulation in Management Science*, 5th edn, John Wiley & Sons, Ltd, Chichester.

Popkov, T. and Garifullin, M. (2006) Multi-approach simulation modeling: challenge of the future, in *Proceedings of the Asia Simulation Conference 2006 on Systems Modeling and Simulation: Theory and Applications* (eds K. Koyamada, S. Tamura and O. Ono), Springer, Tokyo, pp. 103–108.

Randers, L. (1980) *Elements of the System Dynamics Method*, Productivity Press, Cambridge.

Richardson, G. and Pugh, A. (1981) *Introduction to Systems Dynamics Modeling with DYNAMO*, MIT Press, Cambridge, MA.

Robinson, S. (2008) Conceptual modeling for simulation Part I: definition and requirements. *Journal of the Operational Research Society*, **59**, 278–290.

Rohleder, T.R., Bischak, D.P. and Baskin, L.B. (2007) Modeling patient service centers with simulation and system dynamics. *Health Care Management Science*, **10**, 1–12.

Ryan, J. and Heavey, C. (2006) Process modeling for simulation. *Computers in Industry*, **57**(5), 437–450.

Sargent, R.G. (2004) Validation and verification of simulation models. *Proceedings of the 2004 Winter Simulation Conference*, pp. 130–143.

Sterman, J.D. (2000) *Business Dynamics: Systems Thinking and Modeling for a Complex World*, Irwin McGraw-Hill, Boston, MA.

Sweetser, A. (1999) A comparison of system dynamics and discrete event simulation. Proceedings of the 17th International Conference of the System Dynamics Society and 5th Australian & New Zealand Systems Conference.

Tako, A.A. and Robinson, S. (2008) Model building a quantitative comparison. Proceedings of the 26th International Conference of the System Dynamics Society.

Tako, A.A. and Robinson, S. (2009) Comparing discrete-event simulation and system dynamics: users' perceptions. *Journal of the Operational Research Society*, **60**, 296–312.

Tan, B.A., Gubaras, A. and Phojanomongkolkij, N. (2002) Simulation study of Dreyer Care Facility. Proceedings of the Winter Simulation Conference, pp. 1922–1927.

Van Horn, R.L. (1971) Validation of simulation results. *Management Science*, **17**(5), 247–257.

Venkateswaran, J. and Son, Y.-J. (2005) Hybrid system dynamic—discrete event simulation-based architecture for hierarchical production planning. *International Journal of Production Research*, **43**(20), 4397–4429.

Vissers, J. and Wijnaard, J. (1979) The outpatient appointment system: design of a simulation study. *European Journal of Operations Research*, **13**, 459–463.

Wainer, G.A. (2009) *Discrete-Event Modeling and Simulation: A Practitioner's Approach*, Taylor & Francis, Boca Raton, FL.

Walsham, G. (1992) Management science and organisational change: a framework for analysis. *Omega*, **20**, 1–9.

Walter, S.D. (1973) A comparison of appointment schedules in a hospital radiology department. *British Journal of Preventive and Social Medicine*, **27**, 160–167.

Werker, G., Sauré, A., French, J. and Shechter, S. (2009) The use of discrete-event simulation modelling to improve radiation therapy planning processes. *Radiotherapy and Oncology*, **92**(1),76–82.

Wolstenholme, E. (1993) A case study in community care using system thinking. *Journal of the Operational Research Society*, **44**(9), 925–934.

Wolstenholme, E. (1999) A patient flow perspective of U.K. health services: exploring the case for new 'intermediate care' initiatives. *System Dynamics Review*, **15**(3), 263–271.

Worthington, D. (1991) Hospital waiting list management models. *Journal of the Operational Research Society*, **42**(10), 833–843.

6

Theoretical comparison of discrete-event simulation and system dynamics

Sally Brailsford
Southampton Business School, University of Southampton, UK

6.1 Introduction

As described in detail in Chapters 2 and 3, discrete-event simulation (DES) and system dynamics (SD) are two quite different approaches to simulation modelling. DES models systems as networks of queues and activities, where state changes in the system occur at discrete points of time. The objects in the system are distinct individuals, each possessing characteristics that determine what happens to that individual, and the activity durations are sampled for each individual from probability distributions. SD, in its quantitative form, models systems as networks of stocks and flows in which state changes occur continuously. The 'objects' in the system are a continuous quantity, like a fluid, which flows through a system of reservoirs or tanks connected by pipes.

This chapter discusses the differences between these two approaches and whether the choice of simulation methodology – DES or SD – is purely down to the personal preference or expertise of the modeller (I have a hammer, so every problem is a nail), or whether there are identifiable features of certain problems that make one approach intrinsically superior to the other. The chapter begins by briefly recapping the fundamental aspects of both approaches, although the reader is referred to Chapters 2

Discrete-Event Simulation and System Dynamics for Management Decision Making, First Edition.
Edited by Sally Brailsford, Leonid Churilov and Brian Dangerfield.
© 2014 John Wiley & Sons, Ltd. Published 2014 by John Wiley & Sons, Ltd.

and 3 for detailed descriptions. The two approaches are contrasted from a theoretical standpoint, and an illustrative example is then presented in some depth. The chapter concludes with some general guidelines to assist the modeller in making the choice of technique.

6.2   System dynamics

The foundations of SD were laid in the 1950s at MIT by Jay Forrester in his pioneering work on 'industrial dynamics' (Forrester, 1960, 1961). The fundamental principle of SD is that *structure determines behaviour*: in other words, the way that the different components of any system relate to and affect each other structurally determines the emergent behaviour of the system as a whole. Often such emergent behaviour can be counterintuitive, and it is only by analysis of the component parts that the reasons for this unexpected behaviour can be understood.

SD has two distinct aspects, one qualitative and one quantitative. The qualitative aspect involves the construction of causal loop or influence diagrams, which depict graphically the way in which the system elements are related. The aim is to enhance understanding of a problem situation through the structure of the system and the relationships present between relevant variables. Such models have the capability of using descriptive or judgemental data as well as numerical data. The overall emphasis is on policy rather than decisions. SD models are not used for optimisation or prediction.

One of the best known early examples of a qualitative SD application is Eric Wolstenholme's model for community care (Wolstenholme, 1993). The intention behind the UK government's community care legislation in the early 1990s was to reduce public spending on social services. This was to be achieved by transferring responsibility for the home care of the elderly from central government to local social services departments. Thus hospitals would no longer be able to discharge patients on purely medical grounds, but would first have to check that money was available in the local social services budget to support each patient before he or she was discharged. In theory, this would reduce the number of discharges and would reduce overall social services costs.

This intended effect is shown in the upper loop of the influence diagram in Figure 6.1. This is a balancing loop (as it contains one minus sign) and therefore appears to suggest that the new legislation should lead to a stable situation. However, the outer loop of this model is a reinforcing loop as it contains four minus signs. This loop shows that an (unintended) consequence of the Community Care Act would actually be to *increase* social services spending: as hospital discharges decrease, fewer beds become available, thus restricting admissions and increasing the number of sick elderly people out in the community awaiting hospital admission and meanwhile requiring costly care at home. Worse still, since the outer loop is a vicious circle the situation will eventually spiral out of control.

Wolstenholme chose SD because he wished to communicate with politicians and health care planners in order to expose how unintended effects might cause waiting

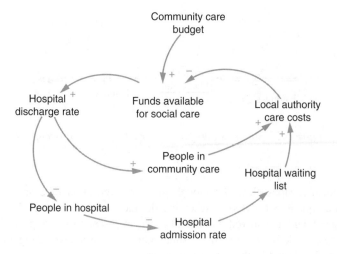

Figure 6.1   Wolstenholme's model for community care (after Wolstenholme, 1993).

lists to increase. Accurately estimating the actual numbers of patients was less important than gaining an understanding of how the system worked. The model was extended to include other types of residential care and was implemented in an interactive gaming mode. This is an excellent example of what Ghaffarzadegan (2010) termed 'powerful small models'. Small models may consist of a few significant stocks and at most seven major feedback loops. These small models can be used to communicate crucial insights to the public, or more generally any stakeholders who do not have in-depth knowledge about the whole system. Furthermore, many small SD models are capable of capturing vital, often counterintuitive insights of a complex problem. These small models enable policy makers to understand complex issues more easily.

Of course in practice the effect of the vicious circle in Figure 6.1 is mitigated by the balancing loop. The two loops counteract each other and the overall net effect cannot be determined merely by inspecting the diagram. To do this we would need to quantify the variables in Figure 6.1, and this is not always straightforward. Even though all the elements in this particular model are potentially measurable, the relationships between them are complex and would require a great deal of data collection and further analysis. Moreover, influence diagrams may include purely qualitative constructs such as 'happiness' and these are obviously hard to quantify numerically.

To develop a quantitative SD model, the influence diagram has to be converted to a stock–flow diagram. Stock–flow models can be conceptualised as a system of water tanks connected by pipes. The 'stuff' in the system is a continuous quantity and may represent money, people, material, product, and so on. Although the flows are theoretically continuous, the underlying equations used in software tools to solve such models are difference equations, usually solved by numerical integration.

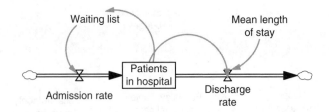

Figure 6.2   Simple stock–flow diagram of hospital occupancy.

Mathematically, stock–flow SD models are a discretisation of a set of ordinary differential equations (ODEs) representing the rates of change of the level of each stock; these ODEs are solved numerically using a discrete time-step d*t*.

A simple stock–flow model for hospital admissions is shown in Figure 6.2. This model was developed in the software Vensim.

The two 'clouds' represent people outside the system. The rectangular box is a stock representing the number of patients currently occupying beds in hospital. The two arrows (pipes) entering and leaving this stock represent admissions and discharges, and the valves on these pipes govern the rates of flow along them. In this model, patients are not really whole people: they are a continuous substance, like water, and therefore at any given point in time a fractional number of people are in hospital. The diagram also shows how influences can be incorporated in stock–flow models. The admissions rate is influenced by (i.e. is a function of) the length of the waiting list and the number of empty beds (not explicitly modelled here, but obviously directly related to the number of patients in the hospital). The discharge rate is a function of the average length of stay and the number of patients in hospital.

6.3   Discrete-event simulation

DES is arguably the most widely used operational research technique in practice. It is used to model systems that can be viewed as a queuing network. Individual objects (*entities*) pass through a series of *activities*, in between which they wait in queues. The rules governing the order in which these activities occur and the conditions for them to take place can be extremely complex. Each individual entity can be given characteristics that determine what happens to that individual in the system. The durations of the activities are usually sampled from probability distribution functions. The modeller has almost unlimited flexibility in the choice of these functions and the logic governing the flow of entities around the system. Indeed this flexibility can sometimes almost be a disadvantage, as it tempts the modeller into creating evermore complex, large models just 'because you can'.

DES models are by definition stochastic in nature and deal with distinct entities, scheduled activities, queues and decision rules. They are simulated in unequal time-steps (when 'something happens') and require large amounts of quantitative, numerical data. The aim of these models is often the comparison of scenarios, prediction or optimising specified performance criteria. They have traditionally been applied at a tactical or operational level. Model validation is an important issue because of the quantitative nature of the results. Reducing the variance of the simulation results can be extremely important, even in an era of powerful computers.

The preliminary or conceptual model in DES is the activity diagram, which can play a similar role to the influence diagram in SD in facilitating discussion between modeller and client to gain understanding of the system. An activity diagram shows the logic of the flow of entities through queues and activities in the form of a linked network of circles and rectangles. An example of a simple activity diagram is shown in Figure 6.3. The circle 'Wait' is a queue, and the rectangle 'Service' is an activity requiring the resource 'Server'. Customer entry and exit points in the system are shown by zigzag lines.

DES software makes frequent use of animation and graphics, and models can be made interactive; all these features are very useful for communication with stakeholders. Modern software can produce graphical output which is almost like watching a movie, and can be very powerful when selling a model to clients or problem owners. DES models produce a vast range of output, often showing the whole distribution of possible outcomes in addition to summary measures. However, each simulation run or iteration only represents one *realisation* of the system (one possible outcome), and highly variable systems require many iterations. Explaining the importance of carrying out multiple iterations is one of the key aspects of teaching simulation to business students, and is one of the potential drawbacks of user-friendly software. The vision of a computer simulation tool on every manager's desktop is alluring, but dangerous, unless the manager understands why decisions made on the basis of one single experiment can be very risky!

Chapter 10 reflects an ever-increasing trend towards user-friendly and cheap DES software. The advantages are obvious – it is easy to develop a rough working model very quickly, no knowledge of programming is required, and communication with clients is facilitated if the modeller can develop and run a demonstration model on his or her laptop during a single meeting. However, as stated above, these packages need to be handled with care; the simulation results need intelligent analysis by people with some knowledge of statistics, as it is easy to draw disastrously wrong conclusions. For some situations, for example in academic research or for very non-standard one-off

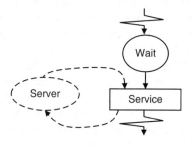

Figure 6.3   Activity diagram for a simple queuing system.

systems, the only solution may be to write the simulation from scratch. This can of course be very expensive and time consuming.

6.4 Summary: The basic differences

SD is essentially a deterministic approach, and purists might argue that it is not strictly correct to call it simulation. It does not consider individual 'entities' or handle variability very effectively, despite efforts by software vendors to introduce probability distributions. The advantage of this is that in comparison with DES, SD models are extremely quick to run as obviously they do not require multiple iterations. SD undeniably lacks the total flexibility of DES, which can use virtually any probability distribution function, or empirical data, to model state dwelling times. DES is capable of modelling great complexity and detail; the downside of this is that DES models can require a vast amount of data. Indeed the data collection part of a project can often be more time consuming than the model-building part. A major disadvantage of DES is that the data requirements, and the need to maintain a next-events list and to carry out long runs or multiple replications to get reliable results, mean that the models are relatively time consuming to develop and run.

Traditionally SD has been used at a higher, more aggregated and strategic level than DES and the data requirements of an SD model are generally much less. In fact, Forrester (1960) described SD models as 'learning laboratories' and, rather more contentiously, Lane (2000) argued that SD models are never more than 40% accurate. Validation of SD models is a tricky and indeed contentious issue given their qualitative nature; it is not possible to apply the standard battery of statistical validation tools that can be applied to a DES model, but simple approaches such as face validity and extreme value tests can still be used, and other methods have been developed (Rodrigues and Bowers, 1996). While SD models are useful for clarifying the complexities of organisational behaviour, their simplified representation of systems, the necessity to aggregate entities, and the use of average flow rates are some of their significant limitations. In contrast, DES models are much more flexible, capturing interactions between entities and detailed characteristics of the system being modelled.

In terms of software, there is undoubtedly a much wider range of DES products compared with SD. This almost certainly reflects the relative popularity of the two approaches in the 1980s and 1990s, and hence the relative maturity of the markets for these tools. DES software, of which there are dozens (if not hundreds) of products on the market, ranges in price from free shareware through to very expensive graphical tools costing five-figure sums. In recent years there has been a trend towards user-friendly SD software that does not require any knowledge of computer programming. Most SD packages have 'drag and drop' interfaces so that the user can select icons for levels, rates, and so on, connect them up, and edit their properties; the software then automatically generates the underlying equations and runs the model, collecting and presenting the output. However, there is a much smaller range of SD software available and it would still be fair to say that this is not yet as sophisticated in terms of graphics and user interfaces as DES software.

Philosophically, there is a profound difference between the two approaches which over the years has led to the development of two entirely separate research communities, with their own journals and conferences (Lane, 2000). DES has been universally recognised as part of the operational research (OR) 'toolkit' since the 1960s, whereas SD, although it has an equally long pedigree, has really only become widely accepted as a mainstream OR technique since the mid-1990s (Brailsford, 2008). DES is a standard part of all OR Masters programmes; SD is of course taught on many programmes, but by no means all. Basically, as Morecroft and Robinson explore in Chapter 9, SD modellers and DES modellers simply see the world in a different way. This in turn influences the way they conceptualise and model problems, which is far more subtle than merely the distinction between discrete and continuous, or deterministic and stochastic. Very crudely put, an SD modeller sees the world as a holistic synthesis of system elements which are dynamically connected, and takes a 'helicopter view' of the world, whereas DES modellers take a microscope to the world and look at the system in detail, paying attention to the variability between individual components. To say that SD is top-down and DES is bottom-up is clearly too much of a simplification, but it does give a flavour of one of the key conceptual differences between them.

As a modeller who uses both approaches in health care, I have observed a fascinating difference in the way that clinicians engage with DES and with SD. Initially, people are more attracted to DES because it accords with their fundamental belief that all patients are different and should be treated as individuals. Moreover, everyone likes pictures, and a DES model showing a floor plan of your own clinic or emergency department, with little people moving around on the computer screen, is very appealing. However, the temptation to include more and more detail in a DES model, partly 'because you can', but also because clinicians know that all sorts of things are really important, means that fairly soon the data requirements become onerous and the modeller has actively to curb the client's enthusiasm and make some very strict assumptions, which can lead to disenchantment with a model which seemed initially to promise so much.

Conversely, an SD model cannot possibly contain such individual detail and so the initial engagement process is very different. There is almost an innate aversion to a modelling approach which asks clinicians to think of patients as a homogenous mass: this is not the way doctors and nurses are trained to think. (Interestingly, public health specialists, who are used to working at the population level, do not seem to have this problem.) However, gradually, the fact that an SD model is able to capture the 'whole system' aspects which people vaguely understand, but have not always truly grasped because they work in silos, wins people over and they become very excited about the power of an SD model to investigate the knock-on effects of things that happen in different parts of the system. At this point, the fact that little people do not move around the screen in an SD model ceases to matter.

6.5 Example: Modelling emergency care in Nottingham

This section presents a case study in which both SD and DES were used. The work was undertaken in 2002 and was originally published in the *Journal of the Operational Research Society*, winning the OR Society's 2004 Goodeve Medal for the best paper published in that journal (Brailsford *et al.*, 2004). Nottingham, a city of about 640 000 inhabitants in the East Midlands of England, was at the time experiencing severe difficulties in coping with increasing numbers of emergency hospital admissions. However, this was just the tip of the iceberg: the real problems lay right across the whole emergency care system in the city. The model was developed as part of a research project led by Professor Valerie Lattimer of the School of Nursing and Midwifery at the University of Southampton. This project, commissioned in 2001 by the (then) Nottingham Health Authority, was itself part of a larger, ongoing project in Nottingham, known as the Emergency Care – On Demand (ECOD) project. The ECOD project was designed to look at the whole health care system, to determine why demand is so high, and to investigate what could be done to alleviate this pressure. The Southampton contribution involved carrying out a system review and providing research support to the ECOD project.

6.5.1 Background

Emergency or unscheduled care can be provided either in hospital or in the community. Many emergency hospital admissions occur as a result of patient visits to a hospital Emergency Department (ED). Patients can also be admitted directly to the wards, usually as a result of a referral by a general practitioner (GP). In both cases, some patients may arrive by ambulance whereas others travel to hospital independently. A third group of emergency patients are admitted directly from outpatient clinics. In the community, unscheduled care is provided in a number of ways. In normal surgery hours, patients may request urgent or same-day GP appointments. After the surgery is closed, patients wishing to see a doctor urgently need to contact an out-of-hours service. Back in 2002, GPs still had 24/7 legal responsibility for all their patients, so this would have been either a large co-operative of local GPs (Nottingham Emergency Medical Services, NEMS), or a commercial deputising service. Other services were (and still are) available, including NHS Direct, a national 24/7 telephone helpline where people can seek medical advice and information. Nottingham had a well-established NHS Direct, which was integrated with NEMS, so that patients simply dialled the NHS Direct number and, if a doctor's visit was required, were transferred directly through to NEMS and given an emergency appointment. Nottingham also had a nurse-run Walk-in Centre which operated extended hours, although not 24/7. Other community services providing health advice or access to the health care system included Social Services, pharmacist shops, the dental services and community mental health teams. Finally, the '999' emergency services – Fire, Police and obviously the Ambulance Service – provided emergency care and access to the NHS system.

The city is served by two acute NHS Hospital Trusts, Queens Medical Centre (QMC) and Nottingham City Hospital (NCH). Both are teaching hospitals, but only QMC had an ED. In 2001, outpatient attendances and inpatient admission rates were approximately three times the national average. At both hospitals there had been an increase in people needing emergency care for the past three years, with a 10% increase at NCH in 2000–2001 compared with 1999–2000. The ED at QMC was one of the busiest in England. All areas of the system were experiencing increasing pressures, manifesting itself in long waiting times for patients, stressed and over-worked staff, hospital wards running close to capacity limits, and fewer elective (planned) admissions as the hospitals struggled to cope with the workload generated by the emergencies.

This problem was by no means confined to Nottingham. Recent reports by the UK Audit Commission (1996, 2001) had highlighted the fact that, despite some improve-ments in a few areas, by and large in England and Wales ED waiting times, both to see a doctor and to be admitted to hospital, had increased steadily since 1996. Patient numbers attending EDs in England and Wales had increased by 1% per annum, whereas the number of nurses had remained roughly the same. However, the total number of doctors had increased by 10% since 1998. This was not, therefore, a simple problem of supply and demand, neither was it a straightforward issue of maximising the throughput of a production system. The Audit Commission report suggested that long waiting times were caused by 'a host of managerial and organisational differ-ences as much by resources and staff levels' (Audit Commission, 2001).

6.5.2   The ECOD project

The ECOD project grew out of an earlier initiative for winter crisis planning in Nottingham, when it became apparent that the 'crisis' was a chronic state of affairs rather than a temporary acute problem. A Steering Committee was set up early in 2001, containing representatives from all the health care providers in Nottingham, and a Project Team formed, chaired by a local GP. The aim was to develop a new Local Services Framework for emergency care, which would form the basis of future strategy in Nottingham. The Southampton team began work in August 2001 and completed the research project in April 2002. The research project actually comprised several strands, involving a literature review, activity data collection and analysis, stakeholder interviews and a patient preference survey. Simulation models were developed to answer the following three questions about the emergency care system:

- How is the system currently configured, and what organisational systems, processes and responsibilities support it?

- What characteristics of demand, demand management and patient flows can be identified from retrospective analysis of activity data, observational data and the views of key informants?

- How should the system be developed to respond to health policy and local needs, and what are the economic implications?

6.5.3   Choice of modelling approach

In this study, we were dealing with a very large, complex system involving a population of over 600 000 potential patients. Furthermore, we considered that although the specific pathways followed by individual patients were of interest, they were of less importance than understanding the major flows of people through the 'front doors' to the NHS, and gaining insight into the general structure of the system and the relationships between its component parts. The problems experienced in the ED, for example, were not principally felt to be due to high variability in casemix or staffing levels, but more to the sheer volume of demand and consequent pressure on resources. Finally, we were less concerned with the waiting times of individual people than with the general flow of patients through the system, in order to identify bottlenecks. Thus SD was chosen as the modelling approach and we used both qualitative and quantitative aspects.

The aim of the qualitative phase was to develop understanding of the ECOD system, not only by the research team, but also by the stakeholders in the system. It was not merely a preliminary stage to the quantitative modelling, but was important in its own right. Many useful insights were gained as a result of the development of the conceptual map and through the interview process. The research team made an initial orientation visit to Nottingham, in which a first-pass 'conceptual map' of the system was drawn up. The aim of this was to list the 'front doors' or access points to the health care system, and then to expand this. The conceptual map was printed on A3 paper and used as the basis of 30 semi-structured interviews with key individuals from all the main health care providers, together with patient representatives. A greatly simplified version of this conceptual map is shown as Figure 6.4. WIC denotes the walk-in centre, NHSD stands for NHS Direct, and OP stands for outpatient. The diamond on the left depicts the patient's (bewildering) choice of service to contact.

The stakeholder participants were selected in consultation with the Project Team and the Steering Committee. During the interviews, participants were asked about their work roles and the capacity they felt they had to influence the interface between their part of the system and other components. This led on to a discussion of the interfaces between components and the factors which might influence patient flows through the system. Participants were asked to draw on the map to show these influences and to annotate or alter the map in any way they felt appropriate. As a result a final agreed version of the map was derived and was later used as the basis for a quantitative computer model of the system using the software Stella.

6.5.4   Quantitative phase

The aim of this phase was to facilitate experimentation with various potential changes in service configurations and demand rates. A stock–flow modelling approach was used, where stocks represented accumulations of patients (e.g. waiting to see a GP, waiting for treatment in the ED, or occupying a bed in an acute admission ward) and the flows were the admission, transfer, treatment and discharge rates. The layout of the computer screen followed that of the conceptual map, so that the top half represented

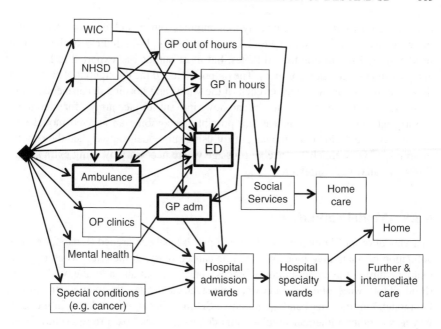

Figure 6.4   Simplified conceptual map of the emergency health care system in Nottingham.

the primary care sector (in-hours and out-of-hours GP surgeries, NHSD and WIC) and the bottom half represented the secondary sector (the two main hospitals). In the middle were the Ambulance Service, Social Services and the ED. We did not attempt to model every single hospital ward, but just the admissions wards, as we were concerned principally with patient flows within the emergency system.

The Stella model was populated with data for the year April 2000 to March 2001, obtained from the various providers in Nottingham. These comprised the patient arrivals, broken down where possible by hour and day, sex and age band, and where appropriate category of urgency; the source of the arrival; and the destination (e.g. emergency hospital admission, discharge home or elsewhere). Hospital length of stay data were derived from the Hospital Episode Statistics provided by the Department of Health (DH, 2002). This enabled flow balance cross-checking to be carried out, although the quality and level of detail of the data were variable. The outflow to B reported by A must equal the inflow from A reported by B. Unfortunately no system-wide data were available for the in-hours GP sector, although we collected prospective data for a single week from four individual practices. We therefore had to rely solely on the hospital data regarding GP admissions, which essentially produced a discontinuity in the model for this particular flow.

As in all stock–flow systems, the contents of each stock or reservoir are updated at regular intervals by solving a set of difference equations representing the inflows and outflows from that stock. The choice of the time-step dt was difficult, given the wide range in activity durations (some only took minutes, others took days or even weeks) but we chose a value of d*t* equal to 2.4 hours (0.1 days, 144 minutes). Stella presents results in the form of graphs and tables, but most output was exported to Excel for analysis and presentation purposes. The output included the throughput of each 'front door' and the occupancy rates of each of the wards and hospital departments. Stella allows the user to break down stocks and flows into subscripted arrays, for example to classify patients by age, but it is not possible to combine an arrayed model with submodels. We decided that the benefits of using submodels outweighed the benefits of arrays, as we were able to account for age where necessary by using extra stocks, flows and auxiliary variables.

6.5.5   Model validation

The validation of SD models is a thorny topic. It has been argued (Lane, 2000) that validation of qualitative models should be carried out with the client as an ongoing dialogue during the model-building process, and is essentially a 'white box' process (Pidd, 2004), where the client knows, understands and trusts the internal structure of the model. On the other hand, quantitative SD models can be validated, in the same way as any other numerical simulation model, by a 'black box' process (Pidd, 2004) where emphasis is not on the model structure, but on the output it produces.

In our case we used both approaches. We developed the model in close collaboration with the Steering Group during frequent visits to Nottingham. In addition to the inflow–outflow balance checking described above, we carried out black box validation by running the model for the period April 2000 to March 2001, using the known arrivals data, and comparing the model output with real-life system performance data which had not been used in the construction of the model. For example, we used the total daily bed occupancy supplied by the hospitals, and compared this with the corresponding model output by aggregating all the individual ward bed occupancies to give confidence that the model was producing sensible output. Similar plots were obtained for other output parameters such as individual ward occupancies.

6.5.6   Scenario testing and model results

The Steering Committee suggested a range of scenarios for testing, based on the comments of the interview participants. For example, it was suggested that GPs may have been admitting some patients as emergencies in order to get investigations carried out more rapidly which could equally well be performed as day cases or even outpatients, because of the lack of suitable facilities. This is a similar behavioural response to that identified in Lane *et al.*'s study in London (Lane *et al.*, 2000). A community Diagnostic and Treatment Centre (DTC) where such tests could be done could therefore prevent many 'unnecessary' admissions.

A planning horizon of five years was used. The scenarios included the 'Doomsday scenario' (maintaining current growth in demand with no additional resources) and a variety of possible alternatives, including:

• 3% year-on-year growth in GP referrals for planned admissions;

• reduced emergency admissions for certain patient groups (e.g. the elderly or people with respiratory disease), for example by the use of a DTC;

• earlier discharge of the elderly to nursing homes;

• the effects of *streaming* in the ED, that is separate resources for certain patient groups.

The key outputs from the system map and Stella model were, initially, the insights gained into different parts of the system by people seeing it as a whole for the first time. Simple influence diagrams (small powerful models) describing parts of the system were found to be a powerful tool in stimulating debate. For example, it could be argued that long waiting times in the ED are not necessarily always a bad thing, in that the expectation of a long wait might discourage 'inappropriate' attenders and lead them to seek help elsewhere, perhaps in the WIC or by phoning NHSD. Thus, inadequately thought-out initiatives designed to reduce waiting times in the ED might actually turn out to be counterproductive.

The first main result from the scenarios was that the system is currently operating dangerously close to capacity. This reinforced the message coming across from many of the stakeholder interviews. The model showed that if growth in emergency admissions continues at the current rate, both hospitals will see a significant decrease in the number of elective admissions within four years. NCH, for example, could expect to see at least a 25% drop in elective admissions by 2005 (from 1100 per month in 2000–2001 to 700 in 2004–2005). The scenario where planned GP admissions were constrained to increase by 3% per annum was even worse, with average bed occupancies exceeding 100% by 2005 (assuming no additional resources).

The model also showed how small changes to one part of the system can have a considerable impact elsewhere in the system. For example, the effect on average total bed occupancy of sending 3% of patients aged over 60 to a DTC instead of admitting them is shown in Table 6.1. If a 3% reduction were maintained year on year for five years, a significant decrease in total occupancy could be achieved. The

Table 6.1 Average percentage occupancy of both hospitals, assuming a sustained year-on-year decrease of 3% in emergency admissions of people aged over 60.

	NCH	QMC
'Status quo'	86.7	84.7
2000–2001	85	83
2001–2002	84	82
2002–2003	83	81
2003–2004	82	80
2004–2005	80	79

bed occupancy target for 2004 set by the government (DH, 2001) was 82% and the current figures for QMC and NCH are 84.7 and 86.7% respectively. Bagust, Place and Posnett (1999) used DES to show that it is risky to have average occupancy figures higher than 85%.

Interventions targeted at patients with specific health problems, such as respiratory conditions or ill-defined diagnoses, did have an effect, though it was not large. Reducing emergency admissions for patients with respiratory problems (by 20% per annum year on year for four years) reduced overall bed occupancy by approximately 2%, a small annual effect. However, the seasonal nature of the reductions in admissions gave increased benefits, as the January peak in occupancy was more significantly reduced relative to other months.

Interventions aimed at preventing 3 or 6% emergency admissions of patients over 60 years of age made a substantial difference in the model. Even without assuming any decrease in average length of stay, bed occupancy in both hospitals was reduced by 1% per annum over the five-year duration. This is to be expected since people in this age group comprise around about half of all emergency admissions.

We evaluated the effect of early discharge for patients admitted as emergencies who were subsequently discharged to nursing homes. Despite the common perception of 'bed-blockers', discharging these patients two days early made hardly any difference to overall occupancy rates, and there appeared to be surprisingly little potential for improvement in this area. We also investigated the effects of discharging from hospital on seven days a week. This showed a small decrease in occupancy, though care needs to be taken in interpreting the model results here, since the admission days for elective patients are currently planned to accommodate weekday discharging. However, some benefit might still be achieved. Overall, though, the model showed that the effects of discharging these people earlier were minimal compared with the effects of keeping them out of hospital in the first place.

6.5.7   The ED model

We were asked to investigate the government's suggestion (DH, 2001) that waiting times in the ED could be reduced by the provision of 'fast track' systems for minor injuries or illnesses. Patients streamed in this way would have their own waiting area and dedicated staff, and would not share resources with other ED patients. Streaming patients appears counterintuitive from a queuing theory perspective, until we take into account the fact that different categories of patients have different acceptable waiting times and hence different targets. Thus, although some patients may have to wait for longer, their waiting time could still be within acceptable limits.

Unfortunately SD does not lend itself to narrowly focused systems involving resource-constrained queuing networks. For problems requiring this level of individual detail, DES is the method of choice. A separate, very simple DES model for the ED was therefore rapidly developed using the software SIMUL8 and was populated with patient arrival and staff resource-level data from the ED at QMC. Activity duration data were derived from the literature (Audit Commission, 1996; Shrimpling, 2002) as there was not time to gather primary data in this study.

Table 6.2   Results from the ED streaming model.

	Performance indicator	Without streaming	Streaming minor cases
Shared doctors	% Utilisation	70	73
Stream doctors	% Utilisation	—	58
Cat. 2 treatment	% Queued less than 10 min	96	83
Cat. 3 treatment	% Queued less than 60 min	99	70
Cat. 4 treatment	% Queued less than 120 min	87	100
Cat. 5 treatment	% Queued less than 240 min	86	99

On arrival in the ED, patients are initially prioritised into five urgency categories, where 1 denotes life-threatening conditions and 5 denotes minor injury or illness. This process is called triage. Category 1 patients are always seen immediately, but lower category patients are seen in priority order as resources permit and may have to wait. We investigated the streaming of minor cases (triage categories 4 and 5). We found that the permanent streaming of minor injuries was not an efficient use of clinical resources. Improvements were observed for the less urgent patients, but these were at the expense of patients in categories 2 and 3. The results for this scenario are shown in Table 6.2.

A flexible system appeared to be required in which streaming is only triggered when waiting times reach a certain threshold. A compromise solution for Nottingham was to dedicate one doctor to the fast track patients, and have a second doctor on standby to join the first doctor if there was a sudden rush of minor cases. Other solutions may involve the use of Emergency Nurse Practitioners to deal with less serious patients, releasing doctors to work with the more serious cases. Further DES modelling could help here, for example in determining the threshold for initiating streaming.

6.5.8   Discussion

Both the qualitative and quantitative aspects of the SD approach proved to be very useful in this project. The conceptual map provided a helpful structure around which to base the stakeholder interviews. Many participants commented on the value of seeing the whole system in its entirety, often for the first time, and on the insights they gained about how other parts of the system related to the part with which they were familiar. Although influence diagrams were not constructed for the entire system, they were used to gain insight into the behaviour of parts of the system.

The Stella model was useful on two levels – first, naturally, for investigating specific scenarios in terms of patient flows and bottlenecks, but second (and perhaps equally importantly) as a device for provoking and facilitating discussion and comment. Interestingly, although the Steering Group were initially fascinated by the computer model and the visual and numerical output, they readily accepted the idea that the model gave an indication of the relative effects of different interventions rather than mathematically precise forecasts or point predictions. They were very keen to suggest alternative scenarios for testing, arising from the findings of earlier runs of the model.

However, the Stella model was not able to address the issues in the ED and it was necessary to build a separate DES model to answer the specific questions about streaming.

This study was conducted in partnership with a health and social care community in Nottingham already committed to the concept of partnership working and the need for a 'whole systems approach' to development. The process and findings of this independent enquiry contributed to sustained local efforts to find better solutions for the benefit of the people of Nottingham, and informed the articulation of a local service framework for emergency care.

6.6   The $64 000 question: Which to choose?

How do you choose which method to use? Are some systems 'naturally' better modelled by SD or by DES – and if so, why? Does the choice of technique simply depend on the personal preference of the modeller, because people tend to stick to what they know best and feel most comfortable with? Or what software they happen to own, or can afford? The answers may depend less on the system being modelled than on the *purpose* of the model: what sort of questions do we want our model to answer?

The main criteria on which a selection could be based are summarised in Table 6.3. Of course, these are not hard and fast rules, and value judgements are involved: for example, how large is 'large'? Many problems will have features of both approaches and it will be necessary to prioritise, and decide which approach tackles the most important issues.

For example, suppose the system under consideration is a hospital outpatients clinic. We might choose SD if we are looking at the clinic in a broad context, and we are interested in its interaction with other parts of the hospital or the community health service; if all patients have roughly the same kind of appointments; if there are large

Table 6.3   Criteria for selection of modelling approach.

	DES	SD
Scope	Operational, tactical	Strategic
Importance of variability	High	Low
Importance of tracking individuals	High	Low
Number of entities	Small	Large
Control	Holding (queues)	Rates (flows)
Relative timescale	Short	Long
Purpose	Decisions: optimisation, prediction and comparison	Policy making: gaining understanding

numbers of patients; if we want qualitative output (e.g. why certain clinics overrun, and how this affects discharges from surgical wards); and if we have a timescale of weeks, months or years. On the other hand, we might choose DES if we are looking at the clinic in a narrow context, and there is little contact with the outside world; if individual patients differ considerably in behaviour, such as length or complexity of appointment; if there are relatively small numbers of patients; if we want quantitative output (patient waiting times, resource utilisation, etc.); if we want to compare scenarios, for example different staffing levels; and if we have a timescale of hours or days.

Two comparisons in the literature are of interest in this context. It should be pointed out that both comparisons were made by aficionados of SD, and moreover that the earlier comparison by Randers (1980) was not a direct comparison between SD and DES, but between SD and quantitative modelling approaches in general. The technological developments in computer hardware and software since 1980, when Randers made his comparison, also affect the validity of this comparison today.

Randers' comparison is shown in Figure 6.5. The nine axes represent the criteria for comparison. Few would disagree with Randers' view that SD outperforms DES in the areas of *insight generating ability*, *mode reproduction ability* (the ability for models to be used in different contexts), *fertility* (the ability to spark off creative new ideas through the modelling process), and *ease of enrichment* (the ease with which the model can be extended or developed). Adherents of DES might argue that today DES models are equally, or even more, *transparent* than SD models because of the powerful graphical and animation facilities provided by modern software. For the same reasons DES enthusiasts might also disagree with Randers that SD models score higher on *descriptive realism*.

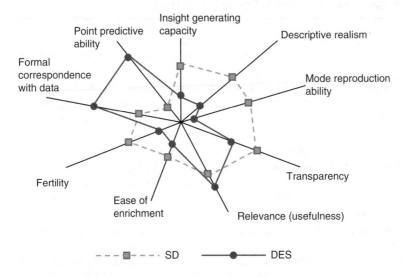

Figure 6.5   Randers' comparison (after Randers, 1980).

The keenest proponent of SD would have to admit that DES is superior in the areas of *formal correspondence with data* and *point predictive ability*, although in fairness SD does not attempt to compete here. In my personal view, both approaches can be equally *useful and relevant* if used appropriately, although it remains true that DES is used more widely than SD.

Lane (2000) presented the comparison shown in Table 6.4. Lane argues that DES models are analytic, in the literal sense that they break a system down into its constituent parts. The complexity of a DES model lies in the detail, whereas SD takes a holistic view and its complexity lies in the dynamic interactions between the elements of the system. Lane also argues that clients find DES models convincing, but that they do not really understand all the underlying mechanics of the model, for example the probability distributions, sampling procedures and simulation methodology – three-phase, process-based. He uses Pidd's analogy of modelling (Pidd, 2004), contrasting a white box, where the client totally understands the internal workings of the model, with a black box, where the client has no idea whatsoever what goes on inside the model. Lane argues that DES is a 'dark grey box' where the client has an inkling of how the model works, whereas SD is a 'fuzzy glass box' where the client has really quite a good grasp of what the model does. I feel that while this may be true of qualitative SD models, it is unlikely to be true of quantitative models; few clients would feel comfortable with numerical integration or the mathematical formulae underlying a large complex model.

In Lane's experience clients find SD models compelling; he believes they are excited by SD models whereas they find DES models more mundane (although convincing). It should be borne in mind that Lane is a proponent of SD, and users of

Table 6.4   Lane's comparison (due to Lane, 2000).

	DES	SD
Perspective	Analytic, emphasis on detail complexity	Holistic, emphasis on dynamic complexity
Resolution	Individual entities, attributes, decisions and events	Homogenised entities, continuous policy pressures and emergent behaviour
Data sources	Numerical with some judgemental elements	Broadly drawn
Problems studied	Operational (?)	Strategic (?)
Model elements	Physical, tangible plus some information	Physical, tangible, judgemental and information links
Human agents	Decisions	Policies
Clients find the model	Opaque, 'dark grey box'; convincing	Transparent, 'fuzzy glass box'; compelling
Outputs	Point predictions, performance measures	Understanding of structural source of behaviour modes

DES might well argue that their clients have found their models both exciting and convincing! However, Lane calls for closer links between the two communities of modellers, which can only be of mutual benefit.

6.7 Conclusion

There is a school of thought that says that it does not matter what modelling approach is used, it is the *process* which counts, and that getting people together round a table to talk about the problem situation is the really important thing. The argument is that in this case the model is of secondary importance and acts only as a catalyst to motivate discussion, or as a boundary object (Carlile, 2002) providing a common ground for different stakeholders to interact; see Chapter 7 for further details on the social role of models as interfaces between stakeholders. While we have sympathy with this view, we would also argue that the approach chosen *does* matter, as it focuses attention on certain aspects of the real-world system which are quite different, according to the approach adopted.

This chapter has attempted to give an overview of DES and SD, to describe the strengths and weaknesses of each approach in (hopefully) an unbiased way, and to offer some guidelines to potential users about the selection of an appropriate technique for a given problem. We echo David Lane's opinion that more communication and discourse between the communities of SD and DES modellers would have great benefit.

References

Audit Commission (1996) *By Accident or Design: Improving A&E Services in England and Wales*, HMSO, London.

Audit Commission (2001) *Accident and Emergency Acute Hospital Portfolio: Review of National Findings*, Audit Commission, London.

Bagust, A., Place, M. and Posnett, J.W. (1999) Dynamics of bed use in accommodating emergency admissions: stochastic simulation model. *British Medical Journal*, **319**, 155–159.

Brailsford, S.C. (2008) System dynamics: what's in it for healthcare simulation modelers. *Proceedings of the 2008 Winter Simulation Conference, Miami, FL, USA, December*, pp. 1478–1483.

Brailsford, S.C., Lattimer, V.A., Tarnaras, P. and Turnbull, J.A. (2004) Emergency and on-demand health care: modelling a large complex system. *Journal of the Operational Research Society*, **55**, 34–42.

Carlile, P. (2002) A pragmatic view of knowledge and boundaries: boundary objects in new product development. *Organization Science*, **13**, 442–455.

Department of Health (2001) Reforming Emergency Care. Downloadable from http://webarchive.nationalarchives.gov.uk/+/www.dh.gov.uk/en/publicationsandstatistics/publications/publicationspolicyandguidance/dh_4008702 (accessed 26 November 2013).

Department of Health (2002) *Hospital Episode Statistics*, Department of Health, London.

Forrester, J.W. (1960) The impact of feedback control concepts on the management sciences, in *Collected Papers of J.W. Forrester* (1975 collection), Wright-Allen Press, Cambridge, MA, pp. 45–60.

Forrester, J.W. (1961) *Industrial Dynamics*, MIT Press, Cambridge, MA (reprinted by Productivity Press 1994 and now available from Pegasus Communications, Waltham, MA).

Ghaffarzadegan, N. (2010) How small system dynamics models can help the public policy process. *System Dynamics Review*, **27**, 22–44.

Lane, D.C. (2000) *You just don't understand me: modes of failure and success in the discourse between system dynamics and discrete event simulation.* LSE OR Dept Working Paper, LSEOR 00-34.

Lane, D.C., Monefeldt, C. and Rosenhead, J.V. (2000) Looking in the wrong place for healthcare improvements: a system dynamics study of an accident and emergency department. *Journal of the Operational Research Society*, **51**, 518–531.

Pidd, M. (2004) *Computer Simulation in Management Science*, 5th edn, John Wiley & Sons, Ltd, Chichester.

Randers, J. (1980) *Elements of the System Dynamics Method*, MIT Press, Cambridge, MA.

Rodrigues, A. and Bowers, J. (1996) System dynamics in project management: a comparative analysis with traditional methods. *System Dynamics Review*, **12**, 121–139.

Shrimpling, M. (2002) Redesigning triage to reduce waiting times. *Emergency Nurse*, **10**, 34–37.

Wolstenholme, E.F. (1993) A case study in community care using systems thinking. *Journal of the Operational Research Society*, **44**, 925–934.

7

Models as interfaces

Steffen Bayer,[1] Tim Bolt,[2] Sally Brailsford[3] and Maria Kapsali[4]
[1]Program in Health Services & Systems Research, Duke-NUS Graduate Medical School Singapore
[2]Faculty of Health Sciences, University of Southampton, UK
[3]Southampton Business School, University of Southampton, UK
[4]Umea School of Business and Economics, Umea University, Sweden

7.1 Introduction: Models at the interfaces or models as interfaces

Simulation models and simulation modelling are used in many different ways. The contexts and objectives of modelling projects vary as much as the approaches and tools used. The system modelled is only part of what determines the modelling process and the modeller often is only one of the stakeholders influencing or being influenced by the model. Others, namely model users such as decision makers or students in a teaching context or participants in a group model building project, also interact directly with the model. Further stakeholders might be influenced indirectly by decisions made based on the model.

In this chapter we pay particular attention to the use of models in group model building projects where a group of domain experts or other stakeholders come together to build a model together with a modelling expert. This chapter draws attention to how models can function as an 'interface' between participants in a modelling project. An interface can be understood as a point of interaction that allows two systems to communicate across a boundary. In everyday use 'interface' is, however, also often

Discrete-Event Simulation and System Dynamics for Management Decision Making, First Edition.
Edited by Sally Brailsford, Leonid Churilov and Brian Dangerfield.
© 2014 John Wiley & Sons, Ltd. Published 2014 by John Wiley & Sons, Ltd.

used as relating to what affords and enables the process of communication across the boundary: interface as the means or mechanism that allows the boundary between subsystems to be permeable. In this chapter we use the metaphor 'interface' loosely in this sense.

The 'interface' metaphor highlights the potential of models (and of the modelling process) to support information transmission in the widest sense between the participants in a group modelling project, but also across the boundary between the model and those engaging with the model. The aim of this chapter is to explore and to illustrate how the interface can shed light on the variety of functions models can have, especially in a group modelling context.

The chapter is motivated by the assumption that reflection on the use of simulation models and the role they are given by stakeholders not only supports the modeller in making a modelling project more effective, more implementable or more insightful for the different stakeholders, but could also help to avoid conflicts and misunderstandings if these stakeholders have a different understanding of the function of simulation models and the role of the modelling process, or if the role of a model evolves over time within a project.

7.2   The social roles of simulation

When participants from different organizational and professional backgrounds come together to develop a simulation model to solve a problem, the emerging model can support them in communicating their diverse experiences and perspectives (which are often expressed in very specific language influenced by their organizational and professional backgrounds). The discussion of relationships and mechanisms to be included in the model, as well as the agreement on decisions about timescales and model boundaries, can give rise to new insights and a shared understanding of the system. Moreover, experiments with a model can also help participants to understand what others are talking about when presenting their perspective of the system. Modelling can support communication and foster understanding in modelling groups with diverse members. Silos, communities of interest and practice, professional and organizational differences, and languages can all make communication between stakeholders difficult. During the process of model building the simulation can take on a variety of social roles.

Boundary objects are artefacts shared between communities of practice, which have their own specific informational codes (Star and Griesemer, 1989; Carlile, 2002; Sapsed and Salter, 2004). Boundary objects can address some of the difficulties in communicating and creating knowledge across (disciplinary and organizational) boundaries. These difficulties include not only the syntactic and semantic challenges of having to overcome differences in language and interpretation, but also those challenges inherent in creating new shared knowledge and dealing with any negative consequences for the participants which might arise from this shared knowledge creation process (Carlile, 2002). Boundary objects can be repositories of knowledge, standardized forms and methods, objects or models, or boundary maps to support interdisciplinary working (Star and Griesemer, 1989). However, while boundary objects can be the basis of negotiation and knowledge exchange, they can also be ineffectual, precisely because their role is at the margins of communities, and their use depends on the frequency of interaction and level of understanding within groups (Sapsed and Salter, 2004).

Simulation can sometimes serve as a boundary object. In a variety of domains, modelling has been shown to be able to support situations where disparate stakeholders need to create new knowledge. In large, complex, transdisciplinary research areas, models can become the facilitators of interdisciplinarity, integrating different knowledge bases (Mattila, 2005). Simulation modelling has been shown to act as a boundary object in engineering planning (Dodgson, Gann and Salter, 2007a), helping to bridge disparate communities involved in innovating, and in particular allowing disparate groups to engage within innovative projects and propose potential solutions to engineering problems (Dodgson, Gann and Salter, 2007b). Models have also been conceptualized as interfaces in policy decision processes drawing on scientific research. In such situations, models can be an interface between scientists and policy makers, making scientific research findings accessible to decision makers (Kolkman, Kok and van der Veen, 2005).

Models can be used to make predictions about outcomes in the real world and allow decision makers to experiment with different courses of action in a safe, quick and low-cost way. However, as has been shown in the case of engineering (Dodgson, Gann and Salter, 2007b), simulation modelling can also help to shape the conversation between stakeholders solving problems together and to foster collaboration. Models can help to achieve this across professional, institutional and other interpersonal or interorganizational boundaries.

By building on the insights of the literature on boundary objects and also its application to group model building (Zagonel, 2002), and drawing on the distinct literature on objects as epistemic and technical objects (Ewenstein and Whyte, 2009), it is possible to distinguish four different social roles of models (Kapsali *et al.*, 2011). Boundary objects and representative objects can both be epistemic and technical objects. Models used as epistemic objects help to create new knowledge either because the system overview provided by the model already triggers new insights for the group, or because the simulation of the model results in new insights. In either case, engagement with the model adds to the total knowledge available, while for models used as technical objects the existing knowledge is communicated among the group members.

The two dimensions of boundary vs representational objects and epistemic vs technical objects therefore allow a stylized classification of four types of model roles to be made (Kapsali *et al.*, 2011). Models which as boundary objects facilitate communication between stakeholders with different knowledge bases can be used to create new knowledge (as epistemic objects) by the stakeholder group, or can be used to make available across the group knowledge which individual members might already possess (as technical objects). Models primarily used to represent a reality which is seen as principally unproblematic can again be used in two different ways: as a micro world or management flight simulator to allow the user to learn; or as a predictive tool to allow the user to draw on the knowledge embodied in the model without necessarily requiring an understanding of the relationships within the system (Kapsali *et al.*, 2011).

While in a typical modelling project these roles will not appear in their pure forms, the roles nevertheless point towards the different ways models are used and the different ways models can act as interfaces. Different stakeholders might have different views of the role of the model: for example, a client might have a predictive tool in mind at the outset, while the modelling process might show that what is required (or maybe in some cases achievable) has to be learned as a group. Over time the role of a model might change: learning as a group might be followed by an expression of knowledge and experimentation, and then by the development of a predictive tool for other users or of a game as a learning environment for students to explore (Kapsali *et al.*, 2011). The discussion of the social roles of models highlights that models have complex and potentially changing roles which go beyond 'prediction' (as a technical and representative object), and in these other roles the transmission of information across boundaries is important.

This interface character can be seen across these social roles (see Table 7.1). As a technical and boundary object a model helps to transmit information between group members by demonstrating already existing knowledge ('express') and by giving group members lacking this knowledge the opportunity to engage with the system ('experiment'). As an epistemic and representative object the model allows those that

Table 7.1   Simulation models as interfaces across social roles.

	Epistemic object (create knowledge)	Technical object (make knowledge available)
Boundary object (facilitate communication across boundaries)	Model allows stakeholder group to learn by *acting as an interface between the group members* as well as an *interface to engage with (preliminary) relationships* captured in the model	Model helps to *transmit information between group members* by demonstrating already existing knowledge and by giving group members lacking this knowledge the *opportunity to engage* with the system
Representative object (represent reality)	Model allows exploration of a filtered and simplified version of the real world. The model as an *interface to engage with the 'real-world' relationships* captured (partly and selectively) in the model	*Access to model outputs* which contain predictions about a constrained subset of the 'real world'

engage with it to create new insights by exploring a filtered and simplified version of the real world. The model can be seen as an interface not between members of the stakeholder group but as an interface to engage with the real-world relationships captured (partly and selectively) in the model. As an epistemic and boundary object a model acts like an interface to both the ways just described: it allows a stakeholder group to learn as a group by acting as an interface between the group members, as well as an interface to engage with the preliminary relationships captured in the model. In this latter case, these relationships captured in the model might be seen as relatively more problematic and more preliminary in their claim to making a statement about the 'real world' than for a model seen as a representative object: essentially these relationships are seen as capturing the mental models of (some members of) the group. In any case, a model is only a snapshot of the whole system: the model necessarily (and importantly for clarity) has to concentrate on what is relevant for a decision, discussion or system behaviour. As Alfred Korzybski memorably expressed, 'The map is not the territory' (Korzybski, 1931). So even a model as representative object can be seen as an interface, filtering aspects of reality and making available outputs which have some predictive power with regard to the expected behaviour of the 'real world'.

7.3   The modelling process

A simulation model might be a product available to others, since developing a simulation model is a process and there are opportunities to learn from both the product and the process. As members of the group learn from each other, they develop and change the model – but also change their own assumptions. Engagement with the model and learning might be a recursive process. Facilitation can play an important role here. Learning can happen throughout the stages of the modelling process and can occur as learning from the model, from modelling and from simulation of the model.

A modelling group can involve very different participants and organizations. Their interaction and later decisions are influenced by explicit and implicit decisions on the interactive use of the model and model building process. The process of engagement with each other is helped by the model acting as an interface to translate between group members and allow learning. Modelling encourages agreement to be reached and that what has been agreed is consistently coded into the model. Clearly, there is a danger that some group members will not fully understand what has been included in the model, or are ignored and overruled, and that therefore disagreement or uncertainty is essentially 'black boxed' (i.e. remain concealed).

During the process of building a simulation model in a group, a shared conceptual model of the issue and the system needs to be developed, which can then be reified or coded as a simulation model (Jonassen, Strobel and Gottdenker, 2005). Engaging with a model allows the members of a simulation group not only to clarify their own understanding, but also to examine the consistency of their own preconceptions through hypothesis testing by simulating the model.

Changes in the model as different versions are built over time reflect changes in the mental models of the members of a model building group. This development is, of course, also shaped and limited by what the specific model building approach allows to be implemented easily: simulation modelling approaches impose their own syntax. The model-based reasoning that the group engage in involves a critical examination of the variables, factors, parts and relationships within their own conceptual models (Jonassen, Strobel and Gottdenker, 2005).

The modelling process begins with a great degree of openness during the creation of the model which then becomes a well-defined artefact that allows experimentation only within boundaries. Simulation models are constrained structures, which limits how they can be interpreted (see also Knuuttila and Voutilainen, 2003), and they are less flexible than a drawing on a blackboard. Simulation models become content rich and increase in rigidity and structure during their development process. This materialization process depends on the learning processes, composition and often the professional background of the group participants, among other factors.

Information about the system will be drawn from the model. Group participants learn about different parts of the system or roles within the systems they are not familiar with. The model therefore allows participants to get an overview of the whole system. Often, there are important insights to be gained from achieving such an overview, in particular in highly fragmented systems where individuals often have constricted roles and no conscious exposure to the wider interrelationships in the system. This might explain some of the reasons why 'whole-systems modelling' has been particularly promoted in healthcare settings where such roles are prominent and complex, and often messy problems involving numerous stakeholders and overlapping priorities have to be considered.

Both the conceptual models of the members of the modelling group and the computer model are necessarily incomplete and fragmentary representations of how the system actually works. As an externalized collective mental model is agreed on and built in the group, the group members will construct, deconstruct and reconstruct their conceptual models drawing on experiences to create sets of structures, factors and variables.

The final computer model is inevitably a compromise between the different conceptual models of the different stakeholders. Other factors also play a role in the conceptual and problem definition phases, such as the definition of needs.

As the model is simulated on the computer, the simulation becomes a source of information: as the model is run it acts as an interface between expected behaviour and the behaviour of the model implied by the assumptions embedded in it. Three sources of learning then build on each other: the modelling, which leads to the model, which can then be simulated; all three are potential sources of insight (see Figure 7.1).

How models can influence and support learning and function as an interface will depend on how the model development process is facilitated and managed; group learning and model scope are influenced, guided and shaped by the management of the model development process.

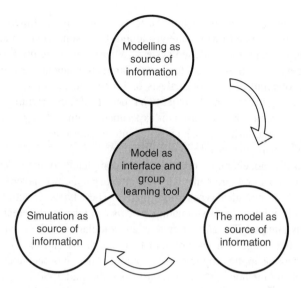

Figure 7.1   Modelling, model and simulation as sources of information.

7.4   The modelling approach

The way models act as interfaces will also depend on the level of abstraction or aggregation in the model. Models need to have the right level of abstraction and aggregation to be useful. From a technical perspective, the amount of complexity and model scope needs to be considered to capture what is relevant for the problem at hand. However, in addition to this technical perspective, what is appropriate for the client and modelling group is also important. These stakeholders might often deal with systems on a more detailed level and might object if some of 'their detail' is not included – even if the modeller argues that it does not affect system behaviour and might obscure some of the clarity of the model. Simplicity in the model itself might, however, not help its communicability if it omits what the audience is looking for.

In the ways of the model as interface described in this chapter, what we consider normally the interface of the model, that is the visual interface, might be significant as it helps to make model outputs and/or relationships within the system accessible to a lay audience. Models can particularly well support 'learning as a group' if they are easily changeable so that suggestions from the model building group and experiments can be rapidly implemented and interactively explored. Such models would typically be simple and visually accessible to stakeholders who might have only limited experience of simulation modelling or understanding of the mathematical under-pinning of models. Frequently in such modelling projects, insights into relationships between variables or parts of the system might be more the focus than precision of the modelling output. Models used to 'predict' might in contrast have the characteristics of being fixed, detailed and precisely focused. While the visual interfaces might still be important, the emphasis might now be more on the visual attractiveness of the output than on the degree to which the visual interfaces support an understanding of the relationships within the modelled system. When models are used to experiment and explore devices such as sliders to change parameters quickly, gaming interfaces or the ability of rapid sensitivity analysis or immediate model simulation can be important. The requirements of the other two types of roles in our framework will fall between these extremes. Models used to 'experiment' with and 'express' knowledge or to 'explore' a system should make insights into relationships easily accessible but need not be so easily changeable in their structure.

While specific models might not correspond completely to these ideal types, and while the exact model requirements will be context specific, we nevertheless believe that this classification of ideal types is informative. These types are actually the vessel in which the different types of simulation models can take up different social roles.

System dynamics (SD) and discrete-event simulation (DES) can be seen to represent the two ends of a spectrum in their emphasis and explanatory power, though both may be applied to the same situations. There has been a discussion and comparison of the methods in the literature since the mid-1990s, the most notable being those by Sweetser (1999), Lane (2000) and Brailsford and Hilton (2001). These themes have been more fully explored by Morecroft and Robinson (2005, 2006) and Tako and Robinson (2009a, 2009b), as well as by modellers subsequently looking at strategies for combining SD and DES in hybrid models.

The differences between the two approaches can be classed into four categories (see Table 7.2):

- the characteristics of the problem/decision under consideration;

- the data requirements and the development process;

- the type of understanding derived; and

- the model output and usability by clients (often based on visual representation).

The social roles of models will affect to a large degree the success of problem solving and decision making and the levels of understanding and interaction among the stakeholders. While it has to be recognized that the actual domains of use of SD and DES might overlap to a wide extent, and modellers can successfully apply SD and DES tools to problems across a spectrum from strategic to tactical (Tako and Robinson, 2009a, 2009b), the characterizations of the SD and DES literature nevertheless allow a hypothesis to be formulated on the 'natural domains' of both modelling approaches. Although the combination of epistemic and boundary object (the top left area in Table 7.1) could be suggested as the natural domain of SD, the combination of technical and representative object seems more the home of DES (the bottom right area in Table 7.1). In this way, consideration of social roles can be added to the traditional criteria for selecting the modelling techniques, leading to a more comprehensive toolbox that will benefit the group decision-making process.

Table 7.2 Key differences between SD and DES modelling in the modelling literature.

Areas	Characteristic	Typical SD use	Typical DES use
Problem/decision type	Decision level	Strategic decisions in systemic and population levels	Operational decisions
	Perspective	Systemic overviews of population level where individual variation is statistically subsumed	Operations level where events impact one another and variations of individuals cumulate or interact
Data requirements and development process	Base data sources	Qualitative to identify system behaviour and find feedback loops; then supported by data to complete stock levels and flow rates	Model build from individual components, putting together entities
	Uncertainty and randomness	Deterministic runs based on provided parameters, feedback loops and delays	Explicit randomness in parameters for each modelled activity and event
Type of understanding derived	Key technical learning	Systemic interactions and feedback effects	Impact of randomness/variation and potential bottlenecks under runs
	Scope of learning	Overall population-level changes for long-term planning efforts	Variation expected for service delivery decisions and contingency planning
Model output and usability by clients	Primary usage mode	Not optimization; understanding influences	Playing with the models; 'what-ifs'
	Representation	System represented as stocks and flows with explicit feedback	System represented as events and queues with implicit feedback effects
	Common user concerns about entities	Lack of individuality among human entities	Probability distributions for each event and entity
	Common user concerns about structure	Continuous, smooth curves and stock accumulation do not match perceptions of users	Rearranging components completely changes interactions

The focus on feedback loops and systemic interactions in SD models more readily supports experimenting with the model in such a way as to create new knowledge about the relationship between system structure and behaviour, and might help to reveal systemic effects of policies and other interventions, while DESs might be very powerful in predicting the impact of randomness on system behaviour.

More empirical work is required to analyse whether this suggested understanding of the natural domains of different modelling approaches corresponds to the actual use of these two approaches and to successful outcomes. Clearly such work will also have to consider the differences between stakeholders in terms of knowledge domains, language used, incentives and social ties, as well as the problem characteristics and the system (e.g. importance of randomness and feedback, relevant level of aggregation, operational vs strategic focus), together with the goal of the planning process or of the modelling engagement.

7.5   Two case studies of modelling projects

In this section we illustrate the discussion with two case studies of group model building projects which show how models can be used as interfaces in the ways discussed in this chapter.

Both projects were done in the healthcare area; they do nevertheless illustrate the diversity of simulation projects. In both projects, modellers met with the expert group on a number of occasions using evidence and knowledge provided by the expert group, together with other sources, to build the model and populate it with data.

In the first project a team of consultants helped to develop a model to understand what the choices might be in providing services and developing public health strategies to reduce a specific type of hospital admission in a locality. This modelling project used SD and involved a diverse stakeholder group drawn from a variety of organizations. A consultant worked on the model between meetings and the (developing and changing) model was presented at every meeting of the stakeholder group, where it was discussed in detail by the group. Over time the model took on (or was given) different roles and became an interface in all the ways indicated in Section 7.2 (see Table 7.3). Especially in the early phase of the project, when it was important to agree on the structure of the system, the model served as a focus for the group to clarify their different understandings of the system and what different group members from different backgrounds actually meant when talking about the system and their experience. The need to capture relationships in a precise form and to a high degree of precision was a challenge for the participants in presenting their views on the relationships. The model helped to transmit information between the group members even before scenarios were explored, or it was used to predict the possible impact of different strategies. The model gave the participants an access point to engage with the mechanisms and real-world relationships captured in the model, and helped them to understand the interdependencies between parts of the system when experiments were suggested and conducted in the group meetings. The model finally produced simulation results as predictions of the likely impact of chosen strategies.

Table 7.3   The model as interface in the first project.

Interface function	Illustrative interview quotes
Interface between the group members and interface to engage with (preliminary) relationships	Discussions about definitions in order to develop language 'because every member in the group had a different definition for every word that you could possibly come up with' (project 1, interview 3) 'I think it helps people discuss . . . yes, it did create a lot of discussion, not necessarily about the model, but about where we were at . . . what we were doing, what needed to be done, so it definitely generated a lot of discussion' (project 1, interview 8) 'the more complicated discussion was, well, what evidence have we got to show the effectiveness of that treatment, affecting the rate of flow from one stock to the next' (project 1, interview 1)
Transmit information between group members	'So at this point they have learnt along the way how it works, what it does and how it works. It's not about how the model works, it's about how the system works, and they don't necessarily see the system' (project 1, interview 3)
Interface to engage with the 'real-world' relationships as captured in model	'was a good learning process . . . to plan scenarios, test scenarios in the model, because that brought it to life' (project 1, interview 1)
Access to model outputs	'I think the main part for me was the capacity that, using the model, you could see what would have the biggest influence and where you needed to target your resources' (project 1, interview 8) 'it was a visual representation that sort of helped me to understand and see the effects' (project 1, interview 4)

In this first project learning occurred in close relationship with the model and throughout the phases of the project. Reflecting on the project, both modellers and stakeholders emphasized the learning aspects of the work. The group were so diverse and came from such different backgrounds and roles that engaging in conversation to define the model helped reveal various insights ('learning from modelling'). The group finally agreed on a structure, even though some members were concerned that aspects of the system they were particularly involved with were not in the model. However, one can clearly see that this model structure contributed to learning ('learning from the model'), albeit to a much smaller extent than learning from the modelling process and also learning from the simulation. There was some degree of overlap of these phases: experimenting with the model and the results from test runs helped decide which elements of the system were relevant for inclusion in the final model. This was not a smooth process and was not without conflict; not all disagreements could be resolved.

The second project was a DES project where a simulation model was built based on discussions with an expert group drawn from two departments providing the same type of service to different parts of a larger healthcare organization. This work investigated the operational efficiency of different practices and studied the impact of changes in the availability of resources on operational performance. The expert group were (in addition to a dedicated data collection exercise) used as a source of data and also involved in decisions about what aspects of the system should be included and what scenarios should be explored. Interestingly the expert group were never shown the actual model on a screen, but instead were presented with a flowchart showing the processes, as well as with results from model experiments that had been run between meetings. The model was, however, not modified and simulations were not run during the group sessions.

In this model there was a lot of shared decision making on what should be included in it and what should be tested with it, but the model itself played only a small role within the group: 'we had a model which obviously will be run through experiments, and . . . we'll discuss experiments and the results of the experiments, but . . . there was no such thing as a coding or a drawing of the schematic representation or a model . . . together'(project 2, interview 1).

The learning process was different from the one in the first project: the stakeholders involved in the project had a good overview of the system and knew each other well, so there was much less to be learned from the modelling process. The expertise of the stakeholders informed the model building and decisions on what was important to include, but the participants themselves gained less from the discussion. However, there were some insights drawn from the detailed data collection which was undertaken to build the model. Learning from modelling and learning from the model were far less important than learning from the simulation: in this project the model served mainly as a tool to provide answers to stakeholders' questions about the consequences of resource use and the policies of operational performance. Interestingly, there was no direct engagement with the model at all: the modellers built the model and experimented with it offline, only reporting the results back to the expert group.

The emphasis of the work was very much on developing as accurate a model as possible so that results could be derived from the simulation; thus the spirit of the modellers was very much 'put this in, try this, try that and then some results' (project 2, interview 2). The group made suggestions to the modeller on what they wanted to be examined with the model: 'The things that we wanted it to model or to test, we put those ideas through but we didn't actually sit and do them' (project 2, interview 3). However, even here, the participants did learn about the need to make their assumptions precise and from the data that was collected for the model: 'And there certainly was things that were perhaps assumptions or unwritten assumptions, unwritten rules that came out of the woodwork' (project 2, interview 4). The meetings were used to discuss the underlying assumptions and outputs as well as to investigate the reasons why the behaviour of the model was different from what was expected: 'I would say that's the most creative we got. It was exploring the reasons why perhaps the result wasn't how we expected' (project 2, interview 5). So there was collective learning in the group about the model and the system, but this was very much about the results of the simulation.

There were a number of reasons why the modellers did not make use of the model in the expert group meetings, such as the size and complexity of the model and the fact that the modellers had used complicated 'workarounds' to implement some features of the process under the constraints imposed by the modelling package and which were difficult to communicate to non-modellers. Showing an animation of a model run was considered unwise as a single run would not be representative of the variance of the stochastic process and therefore very likely to be misleading.

The two projects described here are not representative, but just illustrative, of the different ways models can be used in a model building group. They show the breadth of difference between modelling projects even though both are in healthcare and conducted with the participation of an expert group. They also illustrate typical differences between SD and DES projects. While it is clear that there is flexibility in how these approaches are used, the two projects can be seen as not atypical and illustrate the key differences between the two approaches: the more strategic project used SD, while the more operational project used DES. This operational vs strategic difference is related to the homogeneity vs diversity of the stakeholder group: the more homogeneous stakeholder group in the second project had far less difficulty in understanding each other and far more understanding of the system, so there was much less need of the model as an interface between group members. Moreover, the basic working of the system was relatively clear to everybody in the group, the uncertainty being more in the operational level variance in process times which were gathered in a detailed data collection exercise involving observations. The system structure was also relatively clear to everybody, so there was less scope to learn from the model structure than in the first project. Insights were gained mostly from the detailed data collected and from the simulation runs, but far less from an increased understanding of systemic relationships. For this reason the model in the second project had less of a role as an interface to engage with the 'real-world' relationships as captured in the model. Because the data required was on an individual level and because stochastic effects were important in the performance of the system, these features influenced the building and simulating of the model within the modelling group, as an individual run would be misleading in a DES model but not in a normally deterministic SD model. This reduced the role of the model in the group meetings of this project.

7.6   Summary and conclusions

A model can help a group to communicate, by being consistent in laying down assumptions about how the systems works and helping to show the consequences of those assumptions: modelling, models and simulation can support learning.

To classify the social roles of simulation models, two dimensions can be distinguished: models can be boundary objects or representative objects; and they can be epistemic or technical objects. The metaphor of an interface is relevant for all social roles of a model. The model can act as an interface between the members (boundary object) of a modelling group in a group modelling exercise, as well as in different ways between the group and the system: it can transmit information between group members; engage with 'real-world' relationships as captured in the model; and give access to model outputs. Projects vary in the extent to which models serve as interfaces in these different ways: SD group model building projects typically (though not necessarily) emphasize the model much more as an interface between group members and as an interface to systemic relationships, while DES models are more typical in situations where the model is mainly an interface to the model output as a prediction of the likely consequences of the implementation of different choices in the represented system.

References

Brailsford, S. and Hilton, N. (2001) A comparison of discrete event simulation and system dynamics for modelling healthcare systems. Proceedings of the 26th Meeting of the ORAHS Working Group 2000, Glasgow Caledonian University, Glasgow, Scotland.

Carlile, P.R. (2002) A pragmatic view of knowledge and boundaries: boundary objects in new product development. *Organization Science*, **13**(4), 442–455.

Dodgson, M., Gann, D. and Salter, A. (2007a) In case of fire, please use the elevator: simulation technology and organization in fire engineering. *Organization Science*, **18**(5), 849–864.

Dodgson, M., Gann, D. and Salter, A. (2007b) The impact of modelling and simulation technology on engineering problem solving. *Technology Analysis & Strategic Management*, **19**(4), 471–489.

Ewenstein, B. and Whyte, J.K. (2007) Picture this: visual representations as 'artifacts of knowing'. *Building Research and Information*, **35**, **1**, 81–89.

Jonassen, D.H., Strobel, J. and Gottdenker, J. (2005) Model building for conceptual change. *Interactive Learning Environments*, **13**(1–2), 15–37.

Kapsali, M., Bolt, T., Bayer, S. and Brailsford, S. (2011) The materialization of simulation: a boundary object in the making. HaCIRIC 11 International Conference, Manchester, 26–28 September.

Knuuttila, T. and Voutilainen, A. (2003) A parser as an epistemic artefact: a material view on models. *Philosophy of Science*, **70**(5), 1484–1495.

Kolkman, M.J., Kok, M. and van der Veen, A. (2005) Mental model mapping as a new tool to analyse the use of information in decision-making in integrated water management. *Physics and Chemistry of the Earth*, **30**, 317–332.

Korzybski, A. (1931) A non-Aristotelian system and its necessity for rigor in mathematics and physics. Proceedings of the American Mathematical Society, New Orleans, LA, 28 December.

Lane, D.C. (2000) You just don't understand me: modes of failure and success in the discourse between system dynamics and discrete event simulation. LSE OR Dept Working Paper, LSEOR 00-34.

Mattila, E. (2005) Interdisciplinarity 'in the making': modelling infectious diseases. Working Papers on the Nature of Evidence: How well do facts travel, London School of Economics.

Morecroft, J. and Robinson, S. (2005) Explaining puzzling dynamics: comparing the use of system dynamics and discrete-event simulation. Proceedings of the 23rd International Conference of the System Dynamics Society, Boston, MA.

Morecroft, J. and Robinson, S. (2006) Comparing discrete-event simulation and system dynamics: modelling a fishery, in *Proceedings of the Operational Research Society Simulation Workshop 2006 (SW'06)* (eds J. Garnett *et al.*), Operational Research Society, Birmingham, pp. 137–148.

Sapsed, J. and Salter, A. (2004) Postcards from the edge: local communities, global programs and boundary objects. *Organization Studies*, **25**, 1515–1534.

Star, S. and Griesemer, J. (1989) Institutional ecology, 'translations', and boundary objects: amateurs and professionals in Berkeley's Museum of Vertebrate Zoology. *Social Studies of Science*, **19**, 387–420.

Sweetser, A. (1999) A comparison of system dynamics and discrete event simulation. Proceedings of the 17th International Conference of the System Dynamics Society and 5th Australian & New Zealand Systems Conference, Wellington, New Zealand.

Tako, A.A. and Robinson, S. (2009a) Comparing discrete-event simulation and system dynamics: users' perceptions. *Journal of the Operational Research Society*, **60**(3), 296–312.

Tako, A.A. and Robinson, S. (2009b) Comparing model development in discrete event simulation and system dynamics. Proceedings of the Winter Simulation Conference WSC'09.

Zagonel, A. (2002) Model conceptualization in group model building: a review of the literature exploring the tension between representing reality and negotiating a social order. Proceedings of the 20th International Conference of the System Dynamics Society, Palermo.

8

An empirical study comparing model development in discrete-event simulation and system dynamics*

Antuela Tako and Stewart Robinson
School of Business and Economics, Loughborough University, UK

8.1 Introduction

Simulation is a modelling tool widely used in operational research (OR), where computer models are deployed to understand and experiment with a system (Pidd, 2004). Two of the most established simulation approaches are discrete-event simulation (DES) and system dynamics (SD). They both started and evolved almost simultaneously with the advent of computers (Wolstenholme, 1990; Robinson, 2005), but very little communication existed between these fields (Lane, 2000; Brailsford and Hilton, 2001). This is, however, changing with more DES and SD academics and practitioners showing an interest in the other's world (Morecroft and Robinson, 2005).

---

*This chapter is a slightly adapted version of the published paper: Tako, A. A. and Robinson, S. L. (2010). Model development in discrete-event simulation and system dynamics: an empirical study of expert modellers. *European Journal of Operational Research*, **207**(2), 784–794. Reproduced with permission from Elsevier (License Number: 2878780919919).

Unfortunately there is little assistance for this interest, since work reporting on comparisons of the two simulation approaches is limited. The comparisons that exist are mostly opinion based, derived from the authors' personal views based on their field of expertise. Hence, little understanding exists regarding the differences and similarities between the two simulation approaches. More specifically, this chapter explores the model development process as followed by expert modellers in each field.

Both DES and SD models are simplified representations of a system developed with a view to understanding its performance over time and to identifying potential means of improvement. DES represents individual entities that move through a series of queues and activities at discrete points in time. Models are generally stochastic in nature. DES has been traditionally used in the manufacturing sector, while recently it has been increasingly used in the service sector (Robinson, 2005). Some key DES applications include airports, call centres, fast food restaurants, banks, health care and business processes.

In SD, systems are modelled as a set of stocks and flows, adjusted in pseudo-continuous time. SD models are based on differential equations and are generally deterministic. Feedback, which results from the relationships between the variables in the model, is an important feature in SD models. SD has been applied to a wide range of problems. Applications include economic behaviour, politics, psychology, defence, criminal justice, energy and environmental problems, supply chain manage-ment, biological and health care modelling, project management, educational problems, staff recruitment and also manufacturing (Wolstenholme, 1990).

While the underlying aims of using the two simulation approaches are similar, it is our belief that the approach to modelling is very different among the two groups of modellers. This chapter provides an empirical study that compares the model development process followed in each respective field. The study follows the processes used by 10 expert modellers, five from each field, during a simulation modelling task. Each expert was asked to build a simulation model from a case study on forecasting the UK prison population. The case was specifically designed to be suitable for both DES and SD. Verbal protocol analysis (Ericsson and Simon, 1984) was then used to identify the processes that the modellers followed. This study provides a quantitative analysis that compares the modelling process followed by the DES and SD modellers. Its key contribution is to provide empirical evidence on the comparison of the DES and SD model development process. Its underlying aim is to bring closer the two fields of simulation, with a view to creating a common basis of understanding.

The chapter is outlined as follows. It starts with a review of the existing literature comparing DES and SD, including the comparison of the model deve-lopment process in DES and SD modelling as described in the literature. This is then followed by an explanation of the research approach, describing the case study, the participants, verbal protocol analysis and the coding undertaken. The results of the study, including the quantitative analysis of the protocols of the 10 expert modellers and some observations on how they tackle the case, are then presented. Finally, the main findings and the limitations of the study are discussed.

8.2   Existing work comparing DES and SD modelling

This section reviews the existing literature on the comparison of the two simulation approaches, DES and SD. First, the main comparison studies are briefly considered in chronological order, followed by an account of the views expressed on the model development process in DES and SD.

The first comparison work was that of Meadows (1980) which looked into the epistemological stance taken in SD as a modelling approach and in econometrics (representative of DES). Based on the basic characteristics and the limitations of each paradigm, the author goes on to identify a paradigm conflict resulting from the modellers' perception of the world and the problems concerned, the procedures they use to go about solving them, as well as the validity of model outcomes.

Coyle (1985) approaches the discussion from a SD perspective, while considering ways to model discrete events in an SD environment. His comparison focuses on two aspects: randomness existing in DES modelling and model structure; he claims that open- and closed-loop system representations are developed in DES and SD respectively.

In her doctoral thesis, Mak (1993) developed a prototype of automated conversion software, where she investigates the conversion of DES activity cycle diagrams into SD stock and flow diagrams. DES process flow diagrams, which could be considered as more close to stock and flow diagrams, were not considered in her study.

Baines et al. (1998) provide an experimental study of various modelling techniques, on DES and SD among others, and their ability to evaluate manufacturing strategies. The authors comment on the capability of a modelling technique based on the time taken in building models, flexibility, model credibility and accuracy.

Sweetser (1999) compares DES and SD based on the established modelling practice and the conceptual views of modellers in each area. He maintains that both simulation approaches can be used to understand the way systems behave over time and to compare their performance under different conditions. When comparing the two simulation approaches, Sweetser considers the following aspects: feedback effects and system performance, mental models, systems' view and type of systems represented, the model building process followed and finally validation of DES and SD models. Two DES and SD conceptual models of an imaginary production line are then compared.

Brailsford and Hilton (2001) compare DES and SD in the context of health care modelling. The authors compare the main characteristics and the application of the two approaches, based on two specific health care studies (SD and DES) and on their own experience as modellers. At the end, they provide a list of criteria to assist in the choice between the two simulation approaches, considering problem type, model purpose and client requirements. Theirs is only a tentative list and they admit that the decision is not simple and straightforward.

Lane (2000) provides a comparison of DES and SD, focusing mainly on the conceptual differences. Considering three modes of discourse, Lane maintains that DES and SD can be presented as different or similar based on the position taken (the mode of discourse). In the third discourse a mutual approach is taken. The main comparison aspects considered are: the modeller's perspective on complexity, data sources, problem type, model elements, human agents, model validity and outputs. It should be noted that some of the statements have been contradicted. For example, model outputs in DES do not always represent point predictions (Morecroft and Robinson, 2005). An example of a DES model used to understand the system behaviour can be found in the paper by Robinson (2001).

Morecroft and Robinson (2005) is the first study that undertakes an empirical comparison, using a fishery model. The authors build a step-by-step simulation model, using DES (Robinson) and SD (Morecroft) modelling. However, one could claim the existence of bias, as the two modellers were aware of each other's views on modelling. The authors conclude with a list of differences between DES and SD, regarding system representation and interpretation. They become aware of the different modelling philosophies they take when developing their models, but they still contemplate that there is not a straightforward distinction between the two approaches, but rather it is a result of careful consideration of various criteria.

An empirical study on the comparison of DES and SD from the users' point of view has been carried out by Tako and Robinson (2009). The authors found that users' perceptions of two simple DES and SD models were not significantly different, implying that from the user's point of view the type of simulation approach used makes little difference if any, as long as it is suitable for addressing the problem situation at hand. So far, no study has been identified that provides an unbiased empirical account on the comparison of the DES and SD model development process.

The opinions expressed regarding the comparison of DES and SD are built around three main areas: the practice of model development; modelling philosophy; and the use of respective models. There appears to be a general level of agreement in the opinions stated on the nature of the differences, but also exceptions and contradictions exist. In addition, limited empirical work has been done to support the statements found in the literature. A long list of the views expressed can be compiled. Table 8.1 provides only some of the key differences proposed in the literature.

8.2.1   DES and SD model development process

Considering the modelling process, as suggested in DES and SD textbooks teaching the art of modelling, one can identify similarities between the two approaches, especially in terms of the stages involved. It is clear that the main stages followed are equivalent to generic OR modelling (Hillier and Lieberman, 1990; Oral and Kettani, 1993; Willemain, 1995), which include: problem definition; conceptual modelling; model coding; model validity; model results and experimentation; implementation. For instance, see Figure 8.1 which shows a typical DES modelling process from Robinson (2004).

The main aspects pertaining to the model development process are considered next. First, it is claimed that during the DES model building process emphasis is put on the development of the model on the computer (model coding). In their experimental study, Baines et al. (1998) commented specifically that the time taken in building a DES model was considerably longer compared with a SD model. Furthermore, Artamonov (2002) developed two equivalent DES and SD models of

Table 8.1 Examples of views expressed on the comparison of DES and SD modelling.

Aspects compared	DES	SD	Author(s)
Nature of problems modelled	Tactical/operational	Strategic	Sweetser, 1999; Lane, 2000
Feedback effects	Models open-loop structures – less interested in feedback	Models closed-loop structures based on causal relationships and feedback effects	Coyle, 1985; Sweetser, 1999; Brailsford and Hilton, 2001
System representation	Analytic view	Holistic view	Baines *et al.*, 1998; Lane, 2000
Complexity	Narrow focus with great complexity and detail	Wider focus, general and abstract systems	Lane, 2000
Data inputs	Quantitative, based on concrete processes	Quantitative and qualitative, use of anecdotal data	Sweetser, 1999; Brailsford and Hilton, 2001
Validation	Black-box approach	White-box approach	Lane, 2000
Model results	Provides statistically valid estimates of system performance	Provides a full picture (qualitative and quantitative) of system performance	Meadows, 1980; Mak, 1993

the beer distribution game model (Senge, 1990) and commented on the difficulty involved in coding the model on the computer. He found the development of the model on the computer more difficult in the case of the DES approach, whereas the development of the SD model was less troublesome. Baines *et al.* (1998) attributes this to the fact that DES encourages the construction of a more lifelike representation of the real system compared with other techniques, hence resulting in a more detailed and complex model.

On the other hand, Meadows (1980) highlights that system dynamicists spend most of their modelling time specifying the model structure. Specification of the model structure consists of the representation of the causal relationships that generate the dynamic behaviour of the system. This is equivalent to the development of the conceptual model.

Another feature concerning DES and SD model development is the iterative nature of the modelling process. In DES and SD textbooks, it is highlighted that simulation modelling involves a number of repetitions and iterations (Randers, 1980; Sterman, 2000; Pidd, 2004; Robinson, 2004). Indeed, an iterative modelling process is depicted in Figure 8.1 for both DES and SD. Regardless of the modeller's experience, a number of repetitions occur from the creation of the first model, until

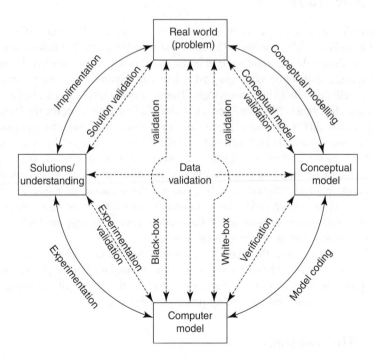

Figure 8.1   The DES modelling process based on (Robinson, 2004, p. 211). Reproduced with permission from the McGraw-Hill Companies.

a better understanding of the real-life system is achieved. So long as the number of iterations remains reasonable, these are in fact quite desirable (Randers, 1980).

8.2.2   Summary

The existing studies comparing DES and SD tackle three main areas of modelling: the practice of model development; modelling philosophy; and the use of models. These studies, however, are mostly opinion based and lack any empirical basis. We focus on the model development process in each field (DES and SD). The key aspects identified consist of the amount of attention paid to the different stages during modelling, the sequence of modelling stages followed, and the pattern of iterations followed. Taking an empirical perspective, we examine in this chapter the statements found in the literature by following the model development process adopted by expert DES and SD modellers. While it is expected that DES and SD modellers pay different levels of attention to different modelling stages, both DES and SD modellers are expected to follow iterative modelling processes. However, we do not have specific expectations about the exact pattern of iteration that the two groups of modellers will follow.

8.3   The study

The overall objective of the study is to compare empirically the stages followed by expert modellers while undertaking a simulation modelling task. We believe that DES and SD modellers think differently during the model development process. Therefore, it is expected that while observing DES and SD experts developing simulation models, these differences become evident. The authors use qualitative textual analysis (Miles and Huberman, 1994) and perform a quantitative analysis of the resulting data to identify the differences and similarities in the model development process. The current chapter compares DES and SD modellers' thinking process by analysing the modelling stages they think about while developing simulation models. The aim is to compare the model building process followed by DES and SD modellers regarding: the attention paid to different modelling stages, the sequence of modelling stages followed, and the pattern of iterations. In order to provide a more complete picture, some observations are also made on the models developed and the DES and SD experts' reactions during the modelling sessions.

This section explains the study undertaken. First, the case study used is briefly described, followed by a brief introduction to the research method employed, namely verbal protocol analysis (VPA). Next, we report on the profile of the participants involved in the study and the coding process carried out.

8.3.1   The case study

A suitable case study for this research needs to be sufficiently simple to enable the development of a model in a short period of time (60–90 minutes). In addition it needs to accommodate the development of models using both simulation techniques so that the specific features of each technique (randomness in DES vs deterministic models in SD, aggregated presentation of entities in SD vs individual representation of entities in DES, etc.) are present.

After considering a number of possible contexts, the prison population problem was selected. The prison population case study, where prisoners enter prison initially as first-time offenders and are then released or return to prison as recidivists, can be represented by simple simulation models using both DES and SD. Both approaches have been previously used for modelling the prison population system. DES models of the prison population have been developed by Kwak, Kuzdrall and Schniederjans (1984), Cox, Harrison and Dightman (1978) and Korporaal *et al.* (2000), while SD models have been developed by Bard (1978) and McKelvie *et al.* (2007); the UK prison population model of Grove, MacLeod and Godfrey (1998) is a flow model analogous to an SD model.

The UK prison population example used in this research is based on Grove, MacLeod and Godfrey (1998). The case study starts with a brief introduction to the prison population problem with particular attention to the issue of overcrowded prisons. Descriptions of the reasons for and impacts of the problem are provided. More specifically, two types of offenders are considered, petty and serious. There are in total 76 000 prisoners in the system, of which 50 000 are petty and 26 000 serious offenders. Offenders enter the system as first-time offenders and receive a sentence depending on the type of offence; on average 3000 petty offenders vs 650 serious offenders enter each year. Petty offenders receive a shorter sentence (on average 5 years vs 20 years for serious offenders). After serving time in prison they are released. A proportion of the released prisoners reoffend and go back to gaol (recidivists) after two years (on average), whereas the rest are rehabilitated. The figures and facts used in the case study are mostly based on reality, but slightly adapted for the purposes of the research. The modellers were provided with a basic conceptual model (Figure 8.2) and some initial data (i.e. the initial number of petty and serious offenders), while other data, namely statistical distributions, were intentionally omitted in order to observe modellers' reactions. The task for participating modellers was to develop a simulation model to be used as a decision-making tool by policy makers. For more details of the case study the reader is referred to Tako and Robinson (2009).

8.3.2 Verbal protocol analysis

VPA is a research method derived from psychology. It requires the subjects to 'think aloud' when making decisions or judgements during a problem-solving exercise. It relies on the participants' generated verbal protocols in order to understand in detail the mechanisms and the internal structure of cognitive processes that take place (Ericsson and Simon, 1984). Therefore, VPA as a process tracing method provides access to the activities that occur between the onset of a stimulus (case study) and the eventual response to it (model building) (Ericsson and Simon, 1984; Todd and Benbasat, 1987). Willemain (1994, 1995) was the first to use VPA in OR to document the thought processes of OR experts while building models.

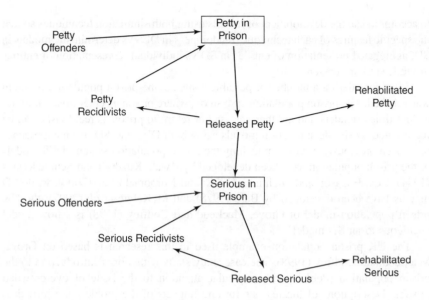

Figure 8.2 A simple diagram depicting a conceptual model of the UK prison population case study.

VPA is considered to be an effective method for the comparison of the DES and SD model building process. It is useful because of the richness of information and the live accounts it provides on the experts' modelling process. Another potential research method would have been to observe real-life simulation projects, using DES and SD. This would, however, mean that only two situations could be used, which would result in a very small sample size. Additionally, for a valid comparison it is necessary to have comparable modelling situations, which would require two potential real-life modelling projects of equivalent problem situations. The potential for getting access to two modelling projects of a similar situation was deemed unfeasible. We also considered running interviews with modellers from the DES and SD fields. Given that the overall aim of this research is to go beyond opinions and to get an empirical view about model development, one can claim that modellers' reflections may not reflect correctly the processes followed during model building and this would therefore not represent a full picture of model building. Hence, interviews were not considered appropriate. VPA on the other hand mitigates the issues related to the aforementioned research methods, as it can capture modellers' thoughts in practical modelling sessions in a controlled experimental environment, using a common stimulus – the case study.

Protocol analysis as a technique has its own limitations. The verbal reports may omit important data (Willemain, 1995) because the experts, being under observation, may not behave as they would normally. The modellers are asked to work alone and this way of modelling may not reflect their usual practice of model building, where they would interact with the client, colleagues, and so on. In addition, there is a risk that participants do not 'verbalise' their actual thoughts, but are only 'explaining'. To overcome this and to ensure that the experts speak their thoughts aloud, short verbalisation exercises, based on Ericsson and Simon (1984), were run at the beginning of each session.

8.3.3   The VPA sessions

The subjects involved in this study were provided with the prison population case study at the start of the VPA session and were asked to build simulation models using their preferred simulation approach. During the modelling process the experts were asked to 'think aloud' as they model. The researcher (Tako) sat in the same room, but social interaction with the subjects was limited. She only intervened in the case when participants stopped talking for more than 20 seconds to tell them to 'keep talking'. The researcher was also answering explanatory questions, providing participants with additional data inputs (if they asked for them) and prompting them to build a model on the computer in the case when they did not do so under their own initiative. The modelling sessions were held in an office environment with each individual participant. The sessions lasted approximately 60–90 minutes. The participants had access to writing paper and a computer with relevant simulation software (e.g. SIMUL8, Vensim, Witness, Powersim, etc.). The expert modellers chose to use the software they used as part of their work and hence were more familiar with. The protocols were recorded on audio tape and then transcribed.

8.3.4   The subjects

The subjects involved in the modelling sessions were 10 simulation experts in DES and SD modelling, five in each area. The sample size of 10 participants is considered reasonable, although a larger sample would, of course, be better. According to Todd and Benbasat (1987), due to the richness of data found in one protocol, VPA samples tend to be small, between 2 and 20.

For reasons of confidentiality participants' names are not revealed. In order to distinguish each participant we use the symbol DES or SD, according to the simulation technique used, followed by a number. So DES modellers are called DES1, DES2, DES3, DES4 and DES5, while SD subjects are SD1, SD2, SD3, SD4 and SD5. All participants use simulation modelling (DES and SD) as part of their work, most of them holding consultant posts in different organisations. The companies they come from are established simulation software companies or consultancy companies based in the UK.

A mixture of backgrounds within each participant group (DES and SD) was sought. All participants have completed either doctorates or masters' degrees in engineering, computer science, OR or hold MBAs. Their experience in modelling ranges from at least 6 years up to 19 years. They have also acquired supplementary simulation training as part of their jobs. They boast an extensive experience of modelling in areas such as the National Health Service, criminal justice, the food and

Table 8.2   List of DES and SD modellers' profiles.

DES modeller	Modelling experience	DES software used	SD modeller	Modelling experience	SD software used
DES1	9 years	Witness	SD1	14 years	Stella/iThink
DES2	4 years	SIMUL8	SD2	16 years	Strategy dynamics
DES3	13 years	Flexim	SD3	5 years	Powersim
DES4	8 years	SIMUL8	SD4	20 years	Stella/iThink
DES5	4 years	Witness	SD5	8 years	Vensim

drinks sector, supply chains, and so on. A list of the expert modellers' experience and the software used in the modelling sessions is provided in Table 8.2.

8.3.5   The coding process

A coding scheme was designed in order to identify what the modellers were thinking about in the context of simulation model building. The coding scheme was devised following the stages of typical DES and SD simulation projects, as in Robinson (2004), Law (2007), Sterman (2000) and Randers (1980). Each modelling topic has been defined in the form of questions corresponding to the modelling stage considered. The modelling topics and their definitions are as follows:

1. Problem structuring: What is the problem? What are the objectives of the simulation task?

2. Conceptual modelling: Is a conceptual diagram drawn? What are the parts of the model? What should be included in the model? How to represent people? What variables are defined?

3. Model coding: What is the modeller entering on the screen? How is the initial condition of the system modelled? What units (time or measuring) are used? Does the modeller refer to documentation? How to model the user interface?

4. Data inputs: Do modellers refer to data inputs? How are the already provided data used? Are modellers interested in randomness? How are missing data derived?

5. Verification and validation: Is the model working as intended? Are the results correct? How is the model tested? Why is the model not working?

6. Model results and experimentation: What are the results of the model? What sort of results is the modeller interested in? What scenarios are run?

7. Implementation: How will the findings be used? What learning is achieved?

The coding process starts with the definition of a coding scheme. Initially, the recordings of each verbal protocol were transcribed and then divided into episodes or 'thought' fragments, where each fragment is the smallest unit of data meaningful to the research context. Each single episode was then coded into one of the seven modelling topics or an 'other' category for verbalisations that were not related to the modelling task. Some episodes, however, referred simultaneously to two modelling topics and were, therefore, coded as containing two modelling topics.

Regarding the nature of the coding process followed, a mix of top-down and bottom-up approaches to coding was taken (Ericsson and Simon, 1984; Patrick and James, 2004). A theoretical base was already established (the initially defined modelling topics), which enabled a top-down approach. Throughout the various checks of the coded protocols undertaken, the coding categories were further redefined through a bottom-up approach. An iterative coding process was followed, where the coding scheme was refined and reconsidered as more protocols were coded. Pilot studies were initially carried out to test the case study, use of VPA and the coding scheme. The coding scheme was refined during the pilots and to some extent during the coding of the 10 protocols.

The transcripts were coded manually using a standard word processor. According to Willemain (1995), the coding process requires attention to the context a phrase is used in and, therefore, subjectivity in the interpretation of the scripts is unavoidable. In order to deal with subjectivity, multiple independent codings were undertaken in two phases. In the first stage, one of the researchers (Tako) coded the transcripts twice with a gap of three months between coding. Overall, a 93% agreement between the two sets of coding was achieved, which was considered acceptable. The differences were examined and a combined coding was reached. Next, the coded transcripts with the combined codes were further blind-checked by a third party, knowledgeable in OR modelling and simulation. In the cases where the coding did not agree, the researcher who undertook the coding and the third party discussed the differences and re-examined the episodes to arrive at a consensus coding. Overall, a 90% agreement between the two codings was achieved, which was considered satisfactory compared with the minimum 80% match value suggested as acceptable by Chi (1997). A final examination of the coded transcripts was undertaken to check the consistency of the coded episodes. Some more changes were made to the definition of modelling topics, but these were fairly minor. The results from the coded protocols are now presented and discussed.

8.4   Study results

This section presents the results of a quantitative analysis of the 10 coded protocols. The data represent a quantitative description of participants' modelling behaviour, exploring the distribution of attention to modelling topics, the sequence of modelling stages during the model building exercise and the pattern of iterations followed among topics. The findings from each analysis follow.

8.4.1   Attention paid to modelling topics

In order to explore the distribution of attention by modelling topic, the number of words articulated is used as a measure of the amount of verbalisations made by the expert modellers. In turn, this is used to indicate the spread of modellers' attention to the different modelling topics. While Willemain used lines as a measure of verbalisation in his study of modellers' behaviour (Willemain, 1995; Willemain and Powell, 2006), we believe that word counts are a more accurate measure, since they avoid the misrepresentation caused by counting incomplete (half or three-quarter) lines as full lines. This is particularly relevant when measuring the amount of verbalisation in each episode, where the protocol is divided into smaller units, including more incomplete lines. The average number of words articulated in the DES and SD protocols by modelling topic is compared in order to establish significant differences between the two groups. Figure 8.3 shows the number of words verbalised by modelling topic by the two groups of modellers. The corresponding numerical values for the average number of words and the standard deviation for the two groups of modellers are also provided against each modelling topic. Comparing the total number of words verbalised in the overall DES and SD protocols, an average difference of 1751 words is identified, suggesting that DES modellers verbalise more than SD modellers. The equivalent box plots of the total number of words verbalised by DES and SD modellers in Figure 8.3 (bottom) show that, while the

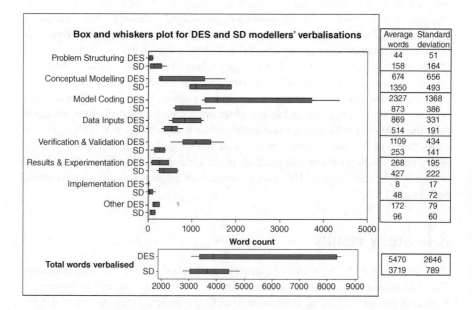

Figure 8.3   Box and whiskers plot of DES and SD modellers' verbalisations by modelling topic. The box shows the 75% range of values and the extreme values as the whiskers.

medians are similar, there is a bigger variation in the total number of words verbalised by DES modellers.

Considering each specific modelling topic, the biggest differences between the DES and SD protocols can be identified with regard to model coding, verification and validation and conceptual modelling (Figure 8.3). This suggests that DES modellers spend more effort in coding the model on the computer and testing it, while SD modellers spend more effort in conceptualising the mental model.

In order to test the significance of the differences identified, the Kolmogorov–Smirnov test, a non-parametric test, is used to compare two independent samples when it is believed that the hypothesis of normality does not hold (Sheskin, 2007). In this case, only five data points (word count for each modeller) are collected from the two groups of modellers (DES and SD). Due to the small sample size and the fact that count data are inherently not normal, it is considered that the assumption of normality is violated. The null hypothesis for the Kolmogorov–Smirnov test assumes that the verbalisations of the DES modellers follow the same distribution as the verbalisations of the SD modellers. The alternative hypothesis is that the data do not come from the same distribution. This test compares the cumulative probability distributions of the number of words verbalised by the modellers in the DES and SD groups for each modelling topic.

The statistical tests performed indicate significant differences, at a 10% level, in the amount of DES and SD modellers' verbalisations for the three modelling topics: conceptual modelling, model coding, and verification and validation (Table 8.3). This suggests that DES modellers verbalise more with respect to model coding and verification and validation, and thus spend more effort on these modelling topics compared with SD modellers. However, SD modellers verbalise more on conceptual modelling. In addition, the total verbalisations of the two groups of modellers are not found to be significantly different. Furthermore, the non-parametric Mann–Whitney test is used to compare the distributions of the verbalisations by modelling topic for

Table 8.3   The results of the Kolmogorov–Smirnov test comparing the DES and SD modellers' verbalisations for seven modelling topics and the total protocols at a 10% level of significance. The significant differences are highlighted, based on the comparison of the greater vertical distance M with the critical value of 0.8 (Sheskin, 2007).

Modelling topic	M	Differences in verbalisations?
Problem structuring	0.6	No (p-value $= 0.329$)
Conceptual modelling	0.8	Yes (p-value $= 0.082$)
Model coding	0.8	Yes (p-value $= 0.082$)
Data inputs	0.6	No (p-value $= 0.329$)
Verification and validation	1.0	Yes (p-value $= 0.013$)
Results and experimentation	0.4	No (p-value $= 0.819$)
Implementation	0.4	No (p-value $= 0.819$)
Total protocol	0.4	No (p-value $= 0.819$)

the two groups of modellers. The test statistic W has a value of 70 for a p-value of 0.8748. At a 5% level of significance, this test suggests that the data for the two groups of modellers represent different distributions and hence that their verbalisations differ significantly.

8.4.2   The sequence of modelling stages

This section focuses on the progression of modellers' attention during a simulation model development task using timeline plots. These plots show when modellers think about each modelling topic during the simulation modelling task (Wille-main, 1995; Willemain and Powell, 2006). A timeline plot is created for each of the 10 verbal protocols.

As examples, Figures 8.4 and 8.5 show timeline plots for DES1 and SD1 respectively. The plots consist of matched sets of seven timelines showing which of the seven modelling topics the modeller attends to throughout the duration of the modelling exercise. The vertical axis takes three values: 1 when the specific modelling topic is attended to by the modeller; 0.5 when the modelling topic and another have been attended to at the same time; and 0 when the modelling topic is not mentioned. The horizontal axis represents the proportion of the verbal protocol, from 0 to 100% of the number of words. The proportion of the verbal protocol is counted as the fraction of the cumulative number of words for each consecutive episode over the total number of words in that protocol, expressed as a percentage.

The timeline plots (Figures 8.4 and 8.5) are representative of most of those generated with the exception of DES3 and SD5. The former modeller did not complete the model due to difficulties encountered with the large number of attributes

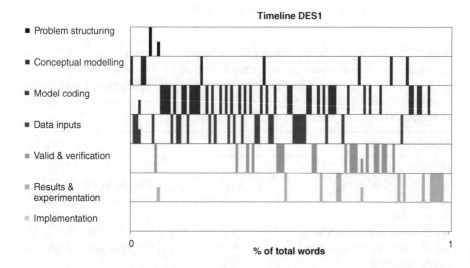

Figure 8.4   Timeline plot for DES1.

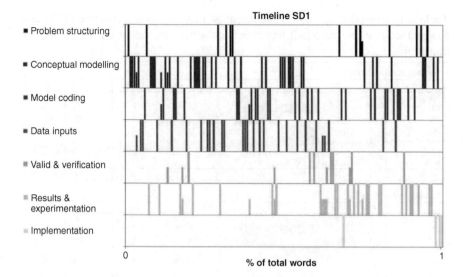

Figure 8.5   Timeline plot for SD1.

and population size. The latter was reluctant to build a model on the computer and so attended to model coding only at the end of the protocol, after being prompted by the researcher.

Observing the DES and SD timeline plots, it is clear that modellers frequently switched their attention to the topics. Similar patterns of behaviour were observed by Willemain (1995), where expert modellers were asked to build models of a generic OR problem. Looking at the overall tendencies in the DES and SD timeline plots, it appears that the DES protocols might follow a more linear progression in the sequence of modelling topics. Linearity of thinking implies that modellers' thinking progresses from the first modelling topics at the top left of the graph to the next one down towards the bottom right corner of the graph, with modelling topics concentrated in the centre of the plot. Meanwhile, in the SD protocols, modellers' attention appears to be more scattered throughout the model building session (Figure 8.5). The transition of attention between modelling topics is further explored in the next subsection.

8.4.3   Pattern of iterations among topics

In this subsection the iterations among modelling topics are explored using transition matrices, with a view to further understanding the pattern of iterations followed by DES and SD modellers. A transition matrix represents the cross-tabulation of the sequence of attention between successive pairs of episodes in a protocol. The total number of transitions occurring in the combined DES and SD protocols is displayed in Table 8.4. It can be observed that DES and SD modellers switched their attention from one topic to another almost to the same extent (505 times for DES modellers and 507

Table 8.4   Comparative view of the transition matrices for the combined DES and SD protocols, where each cell has been highlighted depending on the number of transitions.

Transition matrix - Total DES

	PS	CM	MC	DI	V&V	R&E	Impl	Totals
PS	0	1	2	4	0	3	0	10
CM	2	0	42	21	6	6	0	77
MC	1	37	0	75	62	10	0	185
DI	0	21	70	0	15	2	0	108
V&V	2	11	58	9	0	10	0	90
R&E	3	4	12	5	7	0	2	33
Impl	1	0	1	0	0	0	0	2
Totals	9	74	185	114	90	31	2	505

Transition matrix - Total SD

	PS	CM	MC	DI	V&V	R&E	Impl	Totals	
PS	0	8	1	4	0	8	0	21	
CM	6	0	44	41	7	16	2	116	
MC	4	37	0	46	22	24	2	135	
DI	3	40	50	0	8	10	0	111	0
V&V	1	10	17	11	0	9	0	48	1–10
R&E	5	21	21	13	11	0	1	72	11–20
Impl	0	1	2	0	0	1	0	4	21–40
Totals	19	117	135	115	48	68	5	507	40+

times for SD modellers). In order to explore the dominance of the modellers' thinking, the cells in the transition matrices have been highlighted according to the number of transitions counted. The darker shades represent the transitions that occur most frequently.

The main observations made based on the DES and SD transition matrices are as follows:

• Model coding is the topic DES modellers return to most often (185). Similarly, SD modellers return mostly to model coding.

• The modelling topics that DES modellers alternate between most often are conceptual modelling, model coding, data inputs, and verification and validation (shown by the dark grey highlighted cells in the DES transition matrix in Table 8.4). These transitions form the dominant loop in DES modellers' thinking. Among these transitions the highest are the ones between model coding and data inputs, and vice versa.

• SD modellers alternate mostly in a loop between conceptual modelling, model coding and data inputs. These transitions determine the dominant loop in their thinking process (dark grey highlighted cells in the SD transition matrix in Table 8.4).

• Comparing the dominant loops in the DES and SD transition matrices (Table 8.4), the pattern of the transitions for SD modellers follows a more horizontal progression, while a diagonal progression towards the bottom right-hand side of the matrix is observed for DES modellers. This serves as an indication that DES modellers' thinking process progresses more linearly compared with that of SD modellers.

The indication of linearity identified by comparing the dominant loops in DES and SD modellers' thinking is further verified using the total number of transitions of attention for the parallel linear strips in the two transition matrices. Each parallel strip includes the diagonal row of cells going from the top left to bottom right of the transition matrices (Table 8.4). The cells in each parallel strip have been highlighted in different shades for the DES and SD matrix separately (part (a), Table 8.5). So

Table 8.5   Total number of transitions per parallel strip in the DES and SD transition matrices.

(a)

Transition matrix - Total DES

	PS	CM	MC	DI	V&V	R&E	Impl
PS	0	1	2	4	0	3	0
CM	2	0	42	21	6	6	0
MC	1	37	0	75	62	10	0
DI	0	21	70	0	15	2	0
V&V	2	11	58	9	0	10	0
R&E	3	4	12	5	7	0	2
Impl	1	0	1	0	0	0	0
Totals	9	74	185	114	90	31	2

Topic transition - Total SD

	PS	CM	MC	DI	V&V	R&E	Impl
PS	0	8	1	4	0	8	0
CM	6	0	44	41	7	16	2
MC	4	37	0	46	22	24	2
DI	3	40	50	0	8	10	0
V&V	1	10	17	11	0	9	0
R&E	5	21	21	13	11	0	1
Impl	0	1	2	0	0	1	0
Totals	19	117	135	115	48	68	5

(b)

Parallel strip	DES	SD
	145	116
	125	116
	87	74
	85	74
	20	35
	23	34
	6	18
	7	24
Total	498	491

eight linear strips with different shades have been created, for which the total number of transitions is counted and summed (part (b), Table 8.5). In the case of absolute linear thinking, it is to be expected that all transitions would be concentrated in the central (darkest grey) strip. This would mean that the total number of transitions in the darkest grey strip would be equal to the total number of transitions, that is 505 for DES modellers and 507 for SD modellers. However, this is not the case with any of the DES or SD protocols. Nevertheless, the total number of transitions for all eight linear strips can provide an indication of the extent of linearity involved. The total number of transitions in each corresponding strip is compared for the DES and SD matrices. The highest total of transitions in the most central diagonal strips indicates a more linear modelling process. The highest total of transitions for the strips further away conveys a less linear process. The total number of transitions per linear strip for the DES and SD protocols is shown in Table 8.5(b). The reader should note that the totals shown in Table 8.5(b) differ from the totals in Table 8.4 due to the fact that the cells at the top right and bottom left corners have not been included in a parallel strip.

Comparing the total number of transitions in each parallel strip for the DES and SD modellers (Table 8.5b), it is observed that for DES modellers the higher numbers are at the top of the table, representing the most central strips in Table 8.5(a). This implies that the transition of attention for DES modellers focuses mainly in the most central strips in the matrix, hence representing a relatively more linear process compared with that for SD modellers. However, for SD modellers it can be observed that the higher totals are found at the bottom of the table, representing cells furthest away from the centre strips in Table 8.5(a). This suggests that the SD modellers switched their attention in a more vertical pattern compared with the DES modellers. It should be noted that an almost equal total number of transitions among topics has been found for DES and SD modellers. Based on the comparison of the transitions per linear strip, it can be concluded that DES modellers' attention progresses relatively more linearly among modelling topics than that of SD modellers.

8.5   Observations from the DES and SD expert modellers' behaviour

Understanding the details of the models developed by each expert modeller is not the primary objective of the analysis presented in this chapter. Nor is a detailed analysis of the modellers' behaviour. However, it is considered beneficial to provide a brief overview of the models developed and some key observations (mostly relevant to the model development process), made while the DES and SD expert modellers undertook the modelling exercise. The following paragraphs give the reader a general idea of the data obtained from the VPA sessions and their richness, but these are by all means not exhaustive.

The prison models developed by the DES and SD experts were simple and hence only small differences could be depicted among them. The modellers started from the basic diagram provided (Figure 8.2) and the majority kept close to the brief. The resulting models provided almost similar numerical results for the base scenario, taking into account in some cases mistakes in data inputs or incorrect computations. Most DES modellers produced tidier models more pleasing to the eye compared with the equivalent SD models. This could be due to the fact that more SD modellers added further structures to experiment with various scenarios.

Despite the limited differences observed in the models developed, some differences were observed in DES and SD expert modellers' reactions to tackling the case study. Some key observations that the authors found interesting are now considered, but these are not exhaustive. It was observed that DES and SD modellers considered similar modelling objectives, related to creating a tool that projects the output of interest into the future. SD modellers, however, showed a tendency to consider broader aspects of the problem modelled. They also considered objectives beyond projecting the size of the prison population or solving the problem of prison overcrowding, such as reducing the number of criminal acts or reducing the cost incurred by the prison system to society. Furthermore, SD modellers related the objectives of the model to testing policies, whereas DES modellers rarely related the objectives of the model to a comparison of scenarios for various policies.

It was further observed that the development of a conceptual diagram was not a priority for most DES and SD modellers. Due to the nature of the task, the participants were provided with a basic diagram in order to provide a common starting point. Most expert modellers were happy with this simple diagram (Figure 8.2) and hence did not consider creating a conceptual diagram. Some modellers made some sketches on paper, but these did not resemble any formal DES- or SD-like conceptual diagrams. Most DES and SD expert modellers conceptualised at the same time as coding the model on the computer.

While it is believed that feedback is the basic structure of SD models, it was observed that most SD modellers naturally identified the individual causal relationships and less often the feedback effects present in the model. Most SD modellers considered the effect of one variable on another, which is the basis that leads to the identification of feedback effects. On the other hand, DES modellers did not consider any equivalent structures. While conceptualising, DES modellers were most interested in setting up the sequential flow of events in the model rather than considering the effects among them.

Furthermore, it was observed that the DES modellers go through an analytic thinking process when modelling the prison population model. They did not consider the wider issues involved in the prison system, but focused mostly on the individual parts of the model, without considering the wider environmental or social factors that affect the prison system. In contrast, different aspects of systems' thinking were identified in the SD protocols. The SD modellers thought about the context and the wider environment the prison model is part of. They also considered the interrelationships between variables as well as the effects of prison on society, which did not occur in the DES protocols.

Most of the DES and SD modellers were happy with the structure and nature of the data provided in the case study description. The tendency either to change the structure of the model (SD4 and SD5) or to suggest additional structures or variables (SD1, SD2 and SD3) was most common in the case of SD modellers. DES modellers, on the other hand, followed the structure given in the case study with higher fidelity and suggested the addition of more detailed parts or information in the model, such as the representation of the geographic spread of prison buildings across the UK.

Detailed thinking characterised the DES modellers' protocols. The use of labels and attributes given to each individual in the system, the inclusion of conditional coding and specialised functions in the DES models added to the complexity and, therefore, the difficulties encountered during modelling. On the contrary, in the SD protocols no specific indications of detailed complexity were identified. Most SD modellers were happy to look at aggregate numbers or people in the prison system. DES modellers also commented on the problems encountered with the large population of the prison system, whereas SD modellers did not raise any issues about this.

The DES models were almost entirely built on quantitative data such as length of sentence, number of offenders (petty and serious) entering prison, number of prisoners already in the system, and so on. The SD modellers used both quantitative and qualitative data. References to graphical displays were categorised as qualitative data. One such example was the use of data in the form of a graphical display representing the total prison population over time. Similar attitudes towards missing data were observed among the DES and SD group of modellers. In both groups some individuals required additional data inputs, whereas others were willing to make assumptions. As expected, randomness was an important aspect in DES modelling, whereas no references were made to it by SD modellers.

Both DES and SD modellers were concerned with creating accurate models. Contrary to expectations, SD modellers did not refer to model usefulness as a way of validating the model. Both DES and SD modellers showed an interest in the quantitative and qualitative aspects of the results of the models. DES modellers were naturally thinking about more detailed results. SD modellers were keener on developing scenarios for experimentation with their models.

8.6 Conclusions

This chapter presented an empirical study that compares the modelling process followed by DES and SD modellers, by undertaking a quantitative analysis of the verbal protocols. In summary, the findings of this study consist of the following:

- DES modellers focus significantly more on model coding and verification and validation of the model, whereas SD modellers concentrate more on conceptual modelling.

- DES modellers' thinking process progresses more linearly among modelling topics compared with that of SD modellers.

- Both DES and SD modellers follow an iterative modelling process, but their pattern of iteration differs.

- DES and SD modellers switch their attention frequently between topics, and almost to the same extent (505 times for DES modellers and 507 for SD modellers) during the model building exercise.

- The cyclicality of thinking during the modelling task is more distinctive for SD modellers compared with DES modellers.

- The DES and SD models developed and their outcomes did not differ substantially.

- Differences were found in the DES and SD expert modellers' thinking during modelling, in terms of model objectives, feedback effects, the level of complexity and detail of models, data inputs and experimentation.

The findings supported the views expressed in the literature and were on the whole as expected. As with generic OR modelling, both DES and SD consist of iterative modelling processes. A new insight gained from this analysis was that DES modellers' thinking followed a more linear process, whereas for SD modellers it involved more cyclicality. Based on the study presented in this chapter, it is not possible to identify the underlying reason that causes DES modellers to follow a relatively more linear process. In order to identify whether this is a result of the nature of the modelling approach used or of the modellers' way of thinking, it would be useful to observe DES modellers building SD models and vice versa.

As expected, differences were identified in the attention paid to different modelling stages. The authors believe that this finding is partly a result of the fact that DES modellers naturally tend to pay more attention to model coding and partly because it is inherently harder to code in DES modelling. Clearly, the results are dependent to some extent on the case study used and the modellers selected. Therefore, considerations need to be made about the limitations of the study and the consequent validity of the findings.

Obviously, it should be noted that the findings of this study are based on the researcher's interpretation of participants' verbalisations. Subjectivity is involved in the analysis of the protocols, as well as in the choice of the coding scheme. A different researcher might have reached different conclusions (using a different coding scheme with different definitions). In order to mitigate the problem of subjectivity, the protocols were coded three times, involving a third party in one case. Additionally, the current findings are based on the verbalisations obtained from a specific sample of modellers who were chosen by convenience sampling. A bigger sample size could have provided more representative results, but due to project timescales this was not feasible. In this study only one case study was used. For future research, the use of more case studies could provide more representative results regarding the differences between the two modelling approaches.

Considering the data (verbal protocols) obtained from the modelling sessions implemented, these are derived from artificial laboratory settings, where the modellers at times felt the pressure of time or the pressure of being observed. The task given to the participants was a simple and quite structured task to ensure completion of an exercise in a limited amount of time. These factors have to some extent affected the smaller amount of verbalisations for modelling topics such as problem structuring, results and experimentation, and implementation. Future research could involve less structured tasks or tasks that introduce more experimentation, whereas implementation is more problematic for a laboratory setting.

This chapter compares the behaviour of expert DES and SD modellers when building simulation models. The findings presented ultimately contribute to the comparison of the two simulation approaches. The main contribution of this study lies in its use of empirical data, gained from experimental exercises involving expert modellers themselves, to compare the DES and SD model building processes. The chapter focuses on a quantitative description of expert modellers' thinking process, analysing the processes that DES and SD modellers think about while building simulation models. It is suggested that the modelling processes followed in DES and SD modelling differ. The observations made during the modelling process suggest a number of differences in the approach taken in different stages of modelling.

For further work the analysis presented here could be extended by undertaking an in-depth qualitative analysis of the 10 verbal protocols. This would enable a more detailed examination of the protocols for each modelling topic so that differences and similarities in the underlying thought processes between DES and SD modellers can be identified. This list of differences and similarities can in turn be linked to specific criteria relevant to the problem characteristics that are to be modelled. This can help in developing a set of criteria that can guide the choice between the two modelling approaches. Future work could also concentrate on developing a framework for supporting the choice between DES and SD for a specific modelling study.

Acknowledgements

The authors would like to thank Suchi Collingwood for her help with the coding of the protocols and who patiently undertook the third blind check of the 10 coded protocols.

References

Artamonov, A. (2002) Discrete-event simulation vs. system dynamics: comparison of modelling methods. MSc dissertation. Warwick Business School, University of Warwick, Coventry.

Baines, T.S., Harrison, D.K., Kay, J.M. and Hamblin, D.J. (1998) A consideration of modelling techniques that can be used to evaluate manufacturing strategies. *International Journal of Advanced Manufacturing Technology*, **14**(5), 369–375.

Bard, J.F. (1978) The use of simulation in criminal justice policy evaluation. *Journal of Criminal Justice*, **6**(2), 99–116.

Brailsford, S. and Hilton, N. (2001) A comparison of discrete-event simulation and system dynamics for modelling healthcare systems. Proceedings of the 26th Meeting of the ORAHS Working Group 2000, Glasgow Caledonian University, Glasgow, Scotland, pp. 18–39.

Chi, M. (1997) Quantifying qualitative analyses of verbal data: a practical guide. *Journal of the Learning Sciences*, **6**(3), 271–315.

Cox, G.B., Harrison, P. and Dightman, C.R. (1978) Computer simulation of adult sentencing proposals. *Evaluation and Program Planning*, **1**(4), 297–308.

Coyle, R.G. (1985) Representing discrete-events in system dynamics models: a theoretical application to modelling coal production. *Journal of the Operational Research Society*, **36**(4),307–318.

Ericsson, K.A. and Simon, H.A. (1984) *Protocol Analysis: Verbal Reports as Data*, MIT Press, Cambridge, MA.

Grove, P., MacLeod, J. and Godfrey, D. (1998) Forecasting the prison population. *OR Insight*, **11**(1), 3–9.

Hillier, F.S. and Lieberman, G.J. (1990) *Introduction to Operations Research*, 5th edn, McGraw-Hill, New York.

Korporaal, R., Ridder, A., Kloprogge, P. and Dekker, R. (2000) An analytic model for capacity planning of prisons in the Netherlands. *Journal of the Operational Research Society*, **51**(11), 1228–1237.

Kwak, N.K., Kuzdrall, P.J. and Schniederjans, M.J. (1984) Felony case scheduling policies and continuances – a simulation study. *Socio-Economic Planning Sciences*, **18**(1), 37–43.

Lane, D.C. (2000) You just don't understand me: models of failure and success in the discourse between system dynamics and discrete-event simulation. LSE OR Dept, Working Paper 00.34:26.

Law, A.M. (2007) *Simulation Modeling and Analysis*, 4th edn, McGraw-Hill, Boston, MA.

Mak, H.-Y. (1993) System dynamics and discrete-event simulation modelling. PhD thesis. London School of Economics and Political Science.

McKelvie, D., Hadjipavlou, S., Monk, D. *et al.* (2007) The use of SD methodology to develop services for the assessment and treatment of high risk serious offenders in England and Wales. Proceedings of the 25th International Conference of the System Dynamics Society, Boston, MA.

Meadows, D.H. (1980) The unavoidable a priori, in *Elements of the System Dynamics Methods* (ed. J. Randers.) Productivity Press, Cambridge.

Miles, M.B. and Huberman, A.M. (1994) *Qualitative Data Analysis: An Expanded Sourcebook*, Sage, Thousand Oaks, CA.

Morecroft, J.D.W. and Robinson, S. (2005) Explaining puzzling dynamics: comparing the use of system dynamics and discrete-event simulation. Proceedings of the 23rd International Conference of the System Dynamics Society, Boston, MA.

Oral, M. and Kettani, O. (1993) The facets of the modeling and validation process in operations research. *European Journal of Operational Research*, **66**(2), 216–234.

Patrick, J. and James, N. (2004) Process tracing of complex cognitive work tasks. *Journal of Occupational and Organizational Psychology*, **77**(2), 259–280.

Pidd, M. (2004) *Computer Simulation in Management Science*, 5th edn, John Wiley & Sons, Ltd, Chichester.

Randers, J. (1980) *Elements of the System Dynamics Method*, MIT Press, Cambridge, MA.

Robinson, S. (2001) Soft with a hard centre: discrete-event simulation in facilitation. *Journal of the Operational Research Society*, **52**(8), 905.

Robinson, S. (2004) *Simulation: The Practice of Model Development and Use*, John Wiley & Sons, Ltd, Chichester.

Robinson, S. (2005) Discrete-event simulation: from the pioneers to the present, what next? *Journal of the Operational Research Society*, **56**(6), 619–629.

Senge, P.M. (1990) *The Fifth Discipline: The Art and Practice of the Learning Organisation*, Random House, London.

Sheskin, D.J. (2007) *Handbook of Parametric and Nonparametric Statistical Procedures*, 4th edn, Chapman & Hall/CRC Press, Boca Raton, FL.

Sterman, J. (2000) *Business Dynamics: Systems Thinking and Modeling for a Complex World*, Irwin/McGraw-Hill, Boston, MA.

Sweetser, A. (1999) A comparison of system dynamics and discrete-event simulation. Proceedings of the 17th International Conference of the System Dynamics Society and 5th Australian & New Zealand Systems Conference, Wellington, New Zealand.

Tako, A.A. and Robinson, S. (2009) Comparing discrete-event simulation and system dynamics: users' perceptions. *Journal of the Operational Research Society*, **60**, 296–312.

Todd, P. and Benbasat, I. (1987) Process tracing methods in decision support systems research: exploring the black box. *MIS Quarterly*, **11**(4), 493–512.

Willemain, T.R. (1994) Insights on modeling from a dozen experts. *Operations Research*, **42**(2),213–222.

Willemain, T.R. (1995) Model formulation: what experts think about and when. *Operations Research*, **43**(6), 916–932.

Willemain, T.R. and Powell, S.G. (2006) How novices formulate models. Part II: a quantitative description of behaviour. *Journal of the Operational Research Society*, **58**(10), 1271–1283.

Wolstenholme, E.F. (1990) *System Enquiry: A System Dynamic Approach*, John Wiley & Sons, Ltd, Chichester.

9

Explaining puzzling dynamics: A comparison of system dynamics and discrete-event simulation

John Morecroft[1] and Stewart Robinson[2]
[1]*London Business School, London, UK*
[2]*School of Business and Economics, Loughborough University, UK*

9.1 Introduction

Everyday situations present many examples of puzzling dynamics – performance over time that defies intuition and common sense. You drive for miles at a fast and steady speed on a busy motorway yet sometimes encounter unexpected tailbacks with no apparent cause. You occasionally visit your hairdresser. These visits are on the same day of the week at the same time but you never know in advance whether you will wait one minute or half an hour for a haircut.

One way to investigate such puzzling dynamics is to build a computer simulation model that represents the various interrelated factors and pressures at work in the situation, and then run the model to see whether or not it is capable of generating similar puzzling performance. If a model can, in some meaningful way, mimic observed performance then modellers claim they have an explanation for the phenomenon. In developing a simulation model to investigate puzzling dynamics,

Discrete-Event Simulation and System Dynamics for Management Decision Making, First Edition.
Edited by Sally Brailsford, Leonid Churilov and Brian Dangerfield.
© 2014 John Wiley & Sons, Ltd. Published 2014 by John Wiley & Sons, Ltd.

the analyst can select from a number of approaches, among which two of the most common are discrete-event simulation (DES) (Banks *et al.*, 2001; Pidd, 2004; Robinson, 2004; Law, 2007) and system dynamics (SD) (Forrester, 1961; Richardson and Pugh, 1981; Coyle, 1996; Sterman, 2000; Morecroft, 2007). Both are widely used to examine the performance over time of interconnected systems. The analyst is left to select the appropriate method. It is apparent that most analysts opt for the method with which they are most familiar. Perhaps this tendency is unsurprising given that, in the past, there appear to be very few studies that compare SD and DES, let alone give guidance on which approach might be most appropriate in different circumstances.

Our intention in this chapter is to address the question of which method to use and why, by exploring how SD and DES models help us make sense of puzzling dynamics. We begin with a brief review of existing comparisons of SD and DES in the literature that pre-date this edited volume. Collectively they highlight several important technical and conceptual differences between the approaches. Following this the focus of our research is explained and the case study of erratic fisheries, around which our investigation revolves, is described. A series of SD and DES models are then developed and compared, from which differences in both the representation and interpretation of the fisheries problem are identified. Limitations of the study are discussed before returning to the question of which approach to use and when.

9.2   Existing comparisons of SD and DES

Coyle (1985) discusses how discrete events might be modelled in an SD simulator. In doing so he notes two key differences between the SD and DES approaches. First, there is the tendency for DES models to include randomness. He argues that in SD models stochastic noise can be subsumed into an appropriate delay. The second difference he identifies is that in DES modelling an open-process structure is adopted, while in SD a closed-loop structure is used in which feedback is explicitly identified.

Writing from the perspective of healthcare modellers, Brailsford and Hilton (2000) briefly describe health-related DES and SD studies including an SD model of NHS waiting lists for cardiac surgery and a discrete-event model of AIDS transmission in a localized population. Table 9.1 summarizes the main technical distinctions the authors identify from their experience of modelling in general and from these healthcare studies in particular.

Writing from the perspective of an experienced SD modeller, Lane (2000) identifies conceptual differences between DES and SD in terms of the eight categories shown in Table 9.2. Consider for example the modeller's perspective on problem situations. A DES modeller captures *detail* complexity in the network of activities and queues conceived at a functional operating level. An SD modeller captures *dynamic* complexity in closed feedback loops linking stocks and flows normally conceived at a cross-functional strategic level. Related to the perspective on complexity there is the degree of resolution of models: a close-up resolution in DES that picks out individual entities, attributes, decisions and events versus a distant resolution in SD that captures homogenized entities and continuous policy pressures. There is the type of problem

Table 9.1   Technical differences between DES and SD identified by Brailsford and Hilton (2000).

Discrete-event simulation	System dynamics
Systems (such as healthcare) can be viewed as networks of queues and activities	Systems (such as healthcare) can be viewed as a series of stocks and flows
Objects in a system are distinct individuals (such as patients in a hospital), each possessing characteristics that determine what happens to that individual	Entities (such as patients) are treated as a continuous quantity, rather like a fluid, flowing through reservoirs or tanks connected by pipes
Activity durations are sampled for each individual from probability distributions and the modeller has almost unlimited flexibility in the choice of these functions and can easily specify non-exponential dwelling times	The time spent in each reservoir is modelled as a delay with limited flexibility to specify a dwelling time other than exponential
State changes occur at discrete points of time	State changes are continuous
Models are by definition stochastic in nature	Models are deterministic
Models are simulated in unequal time steps, when 'something happens'	Models are simulated in finely sliced time steps of equal duration

studied, usually operational for DES and more strategic for SD. There is the portrayal of human agents viewed as decision makers in DES (choosing between well-defined options) and as boundedly rational policy implementers in SD (responding to organizational pressures). There are model outputs that in DES take the form of point predictions and detailed performance measures and in SD take the form of simulations that enhance understanding of the structural source of behaviour modes. It should be noted that Robinson (2001) does not agree with this last point in regard to DES, where he argues that DES can be used for developing an understanding of a system's behaviour and does not necessarily have to focus on point prediction.

Further comparisons are found in DES textbooks (though not in SD textbooks until recently). For example, Pidd (2004) comments on the relative level of detail in DES and SD models noting that while discrete-event models concentrate on the state changes and interactions of individual entities, it is normal in SD to operate at a much more aggregate level by concentrating on the rates of change of populations of entities. Moreover, he succinctly observes that: 'In order to model feedback systems for simulation it is important to concentrate on their structure rather than their content. The structure defines how the variables interact, the content is the meaning of those

Table 9.2 Conceptual differences between DES and SD identified by Lane (2000).

	Discrete-event simulation	System dynamics
Perspective	Analytic; emphasis on detail complexity	Holistic; emphasis on dynamic complexity
Resolution of models	Individual entities, attributes, decision and events	Homogenized entities, continuous policy pressures and emergent behaviour
Data sources	Primarily numerical with some judgemental elements	Broadly drawn
Problems studied	Operational	Strategic
Model elements	Physical, tangible and some informational	Physical, tangible, judgemental and information links
Human agents represented in models as	Decision makers	Boundedly rational policy implementers
Clients find the model	Opaque/dark grey box, nevertheless convincing	Transparent/fuzzy glass box, nevertheless compelling
Model outputs	Point predictions and detailed performance measures across a range of parameters, decision rules and scenarios	Understanding of structural source of behaviour modes, location of key performance indicators and effective policy levers

variables for the organization'. Robinson (2004) observes that DES is generally more appropriate when the details of a system need to be modelled, especially when individual items need to be tracked.

Finally, in her doctoral thesis, Mak (1992) investigates how activity cycle diagrams for DES models can be converted into stock and flow representations (used for representing SD models), developing guidelines and software to automate the process. In doing so she identifies a number of differences between the DES and SD approaches. Many of the differences are similar to those already described above. In addition, she notes that SD models explicitly show information feedback while DES models do not, albeit that this information is normally held within the logic of DES models. It is also noted that SD models tend to study the interaction of control policies, exogenous events and feedback structure. DES models tend to be used for 'what-if' experimentation, in which the effects of various options are investigated.

A shortcoming of all these comparisons is that they are written from the perspective of either a specialist in SD or a DES specialist. Until recently a comparison that took a more balanced view did not seem to exist. Tako and Robinson have attempted to address this through three separate studies. The first engaged MBA students with an SD and a DES model of the same problem situation and sought to discover how the two models compared through the eyes of a manager (Tako and Robinson, 2009). They found that there was very little difference from the user's perspective. In a follow-up study, they asked expert SD and DES modellers to build simulations of the same problem (Tako and Robinson, 2010, presented as Chapter 8 of this book). This demonstrated that the modellers followed quite different patterns in the tasks they performed while building the model. The third study is a literature review which compares the use of SD and DES in the supply chain context (Tako and Robinson, 2012). This study demonstrates that there are some different patterns of use with respect to the problems being tackled, but also that there is a significant amount of overlap in the use of SD and DES.

Finally we note that comparisons based on a literature search are confounded by the lack of articles on the underlying philosophy of DES modelling. Whereas SD modellers have a well-formed modelling philosophy (Forrester, 1968; Lane, 1999; Morecroft, 2004; Richardson, 1991; Sterman, 1989), such writings seem to be limited within the DES community, though the situation is changing. Recent interest in conceptual modelling for DES focuses on the process of abstracting a model from a real-world situation, where practice meets philosophy (Robinson et al., 2011). The introduction to DES which forms Chapter 2 of this book also attempts to define the modelling philosophy of a DES modeller.

9.3   Research focus

Rather than focus on technical and conceptual differences, we compare the nature of *explanations and insights* these two approaches have to offer about puzzling dynamics. Our premise is that the modelling style you choose affects the way you represent and interpret phenomena from the real world. Broadly speaking, SD primarily investigates the performance over time of an interconnected system arising from its internal feedback structure. DES primarily investigates the performance over time of an interconnected system subject to internal (e.g. process failure) and external (e.g. environmental conditions) random variability. Either approach can portray realistic situations such as those mentioned above: the movement of traffic on a motorway, the build-up of queues in retail outlets. But what do they tell us? What, if anything, is different about the understanding, explanation and communication of dynamics that arises from such alternative styles of modelling?

These questions stem from our complementary professional backgrounds. Both of us are experienced modellers, one specialising in SD and the other in DES. We share an interest in how people learn from models and simulation, and the relationship between hard and soft OR. We therefore address the shortcoming of previous comparisons by combining expertise from the quite separate SD and DES worlds.

9.4   Erratic fisheries – chance, destiny and limited foresight

For our comparison we developed two models of similar size and detail to represent the dynamics of fisheries. There are several reasons for choosing fisheries. The application is novel (at least among SD–DES comparisons), yet appropriate and important. The problems of overexploitation facing international fisheries are well known, widely reported in the press, and a subject of government policy in many nations. Moreover, fisheries management has been studied by economists (Arnason, 2007; Hardin, 1968), scientists (Roughgarden and Smith, 1996) and operational researchers/modellers (Dudley, 2003; Farber, 1991; Moxnes, 1998; Otto and Struben, 2004). The performance of international fisheries is indeed puzzling. Fish naturally regenerate. They are a renewable resource, in apparently endless supply, providing valuable and healthy food for billions of consumers and a livelihood for millions of fishing communities worldwide. The fishing industry has been in existence since the dawn of civilization and should last for ever. Yet fish stocks around the world are volatile and some are even collapsing. Once-rich fishing grounds such as Canada's Grand Banks now yield no catch at all. Stocks in other areas, such as the English Channel, the North Sea and the Baltic, are in terminal decline.

Figure 9.1 shows typical volatile time series data from real fisheries. The top chart shows the Pacific sardine catch in thousands of tonnes per year over the period 1916 to 1996. The annual catch grew remarkably between 1920 and 1940, starting around 50 000 tonnes and peaking at 700 000 tonnes – a 14-fold increase. Over the next four years to 1944 the catch fell to 500 000 tonnes, stabilized for a few years and then collapsed dramatically to almost zero in 1952. Since then it has never properly recovered. The bottom chart shows a similar story for the North Sea herring catch in the period 1950 to 1998. However, in this case, following a collapse between 1974 and 1979, the fishery did recover in the 1980s and early 1990s with an average annual catch around 600 000 tonnes, similar to the catch in the 1950s and 1960s.

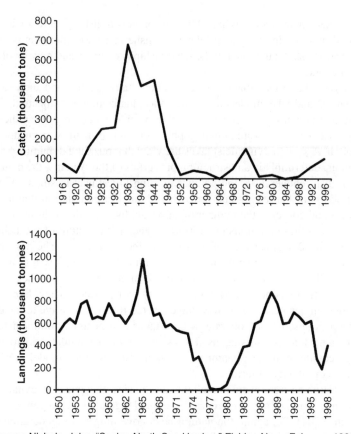

Source: Nichols, John. "Saving North Sea Herring." Fishing News February 1999.

Figure 9.1   Pacific sardine catch (top) and North Sea herring catch (bottom) except from Fish Banks debriefing materials, Meadows et al. (2001). This particular figure was re-printed in Morecroft (2007) and is reproduced by permission of Wiley.

This visible and contemporary problem is amenable to small and transparent models that (despite being small) nevertheless portray dynamic complexity typical of SD and DES. Since neither of us are experts on fisheries we have based our models loosely on a popular fisheries gaming simulator called Fish Banks Ltd (Meadows, Fiddaman and Shannon, 2001) used to teach principles of sustainable development to a wide variety of audiences ranging from politicians, business leaders and government policy advisers to fishing communities and high school students.

   In Fish Banks there is a finite offshore region of ocean containing fish. Fish regenerate as a function of the existing population. The local fishing community buys ships from the shipyard and takes them to sea to harvest fish. The total catch depends on the number of ships, the fish population and other factors such as the weather. In the game, as in real life, the fish population is not known accurately, although it can be estimated. Also, in the game, as in real life, the process of fish regeneration is not fully understood by those in the system (players or fishermen). Regeneration is related to the (unknown) fish population, but the relationship is complex and may involve other external factors.

We have taken this situation as a common starting point and then separately developed SD and DES models using the conceptualization, visualization, formulation and interpretation guidelines of our respective modelling disciplines. From the very outset we were aware of philosophical differences in the way we approached the problem. A system dynamicist takes the view that puzzling dynamics arise from endogenous, deterministic and structural properties of the system – the natural laws governing fish regeneration, the policies guiding investment in ships, the productivity of ships for harvesting and processes of stock accumulation that accurately account for and conserve the number of ships and fish introduced or removed from the system. All these relationships form the feedback structure of the fishery that can be visualized as a network of interlocking feedback loops. The destiny of the fishery (whether it is sustainable and whether future harvest rates will grow, remain stable or collapse) is assumed to be predetermined by the feedback structure, although this future is not known with certainty by fishermen. They, like any other stakeholders in business and society, have limited foresight relative to the dynamically complex system they operate. It only takes two or three nonlinear feedback loops (in a system comprising as few as two stock accumulations) to generate puzzling, counterintuitive dynamics. A simple SD fisheries model comfortably meets this complexity criterion.

In contrast a discrete-event modeller takes the view that puzzling dynamics arise from the interaction of random processes coupled together by endogenous structure (Robinson, 2004). So, again, the natural laws governing fish regeneration and the rules for investment in ships and harvesting are important, but are now assumed to be overlaid by random processes. For example, fish regeneration, though related endogenously to the size of the fish stock, is also affected by a variety of external factors (e.g. the climate and environmental changes), beyond the control of stakeholders, factors that manifest themselves as random variation. Similarly the harvest rate or fish catch is influenced endogenously by the size of the fishing fleet (with more ships you can expect a bigger total catch, other things being equal), but is also affected by random operational factors such as the position of ships at sea, the mood of the crew and the weather. The destiny of the fishery is assumed to be partly and significantly a matter of chance. This future is not known with certainty because it involves random variation. Moreover, the effect of randomness is counterintuitive to fishermen, (as it would be to other stakeholders with limited but normal human foresight), because it involves another kind of dynamic complexity, this time arising from multiple interacting random processes. It takes only two or three interacting random processes, even in a linear system, to generate puzzling dynamics. A simple DES fisheries model comfortably meets this complexity criterion.

It is an empirical matter whether a source of variability in a model is shaped by factors other than those endogenously specified. For example, the fish regeneration rate in an SD fisheries model is expressed as a nonlinear function of the fish population – an endogenous feedback relationship linking the accumulated stock of fish to its rate of increase. A DES model will include a similar relationship, because it fits the facts, though the nonlinearity would not normally receive great attention. Instead attention shifts to the random process that is presumed to overlay population-dependent regeneration. These differences in formulation will be explored more carefully later. But for now consider the philosophical difference. There is scientific evidence for an endogenous formulation controlling regeneration (see for example Townsend, Begon and Harper, 2003, an introductory ecology textbook), which can be described as a 'humped relationship between the net recruitment into a population (births minus deaths) and the size of that population resulting from the effects of intraspecific competition'. There is also evidence of profound external environmental fluctuations, for example an El Niño event 'when warm tropical water from the north reduces the upwelling, and hence the productivity, of the nutrient-rich cold Peruvian current coming from the south'. Either or both deterministic–endogenous and random–external formulations may be appropriate. In a practical fisheries modelling project the facts would obviously be pertinent and modellers, SD or DES, could (with varying degrees of ease) adapt their model and formulations accordingly. However, our point here is to observe that the type of representation and explanation for puzzling dynamics that a modeller seeks depends on the approach adopted. It is natural for an SD modeller to look for an endogenous structural representation that fits the available facts. It is natural for a discrete-event modeller to look for a structural and random process representation that also fits the facts. These unavoidable methodo-logical biases affect what modellers choose to include and how they go about constructing and communicating a model-based argument (see also Meadows (1980) for a related discussion of methodological biases in SD and econometrics). Does this matter? Our in-depth comparison of models and simulations in the next and later sections sheds more light on this question.

9.5 Structure and behaviour in fisheries: A comparison of SD and DES models

The title of this section, 'Structure and behaviour . . . ', is instantly recognizable to a system dynamicist. The reason is that in SD an explanation of puzzling dynamics is deemed to exist when one can show, justify and interpret the interlocking feedback loops (structure) that *cause* the dynamic phenomenon (or behaviour over time) of interest. Growth is caused by reinforcing (positive) feedback. Fluctuations stem from goal-seeking or balancing (negative) feedback involving delayed adjustment. Growth and collapse arise from nonlinear reinforcing and balancing feedback combined. Although discrete-event modellers do not normally think in terms of feedback structure (they rarely if ever visualize feedback loops, even when such loops exist), they are nevertheless aware that a combination of structural relationships and random processes lies behind simulated dynamics. To illustrate this difference in approach to the meaning and use of model structure we now present and compare, step by step, our fisheries models, the corresponding equation formulations, and the simulated behaviour (dynamics) that the structures produce.

Our stepwise analysis begins with a natural fishery in which there are no fishermen, no ships and no harvesting – just a self-regulating fish population governed by biological laws and natural limits to growth. Simulations of both SD and DES models show population growth and saturation, with superimposed random variation in the DES model. We then present a harvested fishery in a series of equilibria, with a fixed number of ships and without randomness or bias in any process. Simulations show the SD and DES models can achieve identical equilibria in terms of fish population and fleet size. We then relax the equilibrium assumptions of our ideal harvested fishery to arrive at more realistic disequilibrium models with potentially volatile population dynamics. But the assumptions we relax are different and conditioned by our contrasting approaches. In the SD model we incorporate pressure for growth in fleet size (a bias of human nature – 'more is better'), presumed to exist in many investment policies. In the DES model we introduce random variation in fish regeneration and the catch. Again we compare simulations. This time the simulations are quite different, although in both cases the fish population is volatile and departs a long way from the ideal equilibrium. Moreover, the style of explanation for these outcomes (how the structure, processes and behaviour are interpreted and presented, and the use made of diagrams, equations and simulations) is distinctive and unique to each approach.

9.5.1   Alternative models of a natural fishery

9.5.1.1   SD model

Figure 9.2 shows the diagram and equations for an SD model of a natural fishery in the format of the popular iThink language (Richmond *et al.*, 2004 and isee systems). This format is pretty much a standard format used for all SD models.

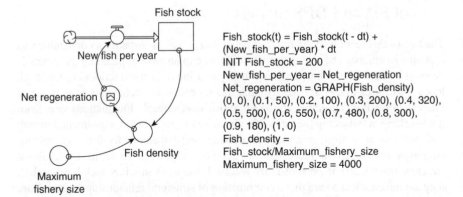

Fish_stock(t) = Fish_stock(t - dt) +
(New_fish_per_year) * dt
INIT Fish_stock = 200
New_fish_per_year = Net_regeneration
Net_regeneration = GRAPH(Fish_density)
(0, 0), (0.1, 50), (0.2, 100), (0.3, 200), (0.4, 320),
(0.5, 500), (0.6, 550), (0.7, 480), (0.8, 300),
(0.9, 180), (1, 0)
Fish_density =
Fish_stock/Maximum_fishery_size
Maximum_fishery_size = 4000

Figure 9.2   Diagram and equations for SD model of a natural fishery.

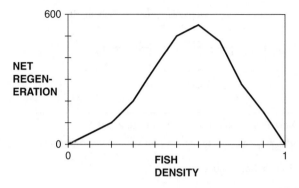

Figure 9.3 Net regeneration as a nonlinear function of fish density.

The fish stock is represented as a level that accumulates the inflow of new fish per year (here the inflow is defined as a net flow of births minus deaths). Initially there are 200 fish in the sea and the maximum fishery size is assumed to be 4000 fish. Incidentally, the initial value and maximum size can be rescaled to be more realistic without changing the resulting dynamics. For example, a fishery starting with a biomass of 20 000 tonnes of a given species and an assumed maximum fishery size of 400 000 tonnes would generate equivalent results.

The structurally important relationships are those that define the endogenous feedback effect of fish stock on net regeneration, which in this model is identically equal to new fish per year. Net regeneration is a nonlinear function of fish density as shown in Figure 9.3. Note that the function is hump-shaped. Normal practice in SD modelling is to avoid the use of functions involving a reversal of gradient because they can sometimes lead to feedback loops with ambiguous polarity (Sterman, 2000, pp. 577–578). In this case we have chosen to retain the function because it is widely used in fisheries models (Townsend, Begon and Harper, 2003) and is identical to the function used in the popular Fish Banks gaming simulator (Meadows, Fiddaman and Shannon, 2001). For those who are interested, it is possible to eliminate the hump-shaped function and to replace it with two monotonic functions by separately modelling fish regeneration and fish deaths (rather than net regeneration). A good example can be found in Dudley (2003).

9.5.1.2 DES model

Figure 9.4 shows the diagram and equations for a DES model of the same fishery. Unlike SD, there is no agreed standard diagramming method for representing DES models. Various approaches are available including activity cycle diagrams (Hills, 1971), process mapping/process flow diagrams (Davis, 2001), Petri nets (Torn, 1981), event graphs (Som and Sargent, 1989) and the Unified Modeling Language (UML) (Richter and März, 2000). Pooley (1991) gives a useful summary of

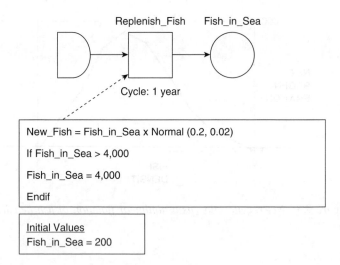

Figure 9.4   Diagram and equations for DES model of a natural fishery.

diagramming techniques for DES. For the purposes of this work, process flow diagrams have been used.

In Figure 9.4 the fish stock is represented as a queue that is fed by the annual process (activity with a cycle of one year) of fish replenishment. New fish are sourced from outside the model by the 'source' on the left of the diagram. As for the SD model, there are assumed to be initially 200 fish in the sea and a maximum of 4000 fish can be sustained.

Fish regeneration is seen as a linear, but random, relation to the number of fish in the sea. Fish grow at an average rate of 20% per year, varying according to a normal distribution with a standard deviation of 2%. This figure is selected as it represents a roughly similar average growth rate as in the SD model. The limit to growth of 4000 is represented as a discrete cut-off which does not allow Fish_in_Sea to exceed this limit.

The simulation was developed using the Witness simulation software (Lanner, 2007). This is one of a number of leading DES software packages.

9.5.1.3   Comparison of SD and DES representations

There are some clear differences in the representations presented by the DES and SD models. The SD model uses stocks and flows while the DES model uses queues and activities. However, the sharpness of this distinction blurs when one recognizes the essential equivalence of stocks and queues, and of flows and activities. Nevertheless, in SD models the contents of a stock accumulation are normally assumed to be perfectly mixed at all times so that all items in the stock have the same probability of exit, independent of their arrival time (Sterman, 2000, pp. 416–417). This perfect mixing is markedly different from the individual entities represented in a discrete-event queue and is consistent with the drive for aggregation in SD in order to portray cross-functional and inter-sector feedback loops.

The feedback structure is explicit within the SD model, but hidden in the equations of the DES representation (the relationship between fish replenishment and the number of fish in the sea). The relationship between fish stocks and fish regeneration in the SD model is nonlinear, but linear in the DES model. As expected, the DES model includes the randomness, which is not present in the SD version. However, it should be noted that SD models are not entirely devoid of random processes, or need not be. For example, it is common in factory and supply chain models to add randomness to demand in order to invoke cyclical dynamics. But system dynamicists do not normally set out to explore surprising bottlenecks and queues (stock accumulations) that stem from interlocking streams of random events. As Forrester (1961) explains in *Industrial Dynamics* (Appendix F on Noise), 'we have chosen to formulate models around the continuous noise-free flows of information, decisions and action. After the noise-free dynamic character of the system is observed, noise is then added to see what randomness contributes to system operation.' An overwhelming number of SD studies in practice are restricted to examination of the noise-free dynamic character of systems, consistent with the philosophical view in system dynamics that enduring feedback structure gives rise to dynamic behaviour.

9.5.1.4   Simulated dynamics of a natural fishery

Our first simulations show the dynamics of a 'natural' fishery starting with an initial population of 200 fish. There are no ships and no investment. Fishermen are not yet part of the system. The SD model (Figure 9.5) shows smooth S-shaped growth due to its nonlinear, deterministic formulations for fish regeneration. Until year 18 the fish stock follows a typical pattern of compounding growth associated with a reinforcing feedback loop. The population grows from 200 to 2500 fish. Fish regeneration (new fish per year) also increases until year 18 as rising fish density enables fish to reproduce more successfully. Thereafter crowding becomes a significant factor according to the nonlinear net regeneration curve in Figure 9.3. The number of new fish per year falls as the population density rises, eventually bringing population growth to a halt as the fish stock approaches its maximum sustainable value of 4000 fish.

The results from the DES model are shown in Figure 9.6. This is the output from a single replication (a run driven with a specific stream of random numbers). If the random number seeds were changed to perform further replications, the exact pattern of growth and fish regeneration would alter. Figure 9.6 shows an increasing growth rate with a discrete cut-off. This is unlike the SD representation in two ways. First, the growth is not as smooth due to the randomness within the regeneration process. This is clearly evidenced by the graph showing new fish. Second, the SD representation shows an asymptotic growth towards the limit of 4000, while the DES model reaches the limit in a discrete step. Both of these differences are a result of the model formulations.

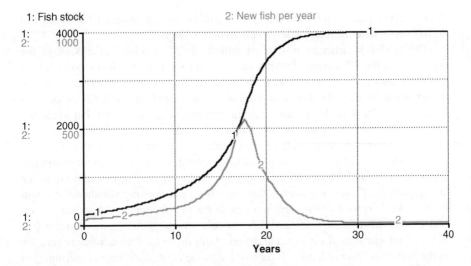

Figure 9.5   Simulation of SD model of a natural fishery with an initial population of 200 fish and maximum fishery size of 4000. From Morecroft (2007). Reproduced by permission of Wiley.

9.5.2   Alternative models of a simple harvested fishery

Imagine that a fleet of ships arrives in the region and sets about harvesting fish. The total catch depends on both the number of ships and their productivity (how many fish each ship catches in a typical year). SD and DES both shed light on the problem of coordinating the size of a fishing fleet (and the catch) with a regenerating fish population. What we know from real fisheries is that satisfactory coordination is difficult to achieve. But why?

9.5.2.1   SD model

In SD we can make a start on this question by investigating the relationship between catch and fish population under a scenario of varying fleet size. A simple harvested fishery is shown in Figure 9.7. All the original relationships of the natural fishery (Figure 9.2) remain intact but now the fish stock is depleted by a harvest rate, equal to the catch and proportional to the number of ships at sea.

9.5.2.2   DES model

Figure 9.8 shows a DES representation of a harvested fishery. The first part of the process flow diagram is the same as for the natural fishery (Figure 9.4). A second activity, 'Catch_Fish', is now added, which represents the catching of fish which are then sent to the sink on the right side of the diagram. The formula for the number of

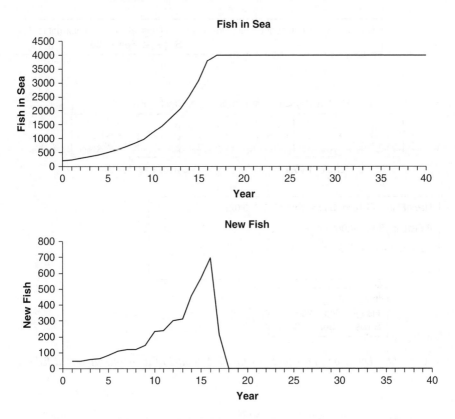

Figure 9.6   Simulation of DES model of a natural fishery with initial population of 200 fish and maximum fishery size of 4000.

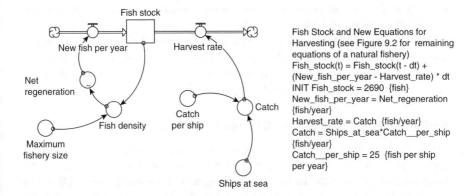

Figure 9.7   Diagram and new equations for SD model of a simple harvested fishery.

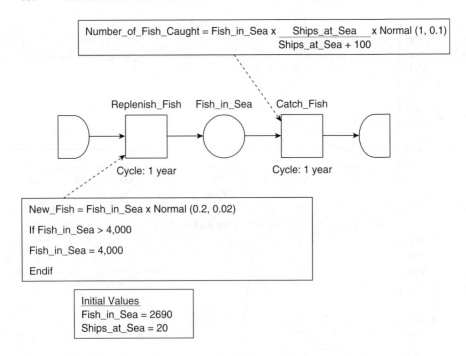

Figure 9.8   Diagram and equations for DES model of a simple harvested fishery.

fish caught consists of two parts. The first sees the catch as an increasing proportion of the fish in the sea, a proportion that increases with the number of ships. The formula is nonlinear, giving a reduced catch per ship with increasing numbers of ships. It is envisaged that as more ships are fishing in the same area their productivity will fall. The second part of the formula adds a random element to the catch. Many factors, such as climate and environmental factors, may affect the catch on any given day and in any year. The variation in annual catch is represented by a normal distribution with a standard deviation of 10% of the mean.

At first sight it may seem that the average fish regeneration rate (20%) is greater than the average catch rate with 20 ships (20/120 = 16.67%). The model is, however, in equilibrium since it first determines the catch and reduces the fish population, before determining the number of fish regenerated.

9.5.2.3   Comparison of SD and DES representations

The differences between the SD and DES models identified for the natural fishery apply equally to this case: the method of representation, the explicitness of the feedback structure and the inclusion of randomness. Interestingly, the DES model represents the catch per ship as a nonlinear function of the number of ships, while for now the SD model represents this as a linear relation.

9.5.2.4   Simulated dynamics of a simple harvested fishery: Equilibrium models

Simulation of the models shows that both are in equilibrium. Each year in the SD model there is a catch of 500 fish by 20 ships from a population of 2690 fish. Similarly, if the randomness is removed from the DES model, the fish population remains at 2690 year on year, since fish regeneration matches the catch exactly.

9.5.2.5   Simulated dynamics of a simple harvested fishery: Non-equilibrium models

To move from an equilibrium to a non-equilibrium model, two different approaches are taken in the SD model and the DES model. In the SD model, non-equilibrium conditions are created by changing the size of the fishing fleet. In the DES model, randomness is added to the model to create a non-equilibrium state. These reflect typical analyses that might be carried out with SD and DES models. The former explores the effects of nonlinearities (by changing the number of ships) and the latter explores the effects of randomness.

Consider a scenario in which the fleet size grows in steps of 10 from 0 to 30 ships. The productivity of these ships is identical. They can each catch 25 fish per year. (Do not dwell on the numerical value – remember that this is an imaginary but scalable world). In typical SD fashion we assume there is no stochastic variation in productivity. Figure 9.9 is a simulation of this stepwise scenario. At the start the fishery is at its maximum size of 4000 fish and the population is in a deterministic equilibrium where births exactly equal deaths. Ten ships arrive in year 4 and for the next 12 years

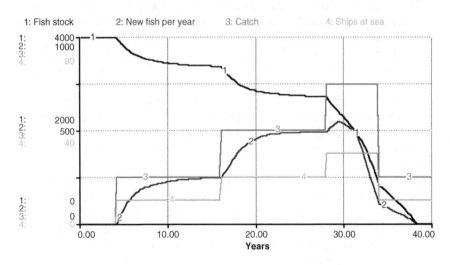

Figure 9.9   Simulation of SD model of a harvested fishery with stepwise changes in fleet size and an initial population of 4000 fish. From Morecroft (2007). Reproduced by permission of Wiley.

they harvest the fishery. The catch rises to 250 fish per year (10×25). As a result the fish stock begins to fall. Then something dynamically interesting happens. Because the fishery is less heavily populated, fish regenerate faster. The lower the fish stock, the lower the fish density and the higher the number of new fish per year, as determined by the values on the right of the nonlinear net regeneration function described earlier. As the years pass the number of new fish added to the population each year approaches ever closer to the harvest rate (and the catch) and so, by the end of year 15, the fish population settles into a sustainable equilibrium.

In year 16 another 10 ships arrive, bringing the fleet size to 20, and they fish for a further 12 years. A similar process unfolds. A doubling of the fleet leads to a doubling of the harvest rate, a fall in the fish population and eventually a compensating rise in the net regeneration of fish.

In year 28 yet another 10 ships arrive, bringing the total to 30. For almost two years it looks as though the fishery will continue its bountiful supply. The catch rises to 750 fish per year (30×25). The population falls below its previous equilibrium (of about 2700 fish) and the number of new fish per year begins to rise. However, by year 30 it is clear something new is happening. While the catch remains at 750, the number of new fish per year begins to fall (for the first time in the entire simulation). As a result the rate of decline of the fish stock begins to accelerate (rather than to moderate). The fishery has passed the peak of the net regeneration curve. It is now operating on the left-hand side of the curve, beyond a critical 'tipping point'. The fish density is now so low that any further reductions cause the regeneration rate to decline rather than rise. There are simply too few fish in the sea to breed at the rate previously achieved. The fish population continues to fall precipitously and is now being over-harvested. At the end of year 33 the population is down to 1000 and still falling.

The model provides one further insight into the dynamics of fish stocks. Imagine, at the start of year 34, the fleet is reduced to only 10 ships, the size it was between years 4 and 16. Back then the fishery had achieved a sustainable equilibrium. Now it is unable to do so. Not surprisingly the catch falls dramatically back to 250 fish per year (10×25) as ships are removed or idled. But net new fish per year is even lower at around 100 fish per year. The fish population and fish density therefore decline and by year 38 the population has collapsed to zero. The message from this compact SD model is that a harvested fishery is dynamically complex, even without randomness, due to stock accumulation, feedback and nonlinearity.

Figure 9.10 shows the results from a simulation of the harvested fishery with the DES model, now with randomness included for the catch and regeneration of fish. Because there is randomness in the model, the simulation has been replicated 10 times, using different random number streams, the results showing the mean of the replications. The use of multiple replications is standard practice in DES modelling for determining the range of outcomes and the mean performance of a system (Law, 2007; Robinson, 2004). The results show the catch and the number of fish in the sea over a 40-year period.

The graphs show the variation in the annual catch with the mean shifting between just above 400 to just below 500. Similarly there is some variation in the fish stock, peaking at about 2800 and falling to about 2600. Such variation is not surprising given

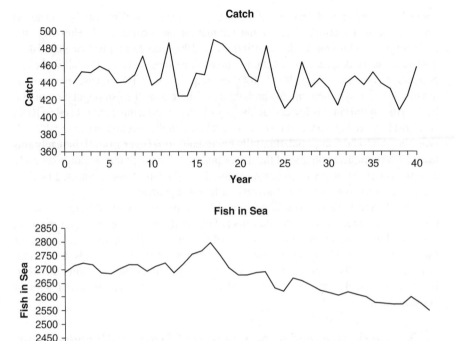

Figure 9.10 Simulation of DES model of a simple harvested fishery: non-equilibrium model.

the randomness in the model. It would be difficult to predict the interconnected effect of variations in fish regeneration and fish catch without a simulation.

Inspecting the graph for the fish in the sea, there appears to be a downward trend. This is surprising as the system is in perfect equilibrium (the catch and regeneration rates are the same) as demonstrated by the results from the equilibrium model; remember that the number of ships is fixed at 20 (Figure 9.9). As such, it would be assumed that the steady-state mean of the model including randomness would remain constant, albeit that there are annual variations. A longer run of the simulation (not shown), however, confirms that there is a downward trend in the fish stock. Over 10 000 years the fish stock falls from 2690 to 300, and after 20 000 years the fish stock disappears altogether. Experiments showed that the greater the randomness in the system, as defined by the standard deviation of the normal distributions, the faster the collapse in fish stocks.

How can these puzzling dynamics occur in a system that is apparently in equilibrium? The reason is most simply explained with reference to an example. Starting with, say, a fish stock of 1000 in any particular year, the probability of there being 10% more or 10% less fish in the next year is the same. This can be concluded because the distributions used in the simulation are symmetrical. However, the probability of returning to the equilibrium of 1000 fish from 1100 or 900 fish is not the same. To decrease from 1100 to 1000 fish requires only a 9.1% fall in the fish population. To increase from 900 to 1000 fish requires an 11.1% increase. Because the distributions are symmetrical, the probability of increasing the fish population of 900 back to the equilibrium is lower than the probability of reducing it from 1100 to 1000. As a result, there is a constant downward pressure on the fish population, with 'bad' years in which the fish population falls being hard to recover from. Ultimately this leads to the complete collapse of the fish population. Although this outcome is purely a numerical effect, it does demonstrate how the inclusion of interconnected random events can lead to a puzzling, and very different, dynamic.

The SD and DES non-equilibrium models are based on quite different assumptions. The SD model investigates the effect of changing the number of ships over time, while the DES model represents the effect of random variation in the regeneration of fish and the catch. The practical, commercial results of the two models are similar. Both lead to a collapse in the fish stock, although the timescales are quite different. Both demonstrate a dynamic that could not easily have been predicted without the simulation.

9.5.3   Alternative models of a harvested fishery with endogenous ship purchasing

Both the SD and DES models show that a fishery can operate for long periods in equilibrium providing there are an appropriate number of ships at sea – not too few and not too many. Here a long period may be hundreds of years, spanning the lifetime of a stable fishing community. The remaining piece of the dynamical puzzle is investment in new ships. Why would fishermen purchase too many ships if, as we have seen, a large fleet can lead to overfishing? And why would they not reduce the fleet size to re-establish a sustainable equilibrium? A comparison of the investment process in SD and DES provides distinctive answers to these questions.

9.5.3.1   SD model

An SD modeller thinks of investment as a collective decision-making process (or policy) representing, in aggregate, the judgements of those people most closely involved (fishermen in this case) and the information sources on which their decisions are based. Such decision-making processes are behavioural in the sense that they capture the broad intention of investment without necessarily assuming decision makers have perfect information or perfect foresight. A typical investment policy has three main parts. There is a goal, a specific condition to be achieved, in this case the desired number of ships in the fleet. There is monitoring of the current state of the system, how many ships are currently in operation. And finally there is corrective action, the purchase or sale of ships, to bring the current state of the system in line with the goal. This overall three-part process is known as 'asset stock adjustment' and is

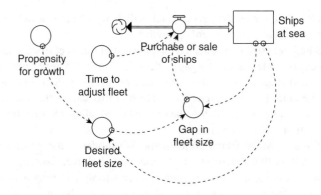

Ships_at_sea(t) = Ships_at_sea(t - dt) + (Purchase_or_sale_of_ships) * dt
INIT Ships_at_sea = 4 {ships}

Purchase_or_sale_of_ships = Gap_in_fleet_size/Time_to_adjust_fleet
{ships/year}
Gap_in_fleet_size = Desired_fleet_size - Ships_at_sea {ships}
Desired_fleet_size = Ships_at_sea * (1 + Propensity_for_growth) {ships}
Propensity_for_growth = ... See later for this important formulation, for now
just assume that normally the propensity for growth is positive and non-zero
Time_to_adjust_fleet = 1 {year}

Figure 9.11   Asset stock adjustment – the decision-making process for the purchase or sale of ships in the fishery.

absolutely central to an information feedback representation of business and social systems (Sterman, 2000). It is a process generalizable across a wide range of investment situations covering inventories, tangible capital goods, human resources and intangible assets. At the heart of the decision making is subtle, purposive (and often judgemental) information processing in which people with responsibility for investment form a view of the appropriate incremental adjustment of important assets.

Figure 9.11 shows asset stock adjustment in the fisheries model. Notice that connections between variables are shown as dashed lines denoting flows of information. The connections are not 'hardwired' as they were for the natural fishery. They are discretionary and reflect the information available and deemed most relevant to investment. The desired fleet size (the goal) depends on the number of ships at sea and the propensity for growth. Specifically the desired fleet size is equal to ships at sea multiplied by a factor (1 + propensity for growth). Here is an important behavioural assumption. Since fishermen do not have the information to decide an optimal fleet size they form their goal more simply with reference to the existing fleet size. We assume that the normal propensity for growth is 0.1, so the desired fleet size is 10% larger than the current fleet size. In other words, fishermen normally and collectively want a bigger fleet than they now have, an attribute of human nature – bigger is better, growth is inherently attractive. As we will see later, the propensity for growth also depends on conditions in the fishery; a poor catch will dampen enthusiasm for a larger fleet, despite an underlying bias towards growth.

Once the desired fleet size is established then the rest of the asset stock adjustment formulation is easy to understand. The gap in fleet size is the difference between the desired fleet size and ships at sea. If there is a large positive gap then conditions for investment are favourable. The purchase or sale of ships closes the gap over an assumed time span of one year, which is the time taken to adjust the fleet (including ordering, construction and delivery).

Perhaps the most crucial formulation in the SD model is the propensity for growth and the factors that determine it. As mentioned above, fishermen do not know the optimal fleet size and so they prefer, more simply and pragmatically, to grow the fleet until there is compelling evidence to stop. The question is what evidence is persuasive enough to curb investment and will this result in a sustainable balance between the fish population and ships at sea? In SD evidence takes the form of information. Persuasive information is both credible and readily available to decision makers. In a real fishery, fishermen know the catch per ship – they themselves bring in the catch and it is vitally important to their livelihood. They may also know, at least roughly, the total catch of all ships in the fishery. Significantly they do not know the fish population or the fish regeneration rate. That is all happening under water. Moreover, they do not believe scientific estimates of low fish stocks unless confirmed by the catch. Such practical considerations suggest that propensity for growth is curbed by low catch rather than by objective evidence of fish stocks. As a result, investment is boundedly rational, sensing only indirectly the true state of the fish population on which the long-term sustainability of the fishery depends.

Figure 9.12 shows one possible SD formulation that captures the essential limited information characteristic of fishermen's boundedly rational decision making. Propensity for growth depends on the normal propensity for growth (set at 0.1 or 10%, to reflect a prevalent view that growth is a good thing) multiplied by the curbing effect of catch per ship. This curbing effect is nonlinear and captures another typical human tendency, which is to ignore bad news until it is really bad. If catch per ship falls from 25 fish per year to 15 per year (a 40% decline) propensity for growth falls from 0.1 to 0.09 (a decline of only 10%). Thereafter the effect becomes much stronger. If the catch per ship falls to 10 fish per year, then propensity for growth falls to zero and fishermen stop purchasing ships. If the catch falls still further then the propensity for growth becomes negative and fishermen sell ships because collectively they sense it is futile to retain a large and unproductive fleet.

Catch per ship measures ships' productivity and is modelled here as a deterministic function of fish density. The scarcer are the fish, then the lower the productivity. But the relationship is nonlinear. For moderate to high fish density (between 0.5 and 1) catch per ship remains close to normal. The assumption is that fishermen do not really notice a difference in the catch if the sea is teeming with fish or only half-teeming with fish, because fish tend to school or cluster. Catch per ship is still 68% of normal when the fish density is only 0.2, or in other words when the fish population is 20% of the maximum sustainable. But thereafter catch per ship falls quickly to

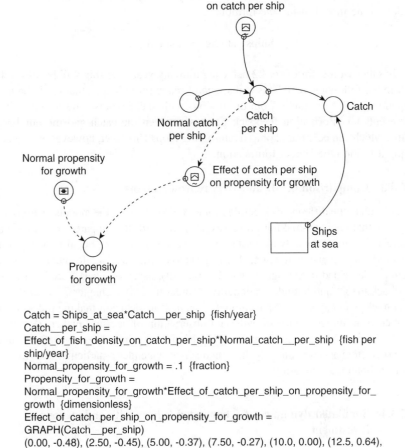

Catch = Ships_at_sea*Catch__per_ship {fish/year}
Catch__per_ship =
Effect_of_fish_density_on_catch_per_ship*Normal_catch__per_ship {fish per ship/year}
Normal_propensity_for_growth = .1 {fraction}
Propensity_for_growth =
Normal_propensity_for_growth*Effect_of_catch_per_ship_on_propensity_for_growth {dimensionless}
Effect_of_catch_per_ship_on_propensity_for_growth =
GRAPH(Catch__per_ship)
(0.00, -0.48), (2.50, -0.45), (5.00, -0.37), (7.50, -0.27), (10.0, 0.00), (12.5, 0.64), (15.0, 0.9), (17.5, 0.995), (20.0, 0.995), (22.5, 1.00), (25.0, 1.00)
Effect_of_fish_density_on_catch_per_ship = GRAPH(Fish_density {dimensionless})
(0.00, 0.00), (0.1, 0.4), (0.2, 0.68), (0.3, 0.8), (0.4, 0.88), (0.5, 0.96), (0.6, 1.00), (0.7, 1.00), (0.8, 1.00), (0.9, 1.00), (1, 1.00)

Figure 9.12   Propensity for growth and catch per ship.

zero as schools of fish become increasingly difficult to find and are hotly contested by rival ships.

9.5.3.2   DES model

In DES modelling there is no equivalent to the 'asset stock adjustment' process in SD modelling. Policies for purchasing and releasing ships would be determined by discussion with relevant stakeholders, as in SD, but without the guiding framework of stock adjustment. In this case a simple policy for adjusting the number of ships is added to the model, using the formula

$$Ships\_at\_Sea = Caught/22$$

In other words, for every 22 fish caught in any year, one ship will be allowed to fish in the following year. This is just one of many formulations that might be used, depending on the findings from discussions with stakeholders. The formula above represents an information feedback process between the catch and the number of ships, which can be thought of as a reinforcing loop. This facet, however, is not made explicit in the DES model formulation.

9.5.3.3   Comparison of SD and DES representations

In addition to the differences identified for earlier versions of the models, there is now the way that decision-making processes are perceived. It is apparent that in SD the modeller is guided not only by discussion with decision makers, but also by principles for modelling decision making. These principles include five formulation fundamentals, described in Sterman (2000, Chapter 13), based on ideas from information feedback theory and bounded rationality (Morecroft, 2007, Chapter 7). In DES such principles do not appear to be in common use. Instead, decision-making processes are derived from direct discussion with and observation of decision makers (Robinson *et al.*, 2005). Uncertainties in making decisions are primarily derived from future unknown stochastic events, which in themselves engender a particular and different form of bounded rationality.

9.5.3.4   Simulated dynamics of a harvested fishery with endogenous investment

The simulations in this section start with both SD and DES models in a sustainable equilibrium. The SD model starts with 10 ships and 3370 fish, resulting in a catch of 250 fish per year (below the maximum sustainable yield to allow room for growth and to investigate boundedly rational misinvestment). The DES model starts with 20 ships and 2690 fish, resulting in an average catch equal to the regeneration of fish (at a theoretical equilibrium of 500 fish per year), to investigate stochastic misinvestment (described later). The equilibrium is then disturbed in contrasting ways typical of each approach.

In the SD model the normal propensity for growth is artificially held at zero at the start of the simulation. A small-is-beautiful mindset has temporarily taken hold. Then in year 10 the normal propensity for growth returns to a value of 0.1, or 10% of the current fleet size. Figure 9.13 shows the results. Equilibrium prevails until year 10. Then the number of ships at sea increases steadily under the influence of an investment policy biased towards growth. For more than 10 years the catch rises. Meanwhile the catch per ship remains steady, suggesting that continued investment is both feasible and desirable. Below the waves conditions are changing, but remember

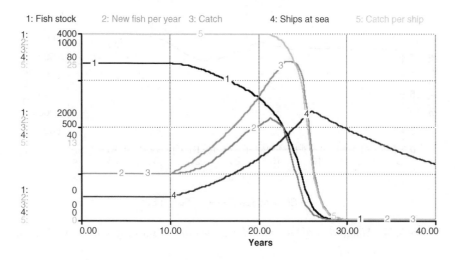

Figure 9.13   Simulation of SD model of a fishery that starts in equilibrium, grows with investment and then unexpectedly collapses. From Morecroft (2007). Reproduced by permission of Wiley.

that these conditions cannot be directly observed by fishermen. The regeneration rate of fish (new fish per year) rises healthily as one would expect in a well-harvested fishery. The fish population falls, but that too is expected in a harvested fishery.

Signs of trouble appear under water in year 21 when, for the first time, regeneration (new fish per year) falls. This reversal of replenishment is a signal that the fishery has passed the tipping point of the nonlinear regeneration curve. The decline in the fish stock begins to accelerate. But interestingly the catch continues to rise for fully three more years, until year 24, and the catch per ship remains close to normal. From the viewpoint of growth-orientated fishermen floating on the surface of the sea, it is business as usual. The fleet continues to grow until year 26 when it reaches a size of 46 ships. By then the catch per ship has fallen to less than one-third of normal (only 8 fish per ship per year instead of 25), sufficient to curb further investment.

By now the hidden fish stock has fallen to a precariously low level of only 300, less than one-tenth of its initial value. With so few fish in the sea the regeneration rate is very low at only 30 new fish per year, well below the catch of around 300 fish per year. Fishermen are now well aware of the underwater crisis and respond accordingly by selling ships. The fleet size falls from a peak of 47 ships in year 26 to 39 ships in year 30. But it is too little action, too late. The boundedly rational investment policy is unable to reduce the fleet quickly enough to halt the decline of the fish stock. By year 30 there are only four fish left in the sea and regeneration has fallen to zero. The fishery has collapsed with a huge excess of relatively new ships owned by fishermen reluctant to sell and still dependent on the fishery for their livelihood. The dismal

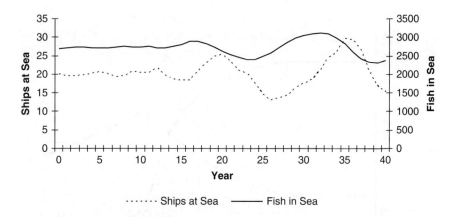

Figure 9.14   Simulation of DES model with investment showing start of oscillation.

dynamics of the Pacific sardine catch in Figure 9.1 have been played out in a purely deterministic nonlinear simulation model.

Figure 9.14 shows the output from the DES model with investment in ships. As before, the results shown are the mean of 10 replications. Over the 40-year period it is apparent that the output is moving into an oscillation that is characteristic of a delayed negative feedback loop. This is confirmed by a longer simulation run (Figure 9.15). The amplitude and frequency of the oscillation vary due to the random nature of the fish catch and fish regeneration.

The oscillation represents a puzzling dynamic. The system is theoretically in equilibrium with the fish catch and regeneration rates set as equal. So what is the cause of the oscillation? Figures 9.14 and 9.15 show, unsurprisingly, that the catch oscillates in line with the number of ships at sea. There is a lag, however, between an increase or

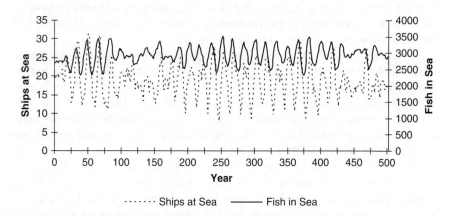

Figure 9.15   Simulation of DES model with investment showing oscillation over 500 years.

decrease in the fish stock and a corresponding increase or decrease in the number of ships. This is to be expected as the purchase and sale of ships is dependent on the previous year's catch.

At the beginning of the simulation the number of ships, the catch and the fish in the sea remain relatively stable, as would be expected in an equilibrium state. In year 17 there is an increase in the number of fish caught. This is presumed to be a combination of two random events, a good year for fishing and a previous good year for regeneration, as evidenced by the catch and fish stock graphs (only the fish stock graph, fish in sea, is shown here). As a result of an increased catch more ships are purchased. As more ships are purchased the catch initially grows, but eventually this affects fish stocks and the catch then falls leading to disinvestment in ships. Once started, the oscillation perpetuates as in any typical delayed negative feedback loop. The oscillation is a result of an initial random variation, to which the ship investment policy reacts, causing the fishery to move into a permanent oscillation of fish stocks, catch and ships. Had there been no random variations, then the oscillation would not have started. Note that from an SD perspective a one-time-step change in either the catch or ships at sea would trigger a similar cyclical behaviour.

What the model demonstrates is a ship investment policy that is too sensitive to random variations (stochastic misinvestment). The decision to grow the shipping fleet is based on the only measurable data concerning likely fish stocks, that is the catch. However, due to random variations in the catch an increased catch does not necessarily indicate a growth in fish stocks. Indeed, it is possible for fish stocks to fall and the catch to rise. In this example there should not have been any investment in ships, as there was no underlying trend suggesting an increase in fish stocks. The fishermen have simply reacted to a random variation in the catch.

Here it is useful to think of the problem in terms of the two types of variation identified in statistical process control theory (Montgomery and Runger, 1994). 'Common' causes of variation are simply random variations that do not indicate any underlying change in the system. 'Special' causes of variation are shifts that do indicate an underlying change, or trend. In quality control the aim is to devise policies that only react to special causes of variation. In the fisheries example we see a typical case of an over-sensitive policy in which a change is made as a result of common causes of variation. Such a reaction can lead to an unexpected outcome; in this case the result is the oscillation evidenced in Figures 9.14 and 9.15.

The SD and DES models represent different policies for investment in the fishing fleet. Both policies are shown to be defective. In the SD model the policy leads to a complete collapse in fish stocks whose deterministic regeneration rate is highly nonlinear. In the DES model the investment policy combined with random variations leads to a boom and bust business cycle in fish stocks whose stochastic regeneration rate is essentially linear, but capped. Interestingly these behaviours closely represent the behaviours of the real fisheries shown in Figure 9.1. The Pacific sardine data showed a collapse in the sardine catch and presumably the sardine stock. This is similar to the output of the SD model. Meanwhile, the North Sea herring data seems to show the beginnings of an oscillation in the catch similar to that predicted by the DES model.

9.6 Summary of findings

The processes of developing SD and DES models of the same problem situation are described above. The models have been developed separately by an expert in each field with a view to understanding the nature of the modelling process and insights gained from the two types of modelling approach. Our premise was that the use of SD or DES affects the representation and interpretation of phenomena from the real world. This premise is borne out in the work described.

Table 9.3 summarizes the key differences in the SD and DES approaches that have emerged as a result of this investigation into the fisheries problem. Although the list of differences is not exhaustive, and not all differences would apply to every modelling situation, nevertheless it is obvious that many of the representational differences are similar to those identified in previous studies comparing SD with DES described at the start of this chapter. But there are new insights too about differences in the interpretation of puzzling dynamics.

Table 9.3 Key differences between SD and DES approaches emerging from fisheries models.

SD 'deterministic complexity'	DES 'interconnected randomness'
Representation	
System represented as stocks and flows	System represented as queues and activities
Feedback explicit	Feedback implicit
Many relationships are nonlinear	Many relationships are linear
Randomness is generally not modelled, or is added later after noise-free simulations	Randomness explicitly modelled
Growth/decay modelled as exponential or S-shaped	Growth/decay represented as random often with discrete steps, for example a cut-off point
Standard recurring modelling structures exist, for example asset stock adjustment process	Standard modelling structures generally do not exist
Standard diagramming format	No agreed standard diagramming format
Interpretation	
Feedback, stock accumulation, delays and nonlinearities are vital to system performance	Feedback, delays and nonlinearities are not emphasized
Randomness is not normally important to system performance	Randomness is a vital element of system performance
Structure leads to system behaviour over time	Randomness leads to system behaviour over time

In general SD is primarily involved with understanding the performance over time of an interconnected system arising from its internal feedback structure. This feedback structure is made explicit in the representation of an SD model and expressed through a series of differential equations that are frequently nonlinear. Randomness is rarely considered and, when included, is simplified and often added later in a study after noise-free simulations have been conducted. Growth and decay processes are normally considered to be exponential or S-shaped in form. A number of standard recurring modelling structures exist (often stemming from assumptions about bounded rationality and managerial decision making) that guide model conceptualization and equation formulation in SD models. The diagramming format for representing SD simulation models (stock and flow diagrams) is seen as a standard, though causal loop diagrams are also widely used qualitatively to depict feedback structure.

DES primarily focuses on the performance over time of an interconnected system subject to internal and external random variation. Feedback structures are included in these models, but they are not made explicit. This use of feedback contradicts the view of Coyle (1985) that DES models are always open-process structures. DES modellers tend to adopt relationships that are linear in form, although not exclusively. Meanwhile, randomness is seen as a vital part of system behaviour and it is explicitly modelled whenever and wherever in the system it is believed to occur. Growth and decay processes are therefore seen as random but limits are often represented as discrete cut-off points. DES modellers do not have standard modelling structures to guide equation formulation, nor is there an agreed diagramming format for representing DES models.

Building on these representational distinctions, SD modellers see feedback, delays and nonlinearities as vital to the performance of a system, while randomness is normally of little or no importance (or is viewed as something that evokes the latent dynamics of feedback structure). DES modellers place the emphasis on randomness with little direct concern for the effects of feedback. As a result SD modellers see feedback structure as the prime source of system behaviour, while DES modellers consider randomness to be the main cause.

What is clear is that SD and DES adopt quite different modelling philosophies. The similarities between the two approaches seem to end at the fact that both are simulation methods aimed at modelling the progress of a system through time. Further to this, and based on our experiences with the fisheries model, we might argue that both adopt an evolutionary approach to model development.

9.7 Limitations of the study

There are some limitations of this study and two specifically need to be highlighted. First, there is a variety of both SD and DES models that could be developed for this problem. We have only developed one example of each. Alternative models may have led to different findings. That said, the fact that these models have been developed by experts of long standing in both fields suggests that they are representative of the typical SD and DES modelling approach. The research strategy adopted here is certainly an improvement on an individual developing both models, where that individual is bound to be more steeped in one modelling approach over the other. This is a drawback of previous comparative research, as identified in Section 9.2.

The second key limitation is that the DES model does not use the full set of DES modelling facilities. In particular, the model does not represent individual fish (instead an entity represents the population of fish) and the time step is fixed (at one year) and does not vary. Modelling individuals and a varying (random) time step are both features of many DES models. It would not be true to say, however, that DES modellers do not adopt the approach taken here. Robinson (2004) describes 'grouping entities' as a standard model simplification method in DES and there are examples of DES models with a constant time step (Bowman *et al.*, 1998). Indeed, DES modellers are keen to stress the importance of building an appropriate model, not one that necessarily uses every feature of the approach (Pidd, 2003). The important feature of this work is that the two models were built for the same problem using the modelling philosophy of both an SD and a DES modeller.

Given these limitations, we believe that it would be beneficial if there were more comparisons of SD and DES that followed a similar approach to that taken here. In this way, through a series of studies, a more complete understanding of the similarities and differences between SD and DES could be derived. Since we first worked on this problem a few such studies have been undertaken and some are reported in this edited volume.

9.8 SD or DES?

Having identified a series of differences between SD and DES modelling, we return to the original question of this chapter: which method to use and why? In the case of fisheries both models seem to offer plausible explanations for the behaviours seen in Figure 9.1. This outcome would suggest that neither method is necessarily superior to the other, but that either may be useful in different circumstances. While each approach represents certain facets of the real world, both approaches also simplify or downplay other facets. Given that both SD and DES modellers would agree that all useful models are simplifications of reality (Meadows, 1980; Pidd, 2003), this selective attention of the two approaches is not in itself a shortcoming. It is, however, important that modellers first recognize what simplifications a modelling approach entails and, second, that they select a modelling approach based on the facets of the problem situation and the attributes of the modelling approach. Table 9.3 should help in determining which attributes each modelling approach includes and which it excludes. Interpreting the important facets of the real world is dependent on the specific problem situation.

Where the facets of the real world and their implications for the problem situation are not clearly understood (which is likely to be the main motivation for modelling) then the advice might be to build both types of models, since both give important and possibly differing insights. Indeed, Renshaw, in writing about the modelling of biological populations, states:

The tragedy is that too few researchers realize that both deterministic [SD] and stochastic [DES] models have important roles to play in the analysis of any particular system. Slavish obedience to one specific approach can lead to disaster . . . So pursuing both approaches simultaneously ensures that we do not become trapped either by deterministic fantasy or unnecessary mathematical detail.

(Renshaw, 1991, p. 2)

Similarly, Koopman (2002) discusses the modelling of a smallpox outbreak with the aim of determining how best to respond. Discrete and continuous simulation models give quite different insights into the benefits of mass vaccination over targeted vaccination.

Perhaps it is time that more SD and DES modellers crossed the divide and considered applying both approaches, or at least considered more carefully situations in which the application of both approaches might yield complementary insights. In fact modellers do sometimes combine both deterministic feedback structure and stochastic inputs. For example, Dudley (2003), a policy adviser to the fishing industry and an expert in fisheries modelling, adds random variation to the recruitment of juvenile fish (i.e. regeneration) in an SD model of a fishery. His model includes not only harvesting but also fishery management. Noise-free simulations show cyclicality in fish stocks and catch. He notes that 'with random variation added to recruitment the system shows significant variability but the overall cyclical pattern of the fishery remains'. This overlaying of randomness in a single rate equation formulation is representative of SD practice (as mentioned earlier). But the practice is quite rare. Moreover, such bolt-on noise tests, applied to individual rate equations, do not necessarily recognize the dynamic significance of *multiple interacting random processes* typical of DES studies.

This step of applying both approaches requires more than simply learning about the modelling techniques and tools involved in either approach. The modeller also needs to adopt, or acknowledge, a completely different modelling philosophy and temporarily to suspend deeply held beliefs about reasons for system behaviour over time. Our experiences with the simple fisheries models have made us more aware, and more accepting, of alternative plausible interpretations for puzzling dynamics. We have moved forward in what Lane (2000) has described as 'mode 3 discourse' between DES and SD, building on a growing appreciation of differences and similarities between the two approaches. Probably each of us will remain anchored in our core disciplines, but we can now see enough of the 'other' discipline to sense where future collaboration might be beneficial.

A sample of ideas for future collaboration is a fitting end to the chapter. We have discovered paradigm differences between DES and SD. One loose yet concise way to communicate these differences is to say that DES illuminates 'interconnected random-ness' whereas SD illuminates 'deterministic complexity'. The real world contains both. Maybe interconnected randomness is most evident in functional/operational problems of the kind frequently tackled by DES modellers, whereas deterministic complexity shows up in cross-functional/strategic problems most often addressed in SD. More likely both SD and DES are capable of addressing operational and strategic issues and it is just a matter of which components you choose for your simulated enterprise and whether or not you believe randomness or feedback structure plays the dominant role in the unfolding future. To take a practical example, the oil producers' microworld (Morecroft, 2007, Chapter 8; Langley, 1995) is an industry-level deterministic SD simulator linking rival aggregate producers of oil to world demand for oil products. Its purpose is to understand global oil market dynamics and long-term oil price. There are five main sectors in the model. There is no reason why these same sectors could not be reconceptualized in a DES model that emphasizes the randomness and turmoil that undoubtedly pervade the oil industry. On the other hand, in the manufacturing heartland of DES exemplified by, for instance, the brick factory model reported by Robinson and Higton (1995), there is surely potential for an SD model of the same factory. However, instead of investigating the stochastic interaction of individual factory machines and manufacturing processes it would emphasize deterministic nonlinear feedback processes in manufacturing control and shed new and complementary light on factory management and brick production. Food for thought . . .

Acknowledgements

This chapter is based on the paper Morecroft, J. and Robinson, S. (2005). Explaining Puzzling Dynamics: Comparing the Use of System Dynamics and Discrete-Event Simulation. *The 23rd International Conference of the System Dynamics Society*, July 17–21, 2005, Boston.

References

Arnason, R. (2007) Fisheries management, in *Handbook on Operations Research in Natural Resources*, vol. 99 (eds A. Weintraub *et al.*), International Series in Operations Research & Management Science, Springer, New York.

Banks, J., Carson, J.S., Nelson, B.L. and Nicol, D.M. (2001) *Discrete-Event System Simulation*, 3rd edn, Prentice Hall, Upper Saddle River, NJ.

Bowman, R.A., Haimowitz, I.J., Mattheyses, R.M., Özge, A.Y. and Phillips, M.C. (1998) Discrete time simulation of an equipment rental business, in *Proceedings of the 1998 Winter Simulation Conference* (eds D.J. Medeiros *et al.*). IEEE Press, Piscataway, NJ, pp. 1505–1512.

Brailsford, S. and Hilton, N. (2001) A comparison of discrete event simulation and system dynamics for modelling healthcare systems. Proceedings of the 26th Meeting of the ORAHS Working Group 2000, Glasgow Caledonian University, Glasgow, Scotland, pp. 18–39.

Coyle, R.G. (1985) Representing discrete events in system dynamics models: a theoretical application to modelling coal production. *Journal of the Operational Research Society*, **36**(4), 307–318.

Coyle, R.G. (1996) *System Dynamics Modelling*, Chapman & Hall, London.

Davis, R. (2001) *Business Process Modelling with ARIS: A Practical Guide*, Springer, Berlin.

Dudley, R.G. (2003) A basis for understanding fisheries management complexities. Proceedings of the International System Dynamics Conference, New York City 2003, manuscript 118, http://www.systemdynamics.org/conferences.

Farber, M.I. (1991) A methodology for simulating the US recreational fishery for billfish. Proceedings of the 1991 Winter Simulation Conference, pp. 832–840.

Forrester, J.W. (1961) *Industrial Dynamics*, MIT Press, Cambridge, MA (reprinted by Pegasus Communications, Waltham, MA).

Forrester, J.W. (1968) Industrial dynamics – after the first decade. *Management Science*, **14**(7), 398–415.

Hardin, G. (1968) The tragedy of the commons. *Science*, **162**, 1243–1248.

Hills, P.R. (1971) *HOCUS*, P-E Group, Egham.

Koopman, J. (2002) Controlling smallpox. *Science*, **298**(5597), 1342–1344.

Lane, D.C. (1999) Social theory and system dynamics practice. *European Journal of Operational Research*, **113**, 501–527.

Lane, D.C. (2000) You just don't understand me: modes of failure and success in the discourse between system dynamics and discrete event simulation. LSE OR Department, Working Paper LSEOR 00-34.

Langley, P. (1995) An experimental study of the impact of online cognitive feedback on performance and learning in an oil producers' microworld. PhD thesis, London Business School.

Lanner (2007) Witness Simulation Software. www.lanner.com (accessed June 2007).

Law, A.M. (2007) *Simulation Modeling and Analysis*, 4th edn, McGraw-Hill, New York.

Mak, H.-Y. (1992) System dynamics and discrete event simulation modelling. PhD thesis. London School of Economics and Political Science.

Meadows, D.H. (1980) The unavoidable a priori, in *Elements of the System Dynamics Method* (ed. J. Randers), MIT Press, Cambridge, MA, pp. 23–57.

Meadows, D.L., Fiddaman, T. and Shannon, D. (2001). *FishBanks, Ltd. A Micro-computer Assisted Group Simulation That Teaches Principles of Sustainable Management of Renewable Natural Resources*, 5th ed., The FishBanks Ltd. game was developed by Professor Dennis Meadows, co-author of "Limits to Growth." The board game kits which include the game software, PowerPoint slide sets for introducing and debriefing the game, instructions for playing the game, the role description, game board, and pieces are sold through the System Dynamics Society http://www.systemdynamics.org/.

Montgomery, D.C. and Runger, G.C. (1994) *Applied Statistics and Probability for Engineers*, John Wiley & Sons, Inc., New York.

Morecroft, J. (2004) Mental models and learning in system dynamics practice, in *Systems Modelling: Theory and Practice* (ed. M. Pidd), John Wiley & Sons, Ltd, Chichester, pp. 101–126.

Morecroft, J. (2007) *Strategic Modelling and Business Dynamics*, John Wiley & Sons, Ltd, Chichester.

Moxnes, E. (1998) Not only the tragedy of the commons: misperceptions of bioeconomics. *Management Science*, **44**(9), 1234–1248.

Otto, P. and Struben, J. (2004) Gloucester fishery: insights from a group modeling intervention. *System Dynamics Review*, **20**(4), 287–312.

Pidd, M. (2003) *Tools for Thinking: Modelling in Management Science*, 2nd edn, John Wiley & Sons, Ltd, Chichester.

Pidd, M. (2004) *Computer Simulation in Management Science*, 5th edn, John Wiley & Sons, Ltd, Chichester.

Pooley, R.J. (1991) Towards a standard for hierarchical process oriented discrete event diagrams. *Transactions of the Society for Computer Simulation*, **8**(1), 1–41.

Renshaw, E. (1991) *Modelling Biological Populations in Space and Time*, Cambridge University Press, Cambridge.

Richardson, G.P. (1991) *Feedback Thought in Social Science and Systems Theory*, University of Pennsylvania Press, Philadelphia.

Richardson, G.P. and Pugh, A.L. (1981) *Introduction to System Dynamics Modeling with DYNAMO*, MIT Press, Cambridge, MA.

Richmond, B., Peterson, S., Chichakly, K., Liu, W. and Wallis, J. (2004) iThink Software, Isee Systems Inc., Lebanon, NH.

Richter, H. and März, L. (2000) Toward a standard process: the use of UML for designing simulation models, in *Proceedings of the 2000 Winter Simulation Conference* (eds J.A. Joines*et al.*), IEEE Press, Piscataway, NJ, pp. 394–398.

Robinson, S. (2001) Soft with a hard centre: discrete-event simulation in facilitation. *Journal of the Operational Research Society*, **52**(8), 905–915.

Robinson, S. (2004) *Simulation: The Practice of Model Development and Use*, John Wiley & Sons, Ltd, Chichester.

Robinson, S., Alifantis, T., Edwards, J.S., Ladbrook, J. and Waller, T. (2005) Knowledge based improvement: simulation and artificial intelligence for identifying and improving human decision-making in an operations system. *Journal of the Operational Research Society*, **56**(8), 912–921.

Robinson, S., Brooks, R., Kotiadis, K. and van der Zee, D.-J. (eds) (2011) *Conceptual Modeling for Discrete-Event Simulation*, CRC Press, Boca Raton, FL.

Robinson, S. and Higton, N. (1995) Computer simulation for quality and reliability engineering. *Quality and Reliability Engineering International*, **11**, 371–377.

Roughgarden, J. and Smith, F. (1996) Why fisheries collapse and what to do about it. *Proceedings of the National Academy of Science*, **93**(10), 5078–5083.

Som, T.K. and Sargent, R.G. (1989) A formal development of event graphs as an aid to structured and efficient simulation programs. *ORSA Journal on Computing*, **1**(2), 107–125.

Sterman, J.D. (1989) Modeling managerial behavior: misperceptions of feedback in a dynamic decision making experiment. *Management Science*, **35**(3), 321–339.

Sterman, J.D. (2000) *Business Dynamics: Systems Thinking and Modeling for a Complex World*, Irwin/McGraw-Hill, Boston, MA.

Tako, A.A. and Robinson, S. (2009) Comparing discrete-event simulation and system dynamics: users' perceptions. *Journal of the Operational Research Society*, **60**(3), 296–312.

Tako, A.A. and Robinson, S. (2010) Model development in discrete-event simulation and system dynamics: an empirical study of expert modellers. *European Journal of Operational Research*, **207**, 784–794.

Tako, A.A. and Robinson, S. (2012) The application of discrete event simulation and system dynamics in the supply chain context. *Decision Support Systems*, **52**(4), 802–815.

Torn, A.A. (1981) Simulation graphs: a general tool for modeling simulation designs. *Simulation*, **37**(6), 187–194.

Townsend, C.R., Begon, M. and Harper, J.L. (2003) *Essentials of Ecology*, 2nd edn, Blackwell, Oxford.

10

DES view on simulation modelling: SIMUL8

Mark Elder
SIMUL8 Corporation, Glasgow, UK

10.1 Introduction

This chapter presents a particular perspective on conducting discrete-event simulation (DES) projects with software and gives the reader enough information about the use of one particular software package to get started with simple, practical, simulation building and running.

Although it is only one perspective, the author has, over too many years to remember, been involved with the creation of about eight different DES software products, so the perspective is a little wider than one product and this chapter is intended to be a perspective on DES with software, not on software products.

Most, if not all, DES software products nowadays do much more than run simulations. In addition to explaining how to use DES software, this chapter also explains what the different parts do and why the products have evolved to where they are now.

Of course the software for DES is just a small part of doing DES projects, so this chapter first explains how to use software in the context of a project. It concludes with a look at the problems still to be solved by DES software.

Discrete-Event Simulation and System Dynamics for Management Decision Making, First Edition.
Edited by Sally Brailsford, Leonid Churilov and Brian Dangerfield.
© 2014 John Wiley & Sons, Ltd. Published 2014 by John Wiley & Sons, Ltd.

10.2   How software fits into the project

The only reason for building and running simulations is because of a need to understand something about a process, probably because of the desire to improve that process but without yet knowing how.

This means the objective is not directly to build a simulation, but to help create understanding and knowledge so the clients of the work then know enough to be able to change the process they manage. This might seem a really obvious point. However, it is always surprising how often builders of simulations think the simulation itself is the main objective. It is all too easy to lose oneself in the technical niceties and details of a simulation rather than directly helping the client.

What does this mean in practice? In a DES project it means working with the client most of the time (Elder, 1992). If one really needs to go off to a back room to do some intricate work on the innards of a simulation model then this should only be done for a couple of days at most. There are many reasons for this, in addition to the obvious: it is worthwhile to maintain visible contact with the clients in case they think you are not working hard enough on their project! Here are some of the more important reasons:

- The issues relevant to a business decision do change. If a business decision maker is left alone for a couple of days, the issues will have evolved by the time the consultant building the simulation returns. So the work done on the simulation needs to include these changes. Indeed the simulation, if effectively used, can be the prime driver for this evolution of the decision maker's thinking.

- Most business decisions need to be taken quickly. There are exceptions – especially when the client's project is part of something larger where there might be a few weeks before the results of a simulation project are required to feed into some other part of the project. But for the majority of occasions when simulation is useful, the decision makers need to know the answers to the questions before they can move forward, and if building the simulation takes weeks it is likely the clients will make decisions using the seat of their pants and not the simulation.

- The scope of what needs to be simulated changes with the progress of the simulation project. For example, one may start by building a strategic model looking at an entire system at a very global level of detail. This will help the clients understand where they need to focus their investigation. Then a detailed simulation model, or several models, are required, focusing on exactly what happens in specific parts of the process. These might feed back changes to the high-level model, or indicate that a new investigation is required at an even more detailed level. It is the answers returned by the simulations that drive this investigative process. If the clients are not involved, they will not understand the need for multiple models.

- The journey is often more important than the destination: if the simulation results are seen as the destination then the process of building and refining the simulation model (or models) is the journey. It is this journey that leads to most of the discoveries along the way and so the client needs to be involved in it. If left in the dark while the simulation builder makes all the discoveries then the client simply will not understand, or buy into, the conclusions reached. The client will not have experienced the blind alleys, or overcome the same hurdles as the simulation builder. So the client cannot be left out of the loop for long if the project is going to be a success.

- Simulation model builders are often quite technical people. They like data and facts to be clear and unambiguous so it is clear how the simulation should behave. But the real business world is not quite like this. Often, subjective judgements need to be made while trying to find ways to improve a process. Clients sometimes have to 'bend' what they might state as their 'hard constraints' when they start to see (from a simulation) the consequences of the limitations that they have imposed on the simulation builder. If they are involved they will appreciate the need to 'bend' and this might lead to a clever way forward for their business. If the client/decision maker is not involved in the process of building and evolving the simulation model(s) then these kinds of judgements will not be included and the project will be less successful.

- Finally clients do not tell everything upfront because either they are forgetful or they simply do not realize the detail required to build the simulation. It is not good to spend two weeks on building a simulation, and then present a finished model to a client, only to be asked, 'Did I mention that there are three types of product?'

The implication of the above is that simulation activity runs all the way through the project. By this it is particularly meant that it is not a good idea for simulation to be brought in at the last minute, once the designs for change are complete, 'just to check the design is right'. From the above discussion it is clear that simulation helps to get the design right, so it is inefficient to use it only to 'check' a design. Furthermore, if at that late stage simulation shows the design is wrong, the designers may have a problem with their time schedule!

The simulation work should therefore take place in parallel with and integrated into most of the rest of the project on improving or designing a process.

How do you get started? How do you know when to finish? When starting a project there are some very basic questions to ask:

- What do you want to improve?

- What is your measure of success? And then:

- What can you change?

These questions start a conversation that will indicate what will be worth simulating first.

The next step is then to build quickly (and roughly) a small simulation that captures some of what the client has indicated that needs to be included. Do this live in front of the client. Do not worry about complications or data. Press the Run button and let the client see the animation. At that point there had better not be an urgent flight to catch – because the members of the client group will now spend two more hours avidly discussing what really happens in their different departments and starting to understand each other in a way they never have before. And no real data has been entered yet! Indeed the author knows of two different occasions where projects have started like this and there has been no need to do more work because the conversation it generated caused the client group to realize the solution to their problem. On another occasion the client saved over $10 million in that initial session, but the project continued because there were many more savings to be had with several properly built and validated simulations.

As stated above, probably several simulations will be built during the project. When do you stop?

That is easy – stop when the client says, 'OK, thanks, I know what decision I'm going to take.' This is really important. But so many simulation people seem to get this wrong. Do not stop when 'The model is finished.' The model is not the objective of the project. The objective is to get the client to learn enough about the way the process works, or could work, so that the client knows what to do.

10.3   Building a DES

Having stated all the above caveats about simulation projects being rather more than building and running a simulation, it is now time to build one.

The client for the project that will be used as an example is Table Corporation. Table Corp. is trying to get as much product as possible through its small factory with as little resource as possible. Table Corp. can change the number of people and/or the number of machines, but obviously does not want to spend money if it does not have to.

Table Corp. makes metal tables and the process is really simple. It only involves welding the parts together and then painting the tables. The parts are made by an external supplier and delivered to the factory at the expected rate of demand.

Ideally the reader should follow the process in the software, but if that is not possible there is sufficient here to understand what is being done and the impact of doing it.

Any version of SIMUL8 is sufficient to create and run this simulation, for example a student version or an evaluation version, but not the 'runtime viewer', which does not build simulations, but only runs them. The same simulation can be built and run with similar results using most modern simulation software, but the instructions here are for SIMUL8.

Once the software is installed, draw the process flow shown in Figure 10.1 (see Figure 10.2).

Use a start point, two activities and an end point (Figure 10.3).

Figure 10.1   Table Corp.'s process.

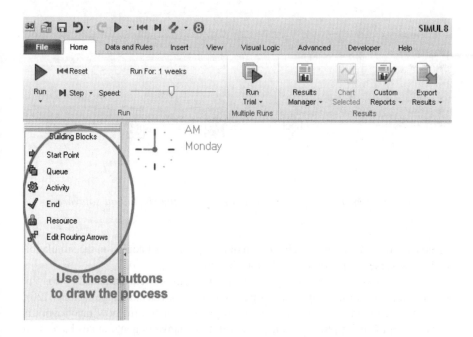

Figure 10.2   Buttons to use to draw the process flow.

Click on the Run button.

What has happened so far is that a first, oversimplified, version of a simulation of Table Corp.'s factory has been built and run in one DES software product. Soon it will be extended and improved, but first a few comments should be made on what the software has done in the background.

Many of the underlying necessities that were detailed in Chapter 2 have been taken care of by the software. For example, deciding what events are going to be handled in the simulation, compiling a table of those events, marshalling them in time, and sampling from random number generators to decide when events should happen, have all been managed automatically based on the process flow that was drawn on the window in the software. Some important detail remains to be added to the process flow, but that will be taken care of and used to modify the underlying events, entities, and so on that are being created and handled by the software.

When the Run button was pressed the simulation software knew that the user had not entered data for many options (e.g. how long the simulation should run), so it

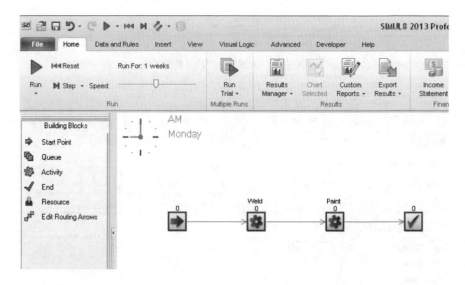

Figure 10.3   Table Corp.'s process drawn in the simulation software.

made reasonable assumptions about each of the options so it could run the simulation and show some animation of the factory in action.

The final position shown in Figure 10.4 shows (a) the current simulated time is 5 p. m. on Friday; (b) how many packs of parts for tables have entered the factory (120); (c) how much work is currently in the weld process (none); (d) how much work is currently in the paint process (none); and (e) how many complete tables have been shipped (120).

There are many more items of data that the simulation software collected as the simulation ran and could be displayed. These are usually important because they reveal how the process is performing and how it could be improved. However, in this case the simulation needs to do some more work before it is worth looking at that data.

Figure 10.4   The simulation at 5 p.m. on Friday.

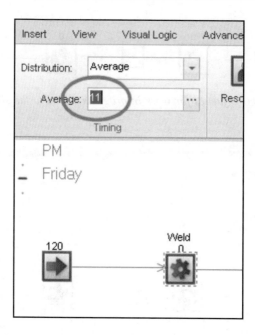

Figure 10.5   Change the data when required.

What typically happens next is the client saying something like 'that does not look quite right to me' and starts talking about how the process map does not exactly reflect what really happens in the factory, and there is some detail missing. In this case the welding takes 11 minutes on average and the painting takes 10 minutes. Change Weld to 11 minutes by clicking on it and changing the time to 11 (Figure 10.5). There is no need to change Paint, as the default time happens to be 10 minutes.

Often, before making this type of comment, the client will ask, 'Where is all the work-in-progress?' The average process map, or client description of a process, tends to mention the value-added parts of the process (places where work takes place, like the welding and painting machines) but not the non-productive, though quite real, parts of the process (like queues of work). Projects are sometimes about trying to eliminate these non-productive parts; nevertheless, they should still be included in the simulation because only then can the model be matched to current working and the impact of reducing or removing them evaluated. Sometimes, completely eliminating a queue means a bottleneck machine is starved of work, severely restricting throughput, and thus queue reduction rather than elimination becomes the objective.

In this case the client had not mentioned that work often queues prior to the two activities. In the software remove the arrows into each of the activities, drop some queues in using the buttons on the left of SIMUL8 and connect up new arrows as shown in Figure 10.6.

Run the simulation and look at the difference between Figure 10.4 and Figure 10.7.

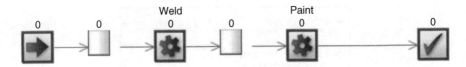

Figure 10.6   Queues for work-in-progress added to the simulation.

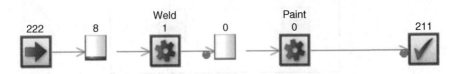

Figure 10.7   Friday 5 p.m. with capacity for work-in-progress and slower welding.

Despite the weld machine being slower (it takes an average of 11 minutes for each table now, whereas before it took only 10 minutes), the output of the factory is close to double that shown in Figure 10.4. This shows the positive value of some queues in buffering processes that are constraining the work. For example, without queues the welding machine must stop and do nothing if it has finished work, but simultaneously the paint machine is busy and cannot accept its welded table.

We are getting near an accurate simulation of the factory now (more later on how to make sure it is as accurate as it needs to be). However, the client then mentions a couple more points that are not obvious from the process flow diagram:

1. Although there is only one welding process that tables go through, there are two welding machines to choose between, giving extra capacity to welding.

2. Welding and painting are not completely automatic; an operator is required at each machine while it is working. There are only two operators to handle the three machines.

People (operators) are normally simulated using 'Resources' in SIMUL8. Resources are facilities that are shared between activities. Activities are not allowed to work unless they have all the resources they need.

Add a resource using the buttons on the left, then name it and set the number of people in the team to 2 as shown in Figure 10.8.

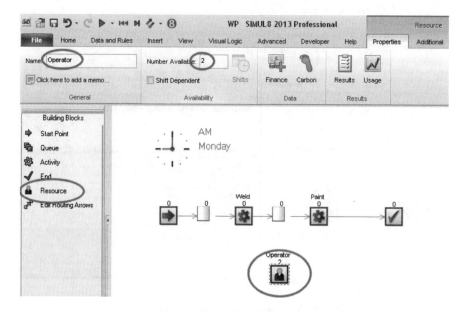

Figure 10.8   Adding resources.

Now tell both 'Weld' and 'Paint' that they need an Operator to be able to work. Click on Weld and use the Resources button and the Add button to add Operators to the list of resources required by Weld (Figure 10.9). Then do the same for Paint.

Now make a second welding machine by copying the first welding machine. Do this just by dragging the first welding machine with the CTRL button on the keyboard held down (Figure 10.10).

Now the client will be happy that the simulation is showing a similar situation to what happens in a typical week in the real factory (Figure 10.11).

Notice that there is quite a large queue for the paint activity. Let us examine some more detailed results. For example, click on the paint activity and then its Results button.

Despite the queue of work, Paint is only working for about 80% of the time (see Figure 10.12):

- What is it doing the rest of the time?

- What can be done about that?

- Try it.

- Is there a way to solve this problem without spending money?

Answering these questions will be left to the reader because they are starting to show some of the benefits of simulation. The act of exploring the simulation results gives rise to ideas on how to improve the process.

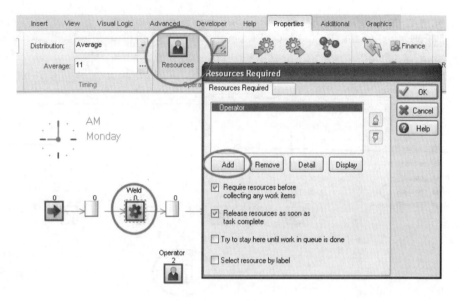

Figure 10.9   Connecting resources to activities.

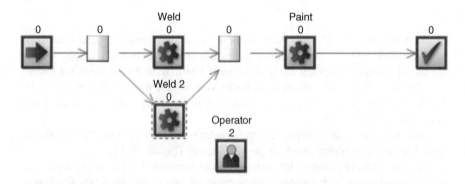

Figure 10.10   Copy the weld machine with CTRL and drag.

10.4   Getting the right results from a DES

There are two important stages in the process of providing results to clients that are easy to forget in the heat of getting the technical work done in a simulation project:

- Ensuring the simulation behaves in the same way as the real process.
- Providing results that account for natural variability.

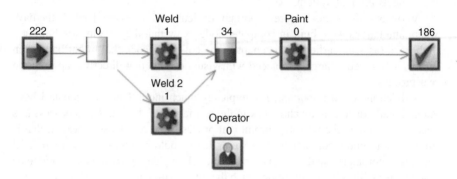

Figure 10.11   Results for two operators working at three machines.

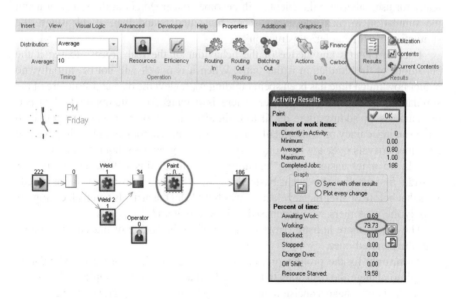

Figure 10.12   Detailed results for paint activity.

10.4.1   Verification and validation

Ensuring the simulation is behaving in the same way as the real process can be seen as two separate activities, although in practice they are often conducted together. 'Verification' is about checking that the simulation has been built in the way that was intended, whereas 'validation' is about checking that the way that was intended is the same as the real system.

Verification was much more important, difficult and necessary back in the days when simulations were built in computer code. Now that they are mostly built at a high level (as seen earlier in this chapter) the computer code that might have been erroneously typed is largely replaced with visual objects that directly map onto the real process.

Validation is still important. It is typically much more than 'is this data I have entered really data from a valid week in the real factory?' To simulate a process it is necessary to know the way the current real process behaves. To some extent that is impossible to know completely. If it were, there would be less need to simulate it! What is important is that the simulation behaves closely enough to the real world to be able to answer the clients' questions such that, if the simulation were 100% accurate, the decision the clients make would be no different.

In a practical setting in the middle of a typical project, the best way to validate (and also verify) is to work with the client(s). Make it very clear that the simulation is not ready for use, otherwise the clients will become concerned about the safety of using the simulation to make decisions. They must understand that, at this stage, the simulation is still 'under construction'.

Use the client's knowledge of the behaviour of the real process to see if the behaviour of the simulation is the same. When it is not (it will not be, initially, no matter how much time has been spent building it), review the data and structure of the simulation, first at a holistic level. There is a danger of attempting to 'correct' inaccuracies by adding more detail to a simulation. A typical example of this is to blame an inaccuracy on some detail that has been omitted, for example using average customer arrivals over a whole day rather than using hour-by-hour data that shows arrivals are much faster during lunchtime. Sometimes this will be necessary, but consider first whether the data is basically right: When were those arrivals measured? Was that before or after the new factory was built? Did the arrivals count all customers or only the customers who purchased at least one item?

The larger, more holistic errors are more often the source of inaccuracies rather than the detailed ones.

Then, only fix the problem/inaccuracy if doing so will improve the simulation enough to change the decisions that would be made. A good example of this was a client who had to make a decision on how many tugboats to purchase to operate a new port. The simulation proved to be inaccurate because it did not include the sea tides that impacted the arrivals of certain types of ship. However, during verification discussions with the client it quickly became obvious that these ship arrivals could not impact the total number of tugs that would be purchased because the technical maximum was four tugs and it was quickly proved that three tugs could hardly ever cope with demand even without the tide-bound ships.

Once a number of issues have been fixed, start the validation process again. Repeat it until the client feels confident enough to make decisions.

10.4.2   Replications

Providing results that account for natural variability is part of what simulation is about, as shown in Chapter 2. In the Table Corp. example used earlier in the current chapter the times taken to weld and paint a table were quoted not as 11 and 10 respectively, but as an *average* of 11 and an *average* of 10. By this it was meant that actual individual times would vary around 11 and 10 but that the averages (taken over a long time) would be those values. Random numbers (as discussed in Chapter 2) are automatically used by DES software to sample times around the specified average.

This is a very important benefit of using DES: that it has behaviour like the real world rather than a smooth, fixed (and unnatural) behaviour that would be shown (for example) in a spreadsheet model that did not include variability.

However, the downside of this is that the behaviour of the simulation can vary from run to run. Actually in most simulation software, rerunning the same simulation will not produce different results because the random numbers are automatically reset and are used again in the same way to make it less confusing for the clients! But if a run of a second week were conducted it would show different results. Try this in SIMUL8 by using the menu under the large Run button (Figure 10.13).

The results are different although of the same character.

To take account of this it is important when making most decisions with DES results to run multiple replications (in SIMUL8 this is known as a 'Trial') and use the range of results from the many runs with different random numbers that are

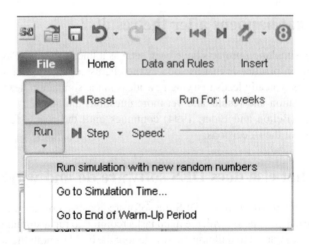

Figure 10.13   Run simulation with new random numbers.

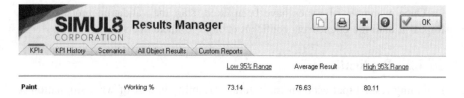

Figure 10.14   Range of results from a series of replications or a 'Trial'.

included in the one 'Trial'. For the interested reader there are more detailed explanations available – see, for example, Sneddon (2013) – but try the following.

Open the detailed results panel for the paint activity in the weld and paint example and click on RIGHT in the 'Working percent' result that was nearly 80% in the earlier discussion. Clicking on RIGHT for most values in SIMUL8 adds the result to the 'Results Manager'. The Results Manager automatically calculates confidence limits to show how accurately the simulation is predicting each of the results it shows.

Now run a trial using the large Trial button on the Home tab. Once the trial is complete the Results Manager will show (Figure 10.14) that this trial has calculated that the real average working percentage of the paint activity is likely to be between 73 and 80%. Often the accuracy of a trial can be improved (reducing the size of the range) by increasing the number of runs in the trial. The only penalty for doing this is the time it takes to conduct this trial.

SIMUL8 includes technology to predict how many runs are required to achieve a given level of accuracy across the results selected as important (Hoad, Robinson and Davies, 2011).

10.5   What happens after the results?

Traditionally simulation projects have been seen as a linear activity (Pidd, 1988) but what happens in practice is that as soon as the clients see the results they ask new questions. This typically leads to trying new ideas in the simulation, leading to a new round of validation and results and yet more questions and ideas.

This loop (Belton and Elder, 1994) continues until the clients feel they know enough to make their decisions.

10.6   What else does DES software do and why?

In the early days of software for DES the software consisted of libraries of routines that could be called to assist with the main tasks required to run a simulation. None of the software was related to building the simulation; the builders built the simulation by coding it themselves in their favourite language. Simulations were also black boxes.

They took input files of data, ran the simulation and output results for a single run, like the numbers used in the Table Corp. example earlier in this chapter. There was nothing visual to see and it was not possible to stop and start the simulation or interactively explore it in any way.

Different software packages evolved along different routes. Many stayed as non-visual black boxes for many years but took some steps forward to reduce the workload on the builder. High-level languages designed specifically for simulation were created, such as GPSS (Gould, 1969) and Siman (Pegden, 1986), while other products stayed with coding in computer languages but became visual and interactive, making it possible to show clients what was happening in the simulation (Fiddy, Bright and Elder, 1982).

Nowadays probably less than 1% of a DES software package comprises the runtime engine (the part devoted to running the simulation), whereas over 50% is related to providing easy options for building the simulation that can be run by the runtime engine. A significant portion of the remainder relates to runtime animation in both 2D and 3D. Almost all DES software also has a portion for automatically optimizing the input variables (e.g. automatically changing the numbers of welding and painting machines, numbers of operators, etc.) to seek an optimal answer for the client.

Ways of displaying results, so that, rather than showing numbers, charts are used to drive understanding that leads to seeing what to improve, have been a key focus of all simulation vendors in recent years.

Another major element of DES software is interfaces to other ways of creating simulations. For example, simulations can now be automatically created from process maps created in standard office software products or CAD packages. Simulations can also be automatically created from direct live connection to corporate data so that they can be built and run to predict the afternoon's performance from the morning's data.

10.7   What next for DES software?

Compared with the early days of simulation, software for DES has achieved a great deal in taking simulation from a backroom activity conducted only by highly trained experts who might take months to build a single simulation to the situation nowadays where, instead, a simulation can be designed, created and run in a small number of hours, often by the people working directly in the process who need the answers.

This is very good news, but not quite good enough.

DES is still more difficult than building spreadsheet models. Admittedly, spreadsheet models do rather less, produce results that fail to account for variability and are difficult to use for communication because they lack animation. However, DES needs to be an order of magnitude easier than it is currently so that improving processes quickly and cheaply via simulations can be as easy as using navigation devices to map new journeys and tablet computers for video editing.

References

Belton, V. and Elder, M.D. (1994) Decision support systems: learning from visual interactive modeling. *Decision Support Systems*, **12**(4–5), 355–364.

Elder, M.D. (1992) Visual interactive modelling: some guidelines for its implementation and some aspects of its potential impact on operational research. PhD thesis. University of Strathclyde, Glasgow.

Fiddy, E., Bright, J.G. and Elder, M.D. (1982) Problem solving by pictures. *Proceedings of the Institute of Mechanical Engineers*, **1982**, 125–138.

Gould, R.L. (1969) GPSS 360: an improved general purpose simulator. *IBM Systems Journal*, **8**, 16–27.

Hoad, K., Robinson, S. and Davies, R. (2011) AutoSimOA: a framework for automated analysis of simulation output. *Journal of Simulation*, **5**, 9–24. doi: 10.1057/jos.2010.22

Pegden, C.D. (1986) *Introduction to SIMAN*, 3rd edn, Systems Modeling Corporation, Sewickley, PA.

Pidd, M. (1988) *Computer Simulation in Management Science*, 2nd edn, John Wiley & Sons, Ltd, Chichester.

Sneddon, F.G. (2013) SIMUL8 online help, http://www.simul8.com/support/help/doku .php?id=model_building_basics:trial (accessed January 2013).

11

Vensim and the development of system dynamics

Lee Jones
Director, Ventana Systems UK, Oxton, Merseyside, UK

11.1 Introduction

This chapter does not dwell on the early history of system dynamics and the software used for implementation, since there are many respectable sources describing the origins and use of the early software programs such as Dynamo, Dysmap and their derivatives. This chapter is written from a personal perspective and includes anecdotal evidence from a number of unnamed but possibly respectable sources. Any errors or omissions are likely to be as a result of failing memory and/or fanciful wishful thinking.

In determining the content of this chapter, a causal loop diagram (CLD) was created as a sort of road map during its initial drafts and was continually updated as the drafts were reviewed and improved. It has been reproduced as Figure 11.1 in the hope that it will help to explain further the concepts described in the chapter, at the risk of exposing the author's potentially flawed thought-processes. It is incomplete, for example it does not include influences on 'actual SD use' other than the size of the pool of practitioners, but it serves to explain the narrative developed within the chapter.

Discrete-Event Simulation and System Dynamics for Management Decision Making, First Edition.
Edited by Sally Brailsford, Leonid Churilov and Brian Dangerfield.
© 2014 John Wiley & Sons, Ltd. Published 2014 by John Wiley & Sons, Ltd.

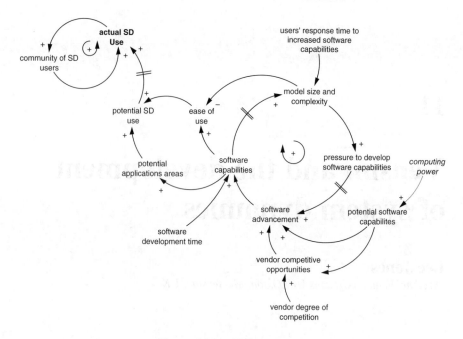

Figure 11.1   Chapter 'map'.

11.2   Coping with complexity: The need for system dynamics

The quote below is from a short story by the British science fiction writer Iain M. Banks. In this story, an advanced alien civilisation sends a small team of scientists to observe 1970s' Earth and the quote is part of a conversation between the star ship 'mind' (a super-intelligent sentient artificial intelligence or AI) and one of its crew. They discuss the merits, or otherwise, of intervening in human life on the planet (making 'contact') and the AI's observation below seems to capture the current state of affairs; system dynamicists are in the process of attempting to simplify complex systems in order to aid understanding. This would not go down too well with the ship 'mind' but, until we have super-intelligent AIs, one has at least to try!

I'm not sure that one approach could encompass the needs of their different systems. The particular stage of communication they're at, combining rapidity and selectivity, usually with something added to the signal and almost always with something missed out, means that what passes for truth often has to travel at the speed of failing memories, changing attitudes and new generations. Even when this form of handicap is recognised all they ever try to do, as a rule, is codify it, manipulate it, tidy it up. Their attempts to filter become part of the noise, and they seem unable to bring any more thought to bear on the matter than that which leads them to try and simplify what can only be understood by coming to terms with its complexity.

(Banks, 2010)

In a 2010 IBM Global Services survey (IBM, 2010) of 1541 chief executive officers in 60 countries and 33 industries (face-to-face conversations), 79% anticipated greater complexity ahead and defined the 'new economic environment as being distinctly different', characterised by being:

- More volatile; deeper/faster cycles, more risk.

- More uncertain; less predictable.

- More complex; multifaceted, interconnected.

- Structurally different; sustained change.

IBM identified three strategies employed by 'standout' organisations:

- Embody creative leadership.

- Reinvent customer relationships.

- Build operating dexterity.

From these strategies a number of recommendations were made and some of the language used in the body of the report is enough to warm the heart of any system dynamics practitioner: 'reach beyond silos', 'encourage experimentation at all levels', 'lead by working together towards a shared vision', 'predict & act, not sense & respond'. Moreover, some recommendations have a remarkable synergy with system dynamics:

- Put complexity to work for your stakeholders:

  o With improved insight into customers, processes and business patterns, drive better real-time decisions and actions throughout the enterprise.

- Take advantage of the benefits of analytics:

  o Identify, quantify and reduce systemic inefficiencies.

  o Elevate analysis from a back-office activity.

- Act quickly:

  o Make decisions when you 'know enough' not when you 'know it all'.

- Push execution speed:

  o Rapid decision making and execution.

- Course-correct as needed:

  o Align metrics with objectives and track results as part of a continuous feedback loop. Modify actions based on what is learned.

There is, however, not one mention of system dynamics (SD), simulation, modelling or systems thinking in the entire report and one would think that if IBM had indeed recommended SD, the CEO readership would have taken notice. How long their interest would have held is up for debate; what would they have found having tasked their internal strategy team to research this apparently new field of analytics? There are a number of published success stories when it comes to the application of SD in the realms of public policy, less so for successful implementation by consultants in business. A lack of transparency and excess of secrecy may account for some, but it is painfully obvious that SD has yet to gain enough traction to become *the* analytical tool of choice, much to the frustration of many in the field. And when SD does have its opportunity to shine, it is oftentimes let down by poor implementation, the product of insufficient investment in time by the potential user and an inability of many practitioners to ensure the successful application of sometimes excellent and innovative models. Once these opportunities are squandered, they affect the immediate future of SD within the specific industry. For example, early use of SD in the UK military was often limited by high-level assumptions that the decision makers could not understand and so the approach became invalid in their eyes. This hit the reputation of SD within UK military circles for years.

SD is simple, open and intuitive. It does not depend on advanced mathematics, it is more powerful than the ubiquitous spreadsheet and it is more capable of addressing problems at the highest level of strategic impact. It is pretty much nailed on as the analytical tool of choice. So why is SD not used more widely? This has perplexed experienced practitioners for decades. But perhaps it should not. In 2001 during an internal marketing symposium for a global IT organisation, the head of marketing stated, with tongue only slightly in cheek, that most of its sales were conducted on the golf course with no company salespeople in sight! The decision to invest millions and sometimes tens of millions of dollars in what was basically a glorified database was being made on the premise that everyone else was doing it. This enormous impact of word of mouth and the fear of being left behind ensured the successful rollout of new and eye-wateringly expensive IT systems across the globe in a matter of a few years with arguably marginal impact on competitiveness. SD could and should benefit from a similar explosion of use but SD appears to fly under the radar, much like a stealth fighter, and is just as invisible.

There are likely many reasons for this paradox but this chapter will focus on one facet of the problem: the software has, until now, been unable to provide the support required to facilitate this leap from understudy to star of the show. There is a need for software vendors to enhance the capabilities of their platforms but in such a way as to make them more acceptable to all stakeholders; the practitioners need greater support in their quest to educate and convince their clients and the clients need greater support in their understanding of the methodology and in the practical application within their current or future business processes.

11.3 Complexity arms race

As recently as the early 1990s, models were being developed through the creation of files containing explicit handwritten equations with minimal syntax checking or automated support. Causal loop diagrams (CLDs) were hand-drawn on paper or, for the sophisticated, created using third-party diagramming tools such as Corel Draw (Corel Corporation, 2013) with manual diagram and model code coordination. Debugging a model involved printing out the code and, with the aid of the mk1 eyeball, scanning hundreds of line of levels, rates, auxiliaries and constants (see the example in Figure 11.2) together with graphical and tabular output. Model variable names were limited to eight or even six characters and arrays (subscripts) were not even in the vocabulary. Consequently, models were simple and progress was slow.

In this simple example, the code required to develop a Cosmic model in 1987 is shown in Figure 11.2 and an equivalent model in Vensim is shown in Figure 11.3. Within the Cosmic software package each line of code would be hand-typed and, with

```
Level and Initialising Value
L PRODSL.K=PRODSL.J+DT*(PRODT.JK-SALES.JK)
N PRODSL=1000


Rates
R PRODT.KL=SDISC.K/TCI
R SALES.KL=MIN(DMD.KL,PRODSL.K/DT)


Auxiliaries
A SDISC.K=(DSL.K-PRODSL.K)
Constants
C DMD=400
C DSL=3000
C TCI=3
```

```
L = Level
N = iNitial value for level
R = Rate
A = Auxiliary
C = Constant
```

Figure 11.2 Example Cosmic code (Coyle, 1988).

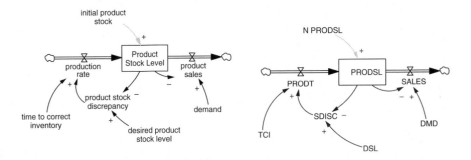

Figure 11.3   (a) Simple Vensim example. (b) Mapping to Cosmic parameter names from Figure 11.2.

the limit to parameter names set to six or eight characters, larger models required a glossary of acronyms in order for the reader to understand the model equations.

If a second product were introduced into the example, the code would need to be manually copied and the parameter names for the additional product updated. For example, PRODSL would become PRODSL1 and PRODSL2 in order to represent two products. In modern software, the use of array structures would make this structure replication a trivial exercise.

In the late 1980s and early 1990s, new software packages such as Stella, Powersim and Vensim were making headway in the market. The new breed of SD software promised advances in productivity only ever dreamt of by users of early SD packages. Using graphical objects linked by arrows and pipes, developers were able to create models without resorting to handwritten code and having to remember the correct syntax. Basic syntax checking and automation enhanced capabilities further and, with the introduction of arrays, productivity improved by an order of magnitude. Simple equation editors became the norm enabling quick access to other model variables, units of measurement and hundreds of functions (Figure 11.4).

Some practitioners were initially sceptical, warning that the use of diagram-to-equation automation would jeopardise the learning experience and lead to fundamental errors in the formulation of models. It is true that early developers, because of

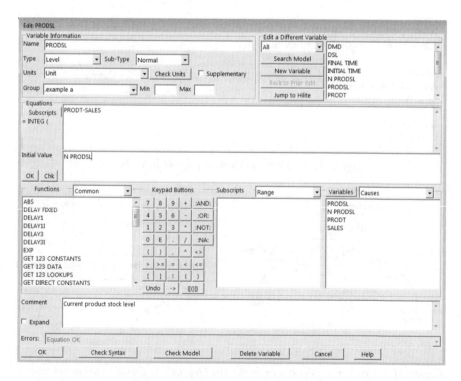

Figure 11.4   Vensim equation editor.

the need to write out the level and rate formulations with explicit reference to time, time step and integration, could not avoid the need for a fundamental understanding of the underlying mathematics of the methodology. It is also true that developers today can 'create' simulation models that will run and produce results but will break all sensible rules, thus rendering the results meaningless.

In the beginning models were, although perhaps dynamically complex, structurally simple, out of necessity. Neither the software nor the hardware could handle anything more complex and so the models and modellers had to be innovative in order to address the issues of the day. Hardware platforms struggled to produce results quickly enough in order to respond to business needs and the turnaround of experimentation was time consuming and cumbersome; productivity was, by today's standard, ridiculously low. Some may argue, however, that having to be 'elegant' in the design of the model, having to really think of the most efficient way to formulate a model in sufficient detail to address the problem and yet still function within the software constraints of the day, enabled better models and better modellers.

11.4   The move to user-led innovation

Enhancements to the SD software packages were heavily influenced by developments in the enabling tools, computing power and software development environments. Vendors' motivations to enhance the tools came from the software developers' insatiable need to innovate and, as their development tools (and they themselves) became more proficient, they applied their new-found skills to the improvement of the software. Those early improvements were software engineer led; optimisation algorithms, arrays and a plethora of under-the-hood enhancements enabled faster, more efficient simulation.

Enhancements were further influenced by the consultancy work undertaken by the vendors themselves, necessitating a tweak here and a tweak there in order to deliver to the customer's needs. Clients' influence on software development strengthened as vendors were able to respond to their specific requirements. For example, in the mid-1990s, end user insistence on documentation deliverables separately documenting stocks, flows, auxiliaries, and so on, led to a request from the modeller in question to the particular software vendor to add this functionality.

Unlike in many software development environments, the competitive imperative was not really a strong motivator for change. New features of one package may or may not show up in another but, if they did, it was invariably some considerable time later (in software timescales). In very recent years, social networking has enabled a much greater user involvement with vendor online forums enabling two-way communication between vendor and user. Conversation within, for example, the Ventana UK online forum (Ventana Systems UK, 2011) has led directly to enhancements or new features within the software. The 'cottage-industry' nature of the vendors also enhances the opportunities for direct, fruitful connection between vendor and user via e-mail, telephone and face-to-face opportunities during the annual SD conference and those organised by the many country chapters. Over the past 20 years, therefore, there has been a marked shift from vendor-led to user-led innovation.

Improvements in ease of use had another positive side-effect: they opened up the world of SD to more people. Gone was the requirement to be a computer programmer and so SD was able to make the move from the back-office computer lab to the executive desktop. The relative lack of success in making this transition has puzzled practitioners for decades; why is SD not the first tool in the business toolbox?

There have been many advances in SD software enabling enhanced efficiency for the modeller and, in turn, delivering more 'bang for the buck' for the end user. It can be argued, however, that the more important enhancements are those enabling SD to become what it was always intended to be: a means to increase understanding and, through learning, allow those with the luxury of being able to advise on or implement decisions to make better informed decisions. In order for decision makers to feel comfortable with SD in an advisory capacity, there exists a growing requirement to increase the 'sex appeal' of the software. Other tools and methods used to advise decision making have developed innovative ways to visualise and summarise data (such as Tableaux and Crystal Reports); the most commonly used 'simulation' or modelling package is Microsoft's Excel, itself vastly improved from the earliest versions, with enhanced graphics and connectivity to other Microsoft Office applications enabling ease of transfer between the most commonly used software tools on the planet. It is increasingly evident that there is a need to improve the 'look and feel' of the SD products and, more importantly, the ease by which output can be analysed, packaged and reported.

The next section will discuss software developments enabling enhanced support to the practitioner and end user (policy maker) in order to utilise efficiently the remarkable IT improvements made in recent years.

11.5   Software support

Advances in IT hardware have enabled developers to increase the capabilities of the SD software and this has resulted in models becoming larger and more complex. As the SD software improves, its potential applications increase, leading to the potential for SD to contribute to decision making across a greater variety of problems. With this complexity and scope explosion, however, comes the need to avoid model development pitfalls introduced as a direct result of this complexity: that is, overly complex models resistant to the level of analysis required to support the understanding of the issues addressed by the modelling effort.

A number of features have been added to SD software to help mitigate the negative impact of increasingly complex models, even though the complexity itself is often introduced by the software improvements themselves! It is useful to discuss the software features within the context outlined by Sterman (2000) (see Figure 11.5), embedding the modelling effort within the dynamics of the system that the modelling effort is attempting to improve. In this way, the impact of improvements to the software can be assessed as they pertain to supporting the modelling effort but also to

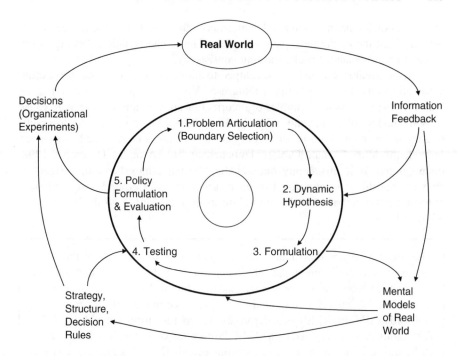

Figure 11.5   Modelling embedded in the dynamics of the system (Sterman, 2000).

the successful application of the SD process in improving the system. Without the understanding and support of the policy makers, SD has no hope of contributing to system improvement in the real world; if SD is inaccessible, difficult to use or simply not capable enough, then it will not be used.

The elicitation and understanding of the problem owners' 'mental models of real world' are central to the successful application of SD. These personally held hypotheses are derived from 'information feedback' received from direct and indirect observations of the 'real world'. One important contribution to be made by the SD methodology is to provide a further signal enhancing, testing and validating these mental models, enabling multiple worldviews to be tested, rejected, negotiated and reformulated. The 'modelling process' should be as transparent as possible to the problem owners – after all, they are ultimately responsible for implementing policy decisions in the real world. In order to support these 'organisational experiments', policy makers need a significant degree of confidence in the modelling process and, with software/hardware improvements enabling increasingly complex representations of reality, a number of software developments have evolved out of necessity.

11.5.1   Apples and oranges (basic model testing)

In order for a model to be useful it must be credible. Credibility must be demonstrated and a number of features have evolved to support the practitioner in this regard. First of all, the practitioner needs to build confidence in the model he or she is creating. At the very least the model must pass a number of basic checks during 'testing', from syntax to mass balance checks and unit consistency.

Software features have been developed to automate or at least support testing, automatic units checking being a feature of Vensim from very early on in its history. There are many examples of incorrect or missing units as the main cause of poor model behaviour and units checking should be second nature to practitioners and a model with no or incomplete units should be regarded as suspect. Indeed, the Mars Climate Orbiter, launched by NASA on 11 December 1998, disintegrated as it mistakenly entered the Martian atmosphere. Investigations showed that this was as a direct result of units errors in ground-based software controlling the insertion of the probe into its correct orbit, with the loss of over $300 million.

The MCO MIB has determined that the root cause for the loss of the MCO spacecraft was the failure to use metric units in the coding of a ground software file, 'Small Forces,' used in trajectory models. Specifically, thruster performance data in English units instead of metric units was used in the software application code titled SM_FORCES (small forces). The output from the SM_FORCES application code as required by a MSOP Project Software Interface Specification (SIS) was to be in metric units of Newton-seconds (N-s). Instead, the data was reported in English units of pound-seconds (lbf-s). The Angular Momentum Desaturation (AMD) file contained the output data from the SM_FORCES software. The SIS, which was not followed, defines both the format and units of the AMD file generated by ground-based computers. Subsequent processing of the data from the AMD file by the navigation software algorithm therefore underestimated the effect on the spacecraft trajectory by a factor of 4.45, which is the required conversion factor from force in pounds to Newtons. An erroneous trajectory was computed using this incorrect data. (NASA, 1999)

11.5.2   Confidence

Confidence in the validity of the model is paramount to its successful use as a tool to inform decision making.

11.5.2.1   Calibration

One leg to the confidence table is the capability to reproduce past performance, that is to simulate a period of time over which there exists good data on the key stocks and flows in the system being modelled, whether the commodity price for an oil production model, sales history for a marketing model, production costs for a manufacturing model or the stock of vehicles in a country. If the model can be shown to replicate closely historical behaviour *for the right reasons* then the user will have greater confidence in the lessons learned from the simulator.

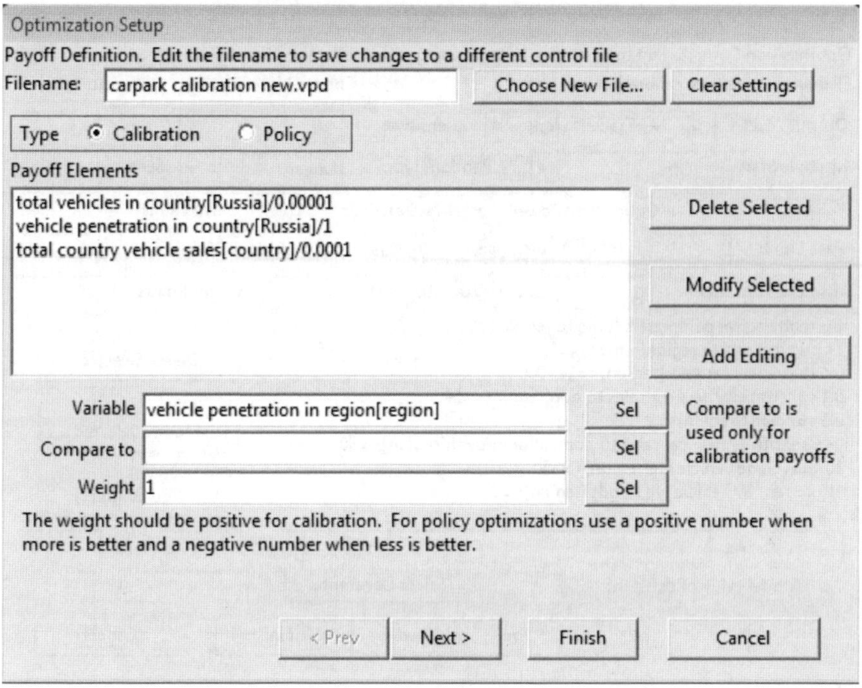

Figure 11.6   Vensim payoff definition for calibration.

Calibration involves finding the values of model constants that make the model generate behaviour curves that best fit the real-world data. Manual calibration is a slow, painstaking process involving manipulation of the input assumptions, the running of the model, and the visual assessment of 'goodness of fit' for a range of performance indicators. Over the years, SD software tools have evolved in order to assist in this process, the most notable being the use of optimisation algorithms.

In the so-called 'calibration optimisation', the payoff is calculated as the accumulated differences between each historical and model-generated data point, the minimisation of which will result in a tendency to select model constant values minimising the difference between the historical data and the results generated by the model over the same historical period.

In the example in Figures 11.6 and 11.7 historical data exists for a number of key model variables, such as 'vehicle penetration in region'. In the first instance, the software is informed of the variables for which historical data exists and, in the second, the range of values for which the selected assumptions may be searched in order to find a best fit.

Once a good fit has been achieved, the software provides a list of the constant values selected during calibration and these can be automatically used as input assumptions for future simulation runs. Figure 11.8 shows calibration output for the

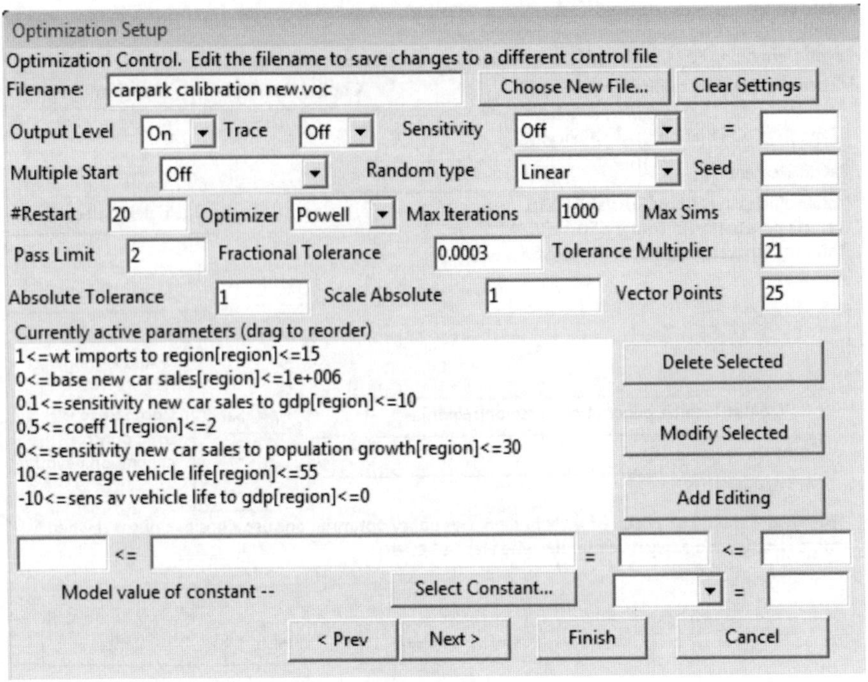

Optimization Setup
Optimization Control. Edit the filename to save changes to a different control file
Filename:
carpark calibration new.voc
Choose New File...
Clear Settings
Output Level
On
Trace
Off
Sensitivity
Off
=
Multiple Start
Off
Random type
Linear
Seed
#Restart
20
Optimizer
Powell
Max Iterations
1000
Max Sims
Pass Limit
2
Fractional Tolerance
0.0003
Tolerance Multiplier
21
Absolute Tolerance
1
Scale Absolute
1
Vector Points
25
Currently active parameters (drag to reorder)
1<=wt imports to region[region]<=15
0<=base new car sales[region]<=1e+006
0.1<=sensitivity new car sales to gdp[region]<=10
0.5<=coeff 1[region]<=2
0<=sensitivity new car sales to population growth[region]<=30
10<=average vehicle life[region]<=55
-10<=sens av vehicle life to gdp[region]<=0
Delete Selected
Modify Selected
Add Editing
<=
=
<=
Model value of constant --
Select Constant...
=
< Prev
Next >
Finish
Cancel

Figure 11.7 Vensim optimisation control for calibration.

vehicle penetration in region
600
450
300
150
0
Vehicle/Thousand Person
2000
2004
2008
2012
2016
2020
2024
Time (Year)
vehicle penetration in region[Northwestern federal district] : carpark calibration
vehicle penetration in region[Northwestern federal district] : Historical and forecast data

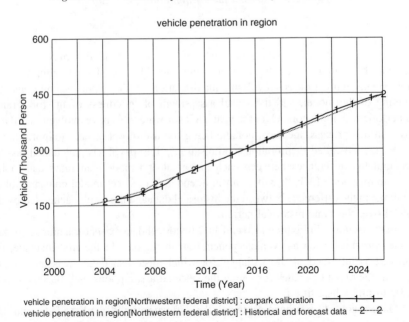

Figure 11.8 Example calibration result.

example model, where the calibration routine has found a good fit for vehicle penetration (in vehicles per thousand of population). Note that, in this case, some of the 'historical' data is actually 'forecast' data obtained from industry analysts.

11.5.2.2 Reality checks

As models are built, there are various checks that must be done against reality. These checks may be explicit and take the form of tests of model behaviour or subsector behaviour under different assumptions, or they may be implicit mental simulations and analyses based on an understanding of models and the modelling process. In either case these checks are very important in ensuring that the models developed can adequately address the problems they are being applied to.

Reality checks provide a straightforward way to express statements that must be true about a model for it to be useful, and the machinery to test a model automatically for conformance with those statements. Reality Check is a technology that adds significantly to the ability to validate and defend models. It can also focus discussion away from specific assumptions made in models onto more solidly held beliefs about the nature of reality.

Models are representations of reality, or our perceptions of reality. In order to validate the usefulness of a model, it is important to determine whether things that are observed in reality also hold true in the model. This validation can be done using formal or informal methods to compare measurements and model behaviour. Comparison can be done by looking at time series data, seeing if conditions correspond to qualitative descriptions, testing sensitivity of assumptions in a model, and deriving reasonable explanations for model-generated behaviour and behaviour patterns.

Another important component in model validation is the detailed consideration of assumptions about structure. Agents should not require information that is not available to them to make decisions. There needs to be strict enforcement of causal connectedness. Material needs to be conserved.

Between the details of structure and the overwhelming richness of behaviour, there is a great deal that can be said about a model that is rarely acted upon. If you were to complete the sentence 'For a model or submodel to be reasonable when I __ it should __' you would see that there are many things that you could do to a model to find problems and build confidence in it.

In most cases, some of the things required for a model to be reasonable are so important that they get tested. In many cases, the things are said not about a complete model but about a small component of the model, or even an equation. In such cases the model builder can draw on experiences and the work of others relating to the behaviour of generic structures and specific formulations.

Ultimately, however, most of the things that need to be true for a model to be reasonable are never tested. Using traditional modelling techniques, the testing process requires cutting out sectors, driving them with different inputs, changing basic structure in selected locations, making lots of simulations, and reviewing the output. Even when this gets done, it is often done on a version of the model that is later revised, and the effect of the revisions not tested. Reality Check equations provide a

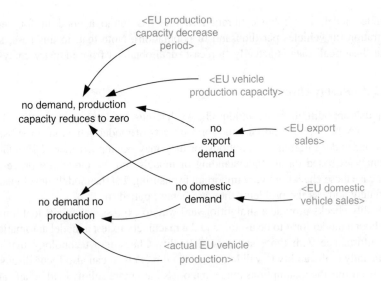

Figure 11.9   Example reality check.

language for specifying what is required for a model to be reasonable, and the machinery to go in and automatically test for conformance to those requirements. The specifications made are not tied to a version of a model. They are separate from the normal model equations and do not interfere with the normal function of the model. Pressing a button shows whether or not the model is in violation of the constraints that reality imposes.

The example in Figure 11.9 illustrates the use of reality checks in a policy model. There are two reality checks shown: 'no demand, no production' and 'no demand, production capacity reduces to zero'. These are named in a fashion indicating the nature of the test inputs and the expected model behaviour. These equations are not part of the normal model structure and are not simulated during normal execution of the model. When a reality check experiment is called, however, each reality check is tested in turn, the software will force the test inputs to be true and it will compare model behaviour with the behaviour expected. For instance, in order to run these reality checks, the software forces 'EU export sales' and 'EU domestic vehicle sales' to be zero for the entire duration of the simulation run, and checks model behaviour, in this case 'EU vehicle production capacity' and 'actual EU vehicle production'. If they conform to the expectation, the check is recorded as 'passed', while a 'failure' indicates a violation of this check.

Such reality checks are developed by those with the greatest knowledge of the real-world system being explored. Failure of any of these checks would need to be investigated and any structural failures in the model rectified. Passing checks designed by the policy makers and/or subject matter experts is a powerful validation of any model.

11.5.2.3 Illuminating model behaviour

Communication between the model end user (decision maker) and model developer (practitioner) is of paramount importance in ensuring models are useful. As previously noted, the software and hardware available in the twenty-first century have enabled increasingly complex representations of systems to be modelled. Increasing model complexity requires increasingly sophisticated analytical tools and methods in order to maintain the necessary level of communication and, therefore, understanding of and confidence in the model. Without the ability to communicate model structure and behaviour, the practitioner is unable to validate the model in the eyes of the policy maker and the project is likely doomed to fail. All SD packages now use diagramming tools in the creation of the model and most support both CLD and stock–flow diagram (SFD) representations. Furthermore, Vensim includes the ability to create multiple 'views' of the model structure serving to emphasise one structural relationship or another, a particular feedback loop or part of an SFD. Combined with the use of symbols, colour, annotations and 'story-telling', the practitioner is able to engage more effectively with the problem owner; this leads to a greater understanding of the modelling process by the latter and of the mental model of the problem owner by the former.

As an example, a manufacturing model, developed as a research tool, is shown in Figure 11.10. Although the project team are able to navigate the model quickly and understand where key influences are likely to be the causes of particular model behaviour, it is clear that those without a similar SD background, or having not been involved in the project development, would need support in order to aid understanding. A first step may be to create another model 'view' and simplify the representation (the diagram in Figure 11.10 represents all equations and influences using an SFD approach) through simply adding a high-level pictorial representation, as shown in Figure 11.11.

Here, the key influences only have been included and much of the intermediate calculation steps omitted for the sake of clarity. The diagram is still, unfortunately, overwhelming for many non-practitioners and so the user may choose to hide all but one small section of the model, the remainder to be progressively unveiled as the practitioner explains each stage of the diagram to the end user (story-telling). The use of font and colour changes further emphasises each section of the model and provides further visual support aiding understanding. In the recent past, this was achieved by copying the model and pasting multiple copies into, for example, Microsoft Power-Point. Manipulation of each copy would enable story-telling as the reader progressed from slide to slide. This is now possible within many SD packages without the need to resort to copy and paste.

'Story-telling' is a powerful method for articulating the model structure as it relates to the behaviour of the system being modelled. Larger models tend to be made more readable through modularisation but decision makers often still have problems when faced with relatively simple modules, or 'plates of spaghetti'. By allowing the practitioner to 'hide' model structure at multiple hide levels, model structure can be revealed in bite-sized, manageable portions. When coupled with explanatory notes

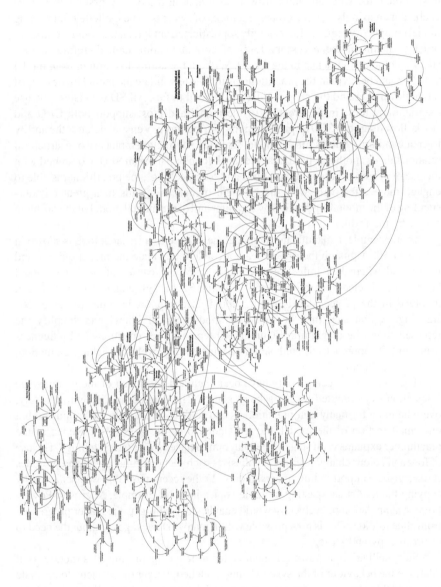

Figure 11.10   Manufacturing model.

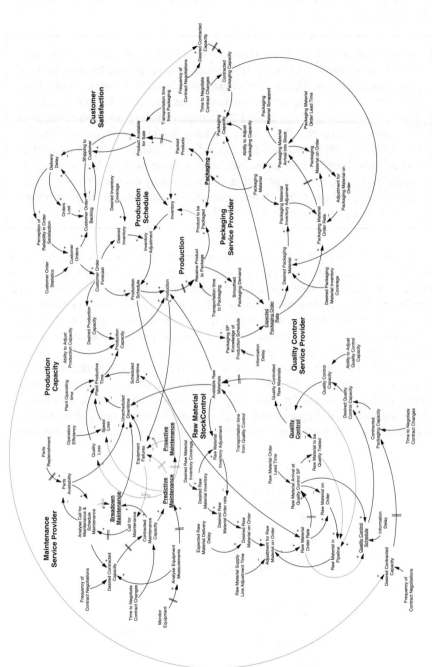

Figure 11.11 *Manufacturing model overview.*

appearing at each successive level, the practitioner is able to communicate more effectively with the end user.

In 2010, the *New York Times* famously published a criticism of Microsoft's PowerPoint software, 'A PowerPoint diagram meant to portray the complexity of American strategy in Afghanistan certainly succeeded in that aim', in which the PowerPoint slide causing such consternation was, in fact, an SD model diagram. This story made headlines around the world, including low-circulation local 'free' papers such as the *Wirral News* and a late-night television comedy show in Germany. Notwithstanding the fact that the article misses the point completely and is, quite frankly, lazy journalism, it does serve to show how a relatively simple model diagram (to a practitioner) is received by the wider community. In reality, the model would have been presented to decision makers in a much more palatable way, using bite-sized story-telling.

Figures 11.12–11.16 show one such sequential 'story-board' where the practitioner is able to describe successfully a large proportion of the model with a series of building blocks beginning with a simple 'call for maintenance' to a maintenance service provider as a result of equipment failure at the manufacturer. In the next step (Figure 11.13), the impact of equipment failures at the manufacturer is explained as a reduction in plant productive time, further complicated by the availability or otherwise of spare parts to affect the repair.

Figure 11.14 introduces the concept of predictive and proactive maintenance. This requires the service provider to monitor the equipment for signs indicating potential failure in the future, and for those parts to be proactively replaced, thus reducing equipment failures and maintaining a higher plant productive time, even accounting for the scheduled downtime as a result of proactive maintenance, as such downtimes

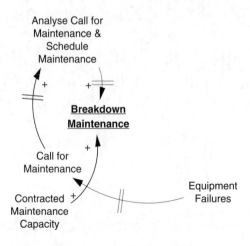

Figure 11.12   Story-telling 1.

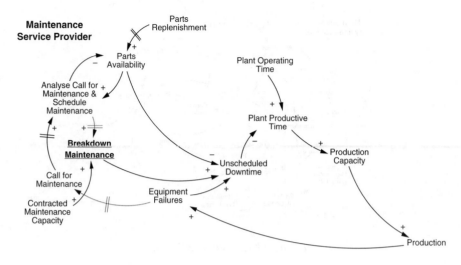

Figure 11.13   Story-telling 2.

are predictable and can often be scheduled at times when the plant would not be operating anyway.

Other impacts on productivity are losses in production hours due to less than optimum operating as a result of worn parts. In Figure 11.15, the concepts of 'speed loss' and 'quality loss' are introduced in order to capture this effect on 'plant productive time'.

Finally, Figure 11.16 includes the contracting issues as they affect the relationship between the manufacturer and its maintenance service provider. It is clear that insufficient contractual support may be the result of improper and infrequent contract

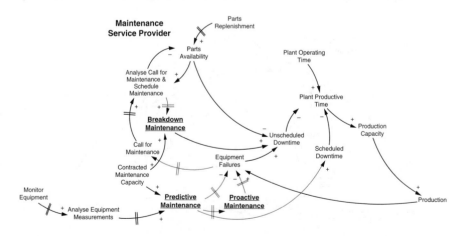

Figure 11.14   Story-telling 3.

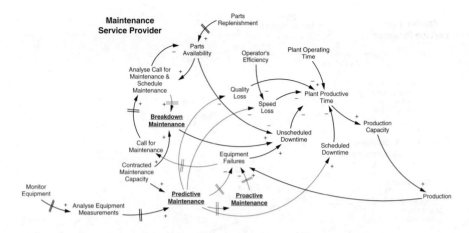

Figure 11.15   Story-telling 4.

reviews, especially if the manufacturing company is growing and, as a result, requesting greater contracted maintenance support.

Once a model is able to demonstrate it passes rudimentary tests, the syntax is correct, material is conserved, units are consistent, and so on, the practitioner will be able to start simulation testing in earnest. Observation of model-generated output and checking of the calculation made by each model equation form the next obvious steps in testing, and SD software analysis tools enable not only the verification that the model equations are calculating as intended, but that the output is valid under a range of assumptions. The ability to demonstrate why the model behaves in a particular way is a powerful means of gaining the confidence of the end user. As models increase in complexity, this ability to trace the cause-and-effect behaviour through layers of model structure becomes more difficult. Causal Tracing^{TM} is a tool specifically

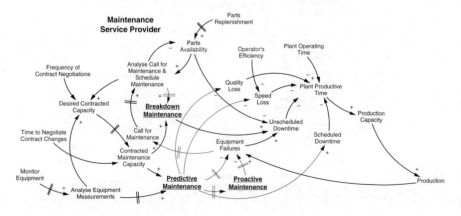

Figure 11.16   Story-telling 5.

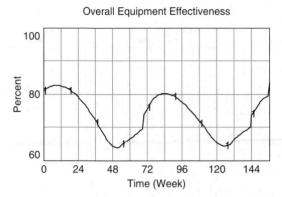

Overall Equipment Effectiveness : Basecase

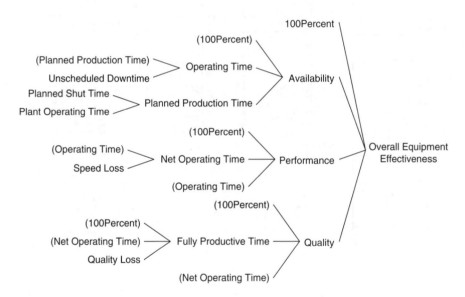

Figure 11.17   An example causal tree.

designed to help the practitioner trace the causes of model behaviour, enabling a more rapid assessment of model output than would otherwise be possible. It is important to be able to interrogate the structure as well as the behaviour of the model, as the former determines the latter, while the latter should behave in a similar fashion to observed real-world behaviour and/or expected future behaviour.

In Figure 11.17, the simulated behaviour of 'overall equipment effectiveness' (OEE) is graphed and managers want to understand the reason for the observed cyclic behaviour. A causal tree for the output in question describes the immediate causal influences on this behaviour, 'availability', 'performance' and 'quality'.

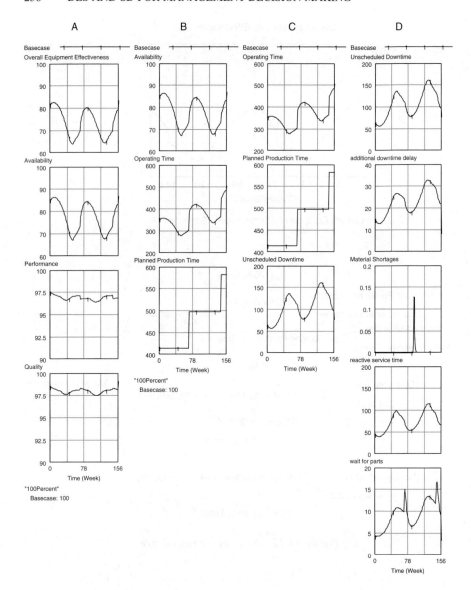

Figure 11.18   'Strip graphs' help trace causes of behaviour.

Immediately, the user can observe the direct structural influences on the output in question and follow this structural causality through multiple model layers. More powerfully, a causal 'strip graph' generates behavioural output for OEE (Figure 11.18, column 'A'). It is immediately clear that the main influence on the OEE behavioural pattern is 'availability' and the user would naturally wish to pursue this avenue of investigation. By simply clicking on 'availability' in the displayed strip graph, a new

Time (Week)	0	1	2	3	4	5	6
"Overall Equipment Effectiveness" and its causes Runs:	Basecase						
Overall Equipment Effectiveness	80.83	81.22	81.56	81.85	82.08	82.27	82.41
"100Percent"	100						
Availability	84.20	84.64	85.03	85.37	85.65	85.89	86.08
Performance	97.48	97.46	97.44	97.42	97.40	97.37	97.35
Quality	98.48	98.46	98.44	98.42	98.40	98.37	98.35
"Availability" and its causes Runs:	Basecase						
Availability	84.20	84.64	85.03	85.37	85.65	85.89	86.08
"100Percent"	100						
Operating Time	348.57	350.42	352.03	353.41	354.59	355.57	356.35
Planned Production Time	414	414	414	414	414	414	414
"Operating Time" and its causes Runs:	Basecase						
Operating Time	348.57	350.42	352.03	353.41	354.59	355.57	356.35
Planned Production Time	414	414	414	414	414	414	414
"Unscheduled Downtime" and its causes Runs:	Basecase						
Unscheduled Downtime	65.43	63.58	61.97	60.59	59.41	58.43	57.65
additional downtime delay	15.38	14.83	14.36	13.97	13.64	13.38	13.17
Material Shortages	0	0	0	0	0	0	0
reactive service time	45.75	44.56	43.42	42.42	41.54	40.81	40.22
wait for parts	4.3	4.202	4.188	4.204	4.224	4.241	4.255

Figure 11.19   Causes Table.

strip graph is displayed showing graphical output for those immediate influences on 'availability' (column 'B').

Once again, the user is able to home in on the main cause of this behaviour and repeat the process (columns 'C' and 'D') and, in this particular case, track down the main cause of the oscillating OEE behaviour (in this complex model the causal tracing has to continue for some levels yet and, in this case, the oscillation was due to the feedback between usage of the manufacturing line and its rate of failure; the higher the number of failures, the less up-time leading to a change in the failure rate and greater up-time).

Causality can also be traced through tabular output of model values, assisting the developer in the verification process (Figure 11.19).

11.5.3   Helping the practitioner do more

One of the key improvements software vendors have made has been to enable the practitioner to achieve more in the limited time usually allotted to problem solving, especially in the business world. Practitioners need to elicit knowledge and experience from experts and capture this information in the form of CLDs and SFDs, formulate and create the model equations, evaluate and clean relevant data for use by the model and present this data in a readable and usable format, verify and validate the model, experiment within a wide range of scenarios, evaluate the sensitivity of the outputs to input uncertainty and communicate the results to the policy maker. Vendors have developed a suite of tools and methods to help the practitioner achieve many of these tasks as quickly and accurately as possible, enabling a reduction in total project cost or, perhaps more importantly, allowing more time to experiment alongside the policy maker and assist in the use of the model to understand the implication of policy options.

Assistance for the verification and validation of the model has been discussed, but there are other ways to improve not only the productivity of the practitioner, but also the communication of model content and output to the model user.

11.5.3.1   Model documentation

Although seemingly of little importance to many model builders, a fully documented model is a must if it is to be peer reviewed and understood. Most of the software

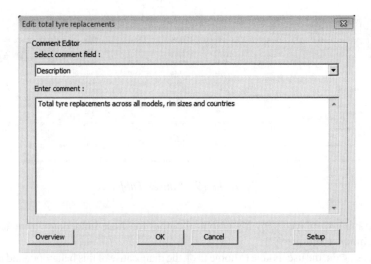

Figure 11.20   'Pop-up' window for documentation.

allows the model builder to enter textual descriptions for each equation and it is good practice to do so as the model is developed. Ventana has developed an advanced documentation add-on for Vensim, allowing model documentation to be created within Microsoft Excel. In Figure 11.22 below we see an example from the earlier example in this chapter. If the model builder has a written description of each variable in the 'comments' field of the equation editor, then this Excel documentation can be created with a few clicks of a mouse button.

Other tools can be added to Vensim by programmers with the necessary skills. For example, many end users would like to see more sophisticated documentation such as 'data source', 'knowledge source', 'telecon information', and so on, and a new application has been created to allow the model builder to enter documentation for a variable in the equation editor but under a range of user-defined fields.

For example, an application is created and set up to be activated upon a double click within the Vensim equation editor; this produces a 'pop-up' window, such as in Figure 11.20. The developer is then able to enter text under a series of user-defined fields.

Fields can be added, deleted and rearranged by the model developer (Figure 11.21) thus enabling a more sophisticated means of documenting each equation. In addition, the macro responsible for the Excel documentation in Figure 11.22 can be extended automatically to populate additional columns for each of the user-defined fields resulting in a fully documented list of variables indicating equation, type, group, causes, uses, units of measurement, subscript ranges, subscript elements and textual information under an unlimited number of user-defined headings.

11.5.3.2   Sensitivity analysis

There are many occasions where uncertainty exists in the value of input assumptions for a model and one method of understanding the impact of this uncertainty is to vary the

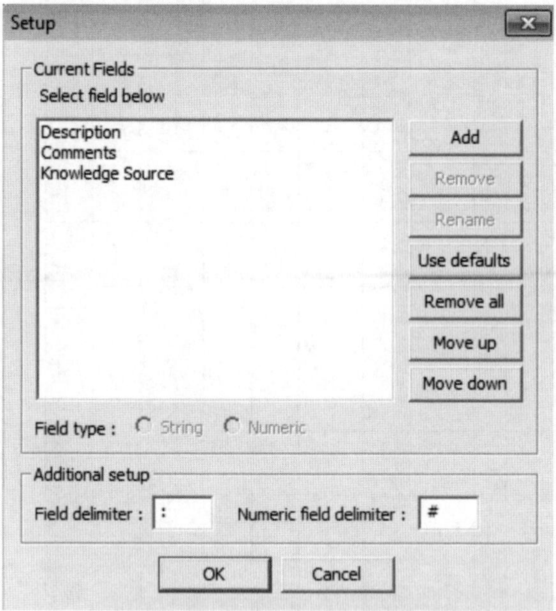

Figure 11.21   User configuration of the documentation tool.

inputs and evaluate the impact on a number of key performance indicators (KPIs). In the past this would have involved time-consuming multiple experiments, although this has become automated in most software tools today, simply by listing the assumptions where uncertainty in the value exists and defining that uncertainty with the use of a random function from which to sample values. During sensitivity analysis, the software will repeat hundreds of simulations, each time sampling for the uncertain assumptions from their distributions, and the results stored as confidence bounds on the KPI outputs.

Figure 11.23 displays sensitivity output for 'vehicle penetration in region', a KPI from a vehicle ownership model trying to understand the possible future use of vehicles in an area of growing vehicle ownership. In this simple example, the developer wanted to assess the impact of uncertainty in a calibration estimate called 'sensitivity new car sales to gdp'. This is added to the list of 'currently active parameters' (see Figure 11.24) and Vensim is instructed to select values from, in this case, a normal distribution function.

11.5.3.3   Policy optimisation

A powerful weapon in the SD software armoury is the use of optimisation. That is, the ability to instruct the software to use the model to maximise a desired outcome, usually profit. By setting a payoff definition file (in much the same way as in the calibration setup at 11.5.2.1) and instructing Vensim to select from a range of values for a number of policy levers, the software will attempt to find the best combination in

Variable Name	Value	Units	Equation & Current Value	Causes	Uses	Subscripts	Subscripts elements	Type	Group	Comment (full)
demand		Unit/Month	demand= 400		product sales			Constant	.example	Current demand for product from stock
desired product stock level		Unit	desired product stock level= 3000		product stock discrepancy			Constant	.example	Product stock desired in inventory
FINAL TIME		Month	FINAL TIME = 36					Constant	Control	The final time for the simulation.
initial product stoc=		Unit	initial product stock= 100		Product Stock Level			Constant	.example	Product stock at the start of the simulation
INITIAL TIME		Month	INITIAL TIME = 0		Time			Constant	Control	The initial time for the simulation.
product sales		Unit/Month	product sales= min(Product Stock Level/TIME STEP ,demand)	Product Stock Level ;demand ;TIME STEP	Product Stock Level			Auxiliary	.example	Current shipment of stock to satisfy demand
product stock discrepancy		Unit	product stock discrepancy= max(desired product stock level-Product Stock Level,0)	Product Stock Level ;desired product stock level	production rate			Auxiliary	.example	:Description: Current value of stock discrepancy
Product Stock Level		Unit	Product Stock Level= INTEG (production rate-product sales, initial product stock)	initial product stock ;product sales ;production rate	product sales ;product stock discrepancy			Level	.example	Current product stock level
production rate		Unit/Month	production rate= product stock discrepancy/time to correct inventory	product stock discrepancy ;time to correct inventory	Product Stock Level			Auxiliary	.example	:Description: Rate of production
SAVEPER		Month	SAVEPER = TIME STEP	TIME STEP				Auxiliary	Control	The frequency with which output is stored.
Time		Month	Time = INTEG(1, INITIAL TIME)	INITIAL TIME				Time Base		Internally defined simulation time.
TIME STEP		Month	TIME STEP = 0.25		product sales ;SAVEPER			Constant	Control	The time step for the simulation.
time to correct inventory		Month	time to correct inventory= 3		production rate			Constant	.example	Period of time over which to attempt to correct inventory

Figure 11.22 Advanced documentation example using Excel macro.

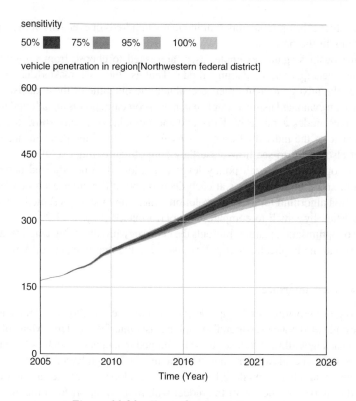

Figure 11.23   Example sensitivity output.

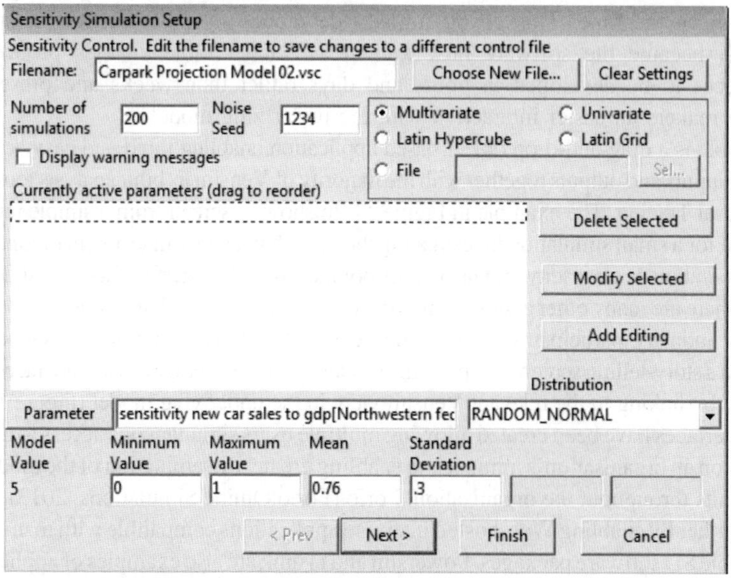

Figure 11.24   Sensitivity Simulation Setup.

order to achieve the optimisation, running many hundreds or even tens of thousands of simulations in the process.

In the business game example, users are asked to choose from a multitude of decisions each quarter, in gaming mode. That is, the user makes decisions and simulates the next quarter in order to evaluate the outcome of those decisions. The decisions are evaluated using a simple balanced scorecard; points are allotted for good performance under a range of KPIs and the cumulative points show the overall performance of the individual user. It is useful to compare the performance of each user but also useful to see how close they come to an optimal solution.

Using optimisation, each policy lever was allowed to be adjusted between its maximum and minimum values at each decision point (i.e. after each quarter). The optimisation algorithm found a best solution which can then be presented to the users as a yardstick by which to compare their performance (Figure 11.25).

Use of optimisation features presents the end user with something tangible: a means to understand the implications of policy change in the improvement of performance.

11.5.3.4   User interface

If there is one innovation surely capable of enhancing the usability of SD models, it has been the ability to create user-friendly graphical user interfaces (GUIs). Many of the SD software packages allow interfaces to be developed in languages such as Visual Basic and C++ and some allow interaction with the model through Excel. The latter has the advantage of familiarity while all enable the developer to create an environment designed to maximise the use of the model with minimal effort from the user.

In the latter half of the 1990s, a move was made to develop a GUI tool for Vensim which would be better looking and easier to use than the in-built capability of the software at the time. This effort, initially developed by practitioners for specific clients, became the software now known as Sable. Sable enables professional interfaces to be developed in hours and days rather than weeks and presents to policy makers an easier interaction with the underlying model.

Sable is a drag-and-drop object-based application enabling rapid and easy access to model inputs and outputs together with the majority of Vensim in-built analysis tools such as Causal Tracing. The example in Figure 11.26 shows a screen from a simple interface created for a small simulation investigating the growth in energy micro-generation, that is the generation of electricity and heat at the point of use. The interface has two slider bars only (there are many other assumptions) and is designed as a quick introduction to SD for energy industry participants who are non-users of SD. The interface developer also created a story-telling screen to explain the model and the scenario the users are faced with, simply by linking to the relevant Vensim view in the underlying model (Figure 11.27).

Interfaces have been created allowing multiple users simultaneous access to a single model on an organisation's intranet and enabling greater dissemination of the model and its results throughout the organisation. Forio (Forio Online Simulations, 2013) goes a step further by enabling Web-hosted interface applications compatible with many of the available SD software packages. Powersim and iThink are also examples of applications enabling interaction with models over the Internet, opening up SD use to the many.

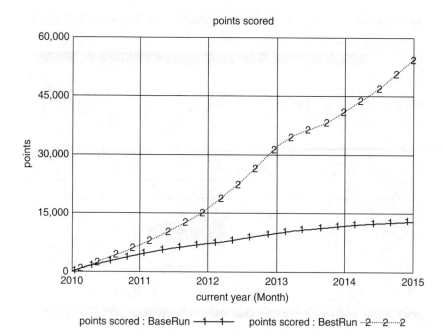

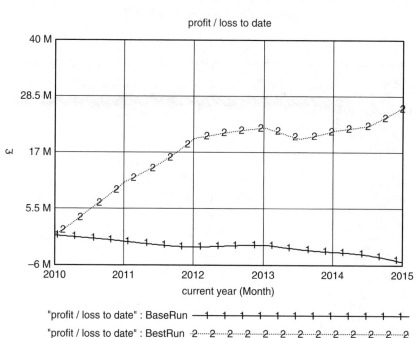

Figure 11.25   Optimisation example.

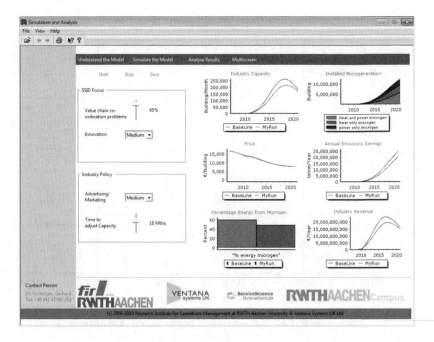

Figure 11.26   Example GUI in Sable.

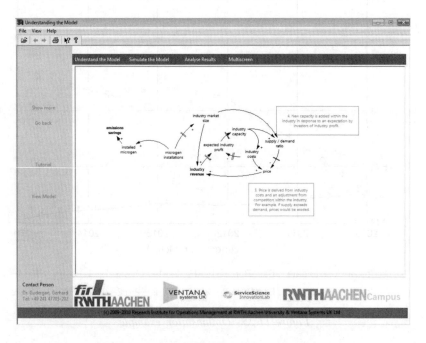

Figure 11.27   Sable-enabled story-telling.

11.6   The future for SD software

The future of SD is a topic often discussed and the debate nearly always seems to centre on ways to expand the use of SD in schools and colleges as a means of 'seeding' the business world and government with systems thinkers. Much effort is expended on this and many demonstrably successful examples are evident. However, we argue here that a critical reason for the disparity between potential and actual application of SD is in its lack of integration with the way in which business and government policy decisions are made. There seems to be too much emphasis on the fun bit, the building of the model, and not enough on the execution of the model in support of real decision making. This failure cannot always be laid at the door of the SD practitioners as it is often outside of their remit to become involved in the decision making or setting of policy, so it follows that the policy makers themselves must become the SD practitioners. Even though SD is 'transparent' and the methodology 'simple', SD tools today are only just becoming accessible in a format usable by everyone.

11.6.1   Innovation

In a plenary session at the 30th International System Dynamics Society (ISDS) Conference a panel was convened to discuss 'Shaping the Future of System Dynamics: Challenges and Opportunities' (System Dynamics Society, 2012). Although meeting with mixed reactions from the audience, Andreas Harbig and Craig Stephens (Greenwood Strategic Advisors, 2013) illustrated this frustration during their Socratic debate and identified the need for SD to become integrated with other analytical tools such as discrete-event simulation (DES), agent modelling, whatever it takes to become practically useful and innovative. The key term used was 'innovation' and it was argued that it will take a radical innovative move to enable SD to reach out as a feasible and widely used solution for business, government and other organisations.

Integrating other software applications with SD is not new. SD software has been 'talking' to Excel and databases for some time now and Vensim can exchange information with almost any other package with a little programming effort. Some of the software packages already handle mixed SD/DES/agent concepts and it is entirely possible to enable an SD simulation, from data entry, through simulation to evaluation of results, entirely within Excel. What may be needed, however, is an industrial-strength amalgam enabling a focus on the practical solutions that business and government policy makers can actually implement. With a focus on products and not just on building models, innovative solutions may then acquire the momentum to generate a huge upsurge of interest in SD.

11.6.2   Communication

Also at the same SD conference, it was suggested we should all 'talk about' SD wherever we are to whoever we are with. This may seem trivial and obvious and it is likely that system dynamicists do indeed talk about SD – with other system dynamicists! Maybe a small number of family or friends have listened patiently to an explanation of causality or feedback from an SD-wise partner while sharing a coffee at the breakfast table or watching the news. Reading a newspaper article or listening to a politician's speech can quickly lead to a red-faced, blood-boiling rant about poor journalism or research and incomprehensible political logic, but it is rare that such encounters lead to true knowledge transfer and so the pool of SD-savvy experienced individuals remains stagnant. Recently, however, there has been the development of viral gaming using the Facebook social network. Millions have tended farms, played poker, built (and destroyed) civilisations and generally interacted with each other through fun (and sometimes educational) games. And even more recently, system dynamicists are starting to exploit this medium as SD software vendors frantically race to enable their software on mobile platforms (iPad, smart phone, etc.) and innovative partnerships emerge between SD practitioners and gaming and graphics experts. One such example is typified by the Facebook game 'Game Change Rio', a collaboration between the developers of a sophisticated SD policy model, an expert gaming company and an SD software vendor (Biovision Foundation, CodeSustainable and Millennium Institute, 2013). The game allows users to make decisions in a host of policy areas affecting the environment, economics and well-being of all Earth's inhabitants. Behind the smart, brilliant interface is an actual SD model running and producing results as the outcome of the user decisions input via the Facebook interface (Figure 11.28).

Perhaps this is the way forward? What if actual real people started to think systemically? What if they started to understand the possibilities through interaction

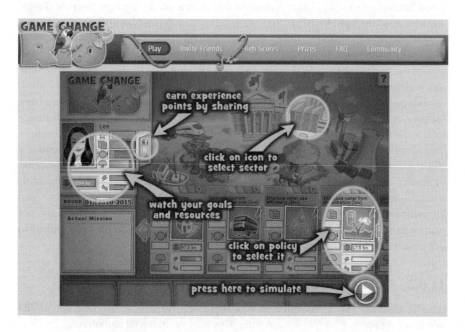

Figure 11.28   Game Change Rio Facebook game using SD.

with insightful games playable from anywhere by anyone, games with an underlying engine designed by SD practitioners for real policy makers? What if, through interaction with such games, the popularity of a systems approach to policy making rises to encompass the majority? Business leaders and politicians would have to sit up and take notice wouldn't they?

References

Banks, I.M. (2010) *The State of the Art*, Hachette, Littlehampton, p. 212.

Biovision Foundation, CodeSustainable and Millennium Institute (2013) Game Change Rio. Available at https://www.facebook.com/gamechangerio (accessed 21 January 2013).

Corel Corporation (2013) CorelDRAW Graphics Suite X6. Available at http://www.corel.com/corel/product/index.jsp?storeKey=gb&pid=prod4260069&trkid=UKSEMGGLGR&gclid=CNas_uqc-bQCFUbKtAodGwwA3Q#tab2&LID=40019623 (accessed 21 January 2013).

Coyle, R.G. (1988) *COSMIC User Manual*, R G Coyle, Salisbury, pp. 1–20.

Forio Online Simulations (2013) Contact. Available at http://forio.com/about-forio/contact-us/ (accessed 21 January 2013).

Greenwood Strategic Advisors (2013) Contact. Available at http://www.greenwood-ag.com/contact.html (accessed 21 January 2013).

IBM (2010) Capitalizing on Complexity: Insights from the Global Chief Executive Officer Study. Available at http://www-935.ibm.com/services/us/ceo/ceostudy2010/index.html (accessed 21 January 2013).

NASA (1999) Mars Climate Orbiter Mishap Investigation Board Phase I Report. Available at http://sunnyday.mit.edu/accidents/MCO_report.pdf (accessed 21 January 2013).

Sterman, J. (2000) *Business Dynamics: Systems thinking and modeling for a complex world*, McGraw-Hill, New York, pp. 83–104.

System Dynamics Society (2012) Proceedings of the 30th International Conference of the System Dynamics Society, 22–26 July 2012, St Gallen, Switzerland. Available at http://www.systemdynamics.org/conferences/2012/index.html (accessed 21 January 2013).

Ventana Systems UK (2011) Forum. Available at http://www.ventanasystems.co.uk/forum (accessed 21 January 2013).

12

Multi-method modelling: AnyLogic

Andrei Borshchev
Managing Director and CEO, The AnyLogic Company, St Petersburg, Russia

The three modelling methods, or paradigms, that exist today are essentially the three different viewpoints the modeller can take when mapping the real-world system to its image in the world of models. The *system dynamics* (SD) paradigm suggests abstracting away from individual objects, thinking in terms of aggregates (stocks, flows) and feedback loops. The *discrete-event (DE) modelling* paradigm adopts a process-oriented approach: the dynamics of the system are represented as a sequence of operations performed over entities. In an *agent-based (AB) model* the modeller describes the system from the point of view of individual objects that may interact with each other and with the environment.

Depending on the simulation project goals, the available data, and the nature of the system being modelled, different problems may call for different methods. Also, sometimes it is not clear at the beginning of the project which abstraction level and which method should be used. The modeller may start with, say, a highly abstract system dynamics model and switch later to a more detailed discrete-event model. Or, if the system is heterogeneous, the different components may be best described by using different methods. For example, in the model of a supply chain that delivers goods to a consumer market, the market may be described in (SD) terms, the retailers, distributors, and producers may be modelled as agents, and the operations inside those supply chain components may be modelled as process flowcharts.

Discrete-Event Simulation and System Dynamics for Management Decision Making, First Edition.
Edited by Sally Brailsford, Leonid Churilov and Brian Dangerfield.
© 2014 John Wiley & Sons, Ltd. Published 2014 by John Wiley & Sons, Ltd.

Frequently, the problem cannot be completely conformed to one modelling paradigm. Using a traditional single-method tool, the modeller inevitably either starts using workarounds (unnatural and cumbersome language constructs), or just leaves part of the problem outside the scope of the model (treats it as exogenous). If the goal is to capture business, economic, and social systems in their interaction, this becomes a serious limitation.

AnyLogic meets this challenge by supporting all three modelling methods on a single, modern object-oriented platform. With AnyLogic, a modeller can choose from a wide range of abstraction levels, can efficiently vary them while working on the model, and can combine different methods into one model.

In this chapter we offer an overview of the most used multi-method model architectures, discuss the technical aspects of linking different methods within one model, and consider three examples of multi-method models, namely:

- consumer market and supply chain

- epidemic and clinic

- product portfolio and investment policy.

The example models are described at a very detailed level so that they can easily be reproduced in AnyLogic development environment.

12.1 Architectures

The number of possible multi-method model architectures is infinite, and many are used in practice. Popular examples are shown in Figure 12.1. We will briefly discuss the problems where these architectures may be useful.

Agents in a SD environment. Think of a demographic model of a city. People work, go to school, own or rent homes, have families, and so on. Different neighbourhoods have different levels of comfort, including infrastructure and ecology, cost of housing, and jobs. People may choose whether to stay or move to a different part of the city, or move out of the city altogether. People are modelled as agents. The dynamics of the city neighbourhoods may be modelled in a SD way, for example, the house prices and the overall attractiveness of the neighbourhood may depend on crowding, and so on. In such a model, agents' decisions depend on the values of the SD variables, and agents, in turn, affect other variables.

The same architecture is used to model the interaction of public policies (SD) with people (agents). Examples are: a government effort to reduce the number of insurgents in society; policies related to drug users or alcoholics.

Agents interacting with a process model. Think of a business where the service system is one of the essential components. It may be a call centre, a set of

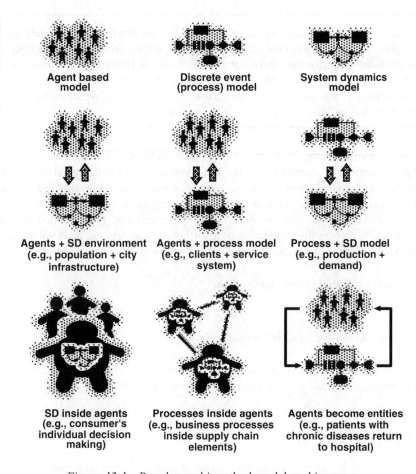

Figure 12.1   Popular multi-method model architectures.

offices, a Web server or an IT infrastructure. As the client base grows, the system load increases. Clients who have different profiles and histories use the system in different ways, and their future behaviour depends on the response. For example, low-quality service may lead to repeated requests, and, as a result, frustrated clients may stop being clients. The service system is naturally modelled in a discrete-event style as a process flowchart where requests are the entities and where operators, tellers, specialists, and servers are the resources. The clients who interact with the system are the agents who have individual usage patterns.

Note that in the previous example the agents can be created directly from the company CRM database and acquire the properties of the real clients. This also applies to the modelling of the company's HR dynamics. You can create an agent for every real employee of the company and place them in the SD environment that describes the company's integral characteristics (the first architecture type).

A process model linked to an SD model. The SD aspect can be used to model the change in the external conditions for an established and ongoing process: demand variation, raw material pricing, skill level, productivity, and other properties of the people who are part of the process.

The same architecture may be used to model manufacturing processes where part of the process is best described by continuous time equations – for example, tanks and pipes, or a large number of small pieces that are better modelled as quantities rather than as individual entities. Typically, however, the rates (time derivatives of stocks) in such systems are piecewise constants, so simulation can be done analytically, without invoking numerical methods.

SD inside agents. Think of a consumer market model where consumers are modelled individually as agents, and the dynamics of consumer decision making are modelled using the SD approach. Stocks may represent the consumers' perception of products, individual awareness, knowledge, experience, and so on. Communication between the consumers is modelled as discrete events of information exchange.

A larger-scale example is interaction of organisations (agents) whose internal dynamics are modelled as stock and flow diagrams.

Processes inside agents. This is widely used in supply chain modelling. Manufacturing and business processes, as well as the internal logistics of suppliers, producers, distributors, and retailers, are modelled using process flowcharts. Each element of the supply chain is at the same time an agent. Experience, memory, supplier choice, emerging network structures, orders, and shipments are modelled at the agent level.

Agents temporarily act as entities in a process. Consider patients with chronic diseases who periodically need to receive treatment in a hospital (sometimes planned, sometimes because of acute phases). During treatment, the patients are modelled as entities in the process. After discharge from the hospital, they do not disappear from the model, but continue to exist as agents with their diseases continuing to progress until they are admitted to hospital again. The event of admission and the type of treatment needed depend on the agent's condition. The treatment type and timeliness affect the future disease dynamics.

There are models where each entity is at the same time an agent exhibiting individual dynamics that continue while the entity is in the process, but are outside the process logic – for example, the sudden deterioration of a patient in a hospital.

12.1.1   The choice of model architecture and methods

AnyLogic, designed as a multi-method object-oriented tool, allows you to create model architectures of any type and complexity, including those previously mentioned. You can develop complex, simple, flat, hierarchical, replicated, static, or dynamically changing structures.

The choice of the model architecture depends on the problem you are solving. The model structure reflects the structure of the system being modelled – not literally, however, but as seen from the problem viewpoint. The choice of modelling method should be governed by the criterion of *naturalness*. Compact, minimalistic, clean, beautiful, easy to understand, and explain – if the internal texture of your model is like that, then you have chosen the right method.

12.2   Technical aspect of combining modelling methods

In this section we will consider the techniques of linking different modelling methods in AnyLogic.

> The very first thing you should know is that all model elements of all methods, be they SD variables, statechart states, entities, process blocks, and even animation shapes or business charts, exist in the "same namespace": any element is accessible from any other element by name (and, sometimes, "path" – the prefix describes the location of the element).

The following examples are all taken from real projects and purged of all unnecessary details. This set, of course, does not cover everything, but it does give a good overview of how you can build interfaces between different methods.

12.2.1   System dynamics → discrete elements

The SD model is a set of continuously changing variables. All other elements in the model work in discrete time (all changes are associated with events). SD itself does not generate any events, so it cannot *actively* have an impact on agents, process flowcharts, or other discrete-time constructs. The only way for the SD part of the model to affect a discrete element is to let that element watch a condition over SD variables, or to use SD variables when making a decision. Figure 12.2 shows some possible constructs.

In case *A*, *SD triggers a statechart transition*. Events (low-level constructs that allow scheduling a one-time or recurrent action) and statechart transitions are frequent elements of agent behaviour. Among other types of triggers, both can be triggered by a condition – a Boolean expression.

> If the model contains dynamic variables, all conditions of events and statechart transitions are evaluated *at each integration step*, which ensures that the event or transition will occur exactly when the (continuously changing) condition becomes true.

A **SD triggers a statechart transition of Condition type**

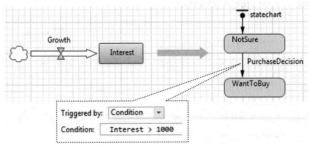

B **SD controls the entity generation in a process flowchart**

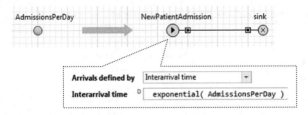

Figure 12.2   SD impacts discrete elements of the model.

In Figure 12.2A the statechart is waiting for the `Interest` to rise higher than a given threshold value. The statechart can be located on the same level as the SD, or in a different active object.

In case *B* the source block `NewPatientAdmission` generates new entities at the rate defined by the dynamic variable `AdmissionsPerDay`. The arrivals are defined in the form of interarrival time and not in the form of rate, because the rate is not re-evaluated during the simulation, whereas the interarrival time is re-evaluated after each new entity.

Note that if the value of the dynamic variable changes *in between* two subsequent event occurrences (or in between two entity arrivals), this will not be "noticed" immediately, but only at the next event occurrence (or next entity arrival).

12.2.2   Discrete elements → system dynamics

For cases *C–E* below, see Figure 12.3.

In case *C, the SD stock triggers a statechart transition, which, in turn, modifies the stock value.* Here, the interface between the SD and the statechart is implemented in the pair condition/action. In the state `WantToBuy`, the statechart tests if there are products in the retailer stock and, if there are, buys one and changes the state to `User`.

C The SD stock triggers a statechart transition, which, in turn, modifies the stock value

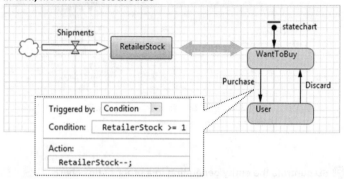

D The SD stock accumulates position properties of a moving agent

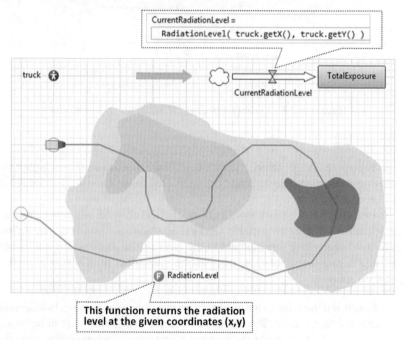

E The SD flow depends on the number of entities in the DE queue

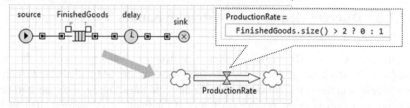

Figure 12.3   Discrete elements of the model impact SD.

You can freely change the values of the SD *stocks* from outside the SD part of the model. This does not interfere with solving the differential equation; the integrator will just start with the new value. However, trying to change the value of a flow or auxiliary variable that has an equation associated with it will not be correct: the assigned value will be immediately overridden by the equation, so assignment will have no effect.

In case *D, the SD stock accumulates the "history" of the agent motion.* This is an interesting example of SD–AB cooperation. The value of the stock `Total-Exposure` is constantly updated as the mobile agent moves through the area contaminated by radiation. The value of the incoming flow `CurrentRadiation-Level` is set to the radiation level at the coordinates of the truck agent. As the truck moves or stays, the stock receives and accumulates a dose of radiation per time unit. Again, in the SD equation, we are referencing the agent and calling its function.

In case *E, referencing DE objects in the SD formula,* the flow `Production Rate` switches between 0 and 1, depending on whether the finished products' inventory (the number of entities in the queue `FinishedGoods` returned by the function `size()`) is greater than 2 or not. Again, one can close the loop by letting the SD part control the production process.

12.2.3   Agent based ↔ discrete event

For cases *F* and *G* below, see Figure 12.4.

In case *F, a server in the DE process model is implemented as an agent.* Consider some complex equipment, such as a robot or a system of bridge cranes. The behaviour of such objects is often best modelled "in agent-based terms" by using events and state-charts. If the equipment is part of the manufacturing process being modelled, you need to build an interface between the process and the agent representing the equipment.

In this example, the statechart is a simplified equipment model. When the statechart comes to the state `Idle`, it checks if there are entities in the queue. If so, it proceeds to the `Working` state and, when the work is finished, unblocks the `hold` object, letting the entity exit the queue. The `hold` object is set up to block itself again after the entity passes through.

The next entity will arrive when the equipment is in the `Idle` state. To notify the statechart, we call the function `onChange()` upon each entity arrival (see the `On enter` action of the queue).

Unlike in models with continuously changing SD elements, in models built of purely discrete elements of events and transitions triggered by a condition, *do not monitor the condition continuously.* The event's condition is evaluated when the event is reset. The transition's condition is evaluated when the statechart comes into the transition's source state. And then the conditions are re-evaluated when something happens to the active object where they are located, or when its `onChange()` function is called.

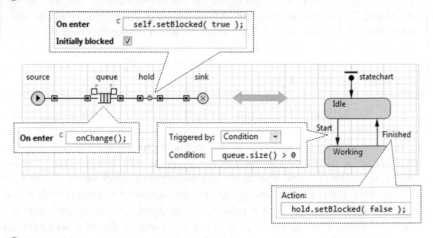

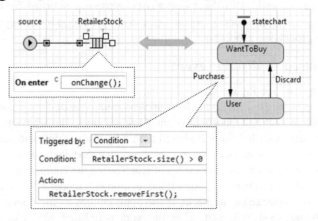

Figure 12.4   Agent-based parts of the model interact with discrete-event parts.

In case *G, the agent removes entities from the DE queue.* Here, the supply chain is modelled using DE constructs; in particular, its end element, the retailer stock, is a `queue` object. The consumers are outside the DE part and are modelled as agents. Whenever a consumer comes to the state `WantToBuy`, it checks the `RetailerStock` and, if it is not empty, removes one product unit. Again, as this is a purely discrete model, we need to ensure that the consumers who are waiting for the product are notified about its arrival – that is why the code `onChange()` is placed in the `On enter` action of the `RetailerStock` queue.

In this simplified version, there is only one consumer whose statechart is located on the same canvas as the supply chain flowchart. In the full version there would be multiple agents–consumers and, instead of calling just `onChange()`, the retailer stock would notify *every consumer* in a loop.

12.3   Example: Consumer market and supply chain

We will model the supply chain and sales of a new product in a consumer market in the absence of competition. The supply chain will include delivery of the raw product to the production facility, production, and the stock of finished products. The QR inventory policy will be used. Consumers are initially unaware of the product; advertising and word of mouth will drive the purchase decisions. The product has a limited lifetime, and 100% of users will be willing to buy a new product to replace the old one.

We will use DE methodology to model the supply chain, and SD methodology, namely, a slightly modified Bass diffusion model (Bass, 1969), to model the market. We will link the two models through the sales events.

12.3.1   The supply chain model

The supply chain flowchart (see Figure 12.5) includes three stocks (the supplier stock of raw material, the stock of raw material at the production site, and the stock of finished products at the same location). Delivery and production are modelled by the two `Delay` objects with limited capacity. The delay time for both has been left at the default value of `triangular(0.5, 1, 1.5)`. To load the supply chain with some initial product quantity we will add this `StartUp code`:

```
Supply.inject( OrderQuantity);
```

If we run this model, at the beginning of the simulation 400 items of the product are produced and accumulate in the `ProductStock`. Nothing else happens in the

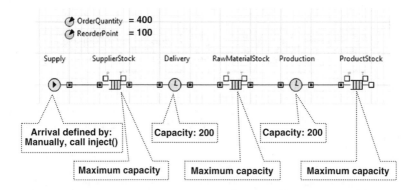

Figure 12.5   The discrete-event model of the supply chain.

model (the `Supply` object is set up to not generate any new entities, unless explicitly asked to do so). The inventory policy is not yet present in our model.

12.3.2   The market model

The market is modelled by an SD stock and flow diagram as shown in Figure 12.6. The SD part is located in the *Main* object – just on the same canvas where the

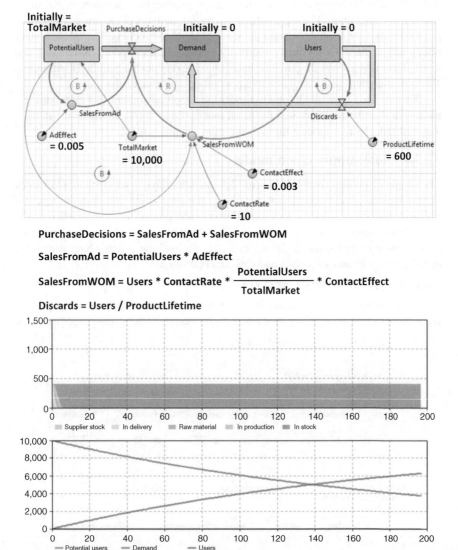

Figure 12.6   The SD model of the market; the dynamics of the unlinked model.

flowchart was created earlier. The difference of our market model from the classical Bass diffusion model with discards (Sterman, 2000) is that the users', or adopters', stock of the classical model is split into two: the `Demand` stock and the actual `Users` stock. The adoption rate in this model is called `PurchaseDecisions`. It brings `PotentialUsers` not directly into the `Users` stock, but into the intermediate stock `Demand`, where they wait for the product to be available. The actual event of sale, that is, the "meeting" of the product and the customer who wants to buy it, will be modelled outside the SD paradigm.

The two pieces of the model are not yet linked. If we run the model, the supply chain will still produce the 400 items and stop, and the potential clients will gradually make their purchase decisions, building up the `Demand` stock. Note that in the current version of the model, the only reason for the potential users to make a purchase decision is advertising. The word-of-mouth effect is not yet working because nobody has actually purchased a single product item. To better view the model dynamics in future experiments, it makes sense to add a couple of charts as shown in Figure 12.6.

12.3.3   Linking the DE and the SD parts

How do we link the supply chain and the market? We want to achieve the following:

- If there is at least one product item in stock and there is at least one client who wants to buy it, the product item should be removed from the `ProductStock` queue, the value of `Demand` should be decremented, and the value of `Users` should be incremented, see Figure 12.7.

We therefore have a condition and an action that should be executed when the condition is true. The AnyLogic construct that does exactly that is the condition-triggered event. The implementation of our scheme in the AnyLogic modelling language is shown in Figure 12.8.

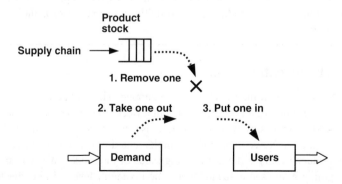

Figure 12.7   Linking the supply chain and the market: the scheme.

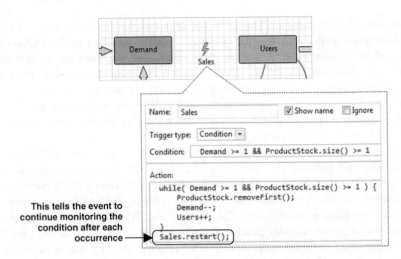

Figure 12.8   Linking the supply chain and the market: the implementation.

A couple of comments on how the condition-triggered events work are in order:

> In the presence of continuous dynamics in the model, the condition of the event is evaluated at each numeric micro-step. Once the condition evaluates to true, the event's action is executed.

We put the sale into a `while` loop, because a possibility exists that two or more product items may become available simultaneously, or the `Demand` stock may grow by more than one unit per numeric step. Therefore, more than one sale can potentially be executed per event occurrence.

> By default, the condition event disables itself after execution. As we want it to continue monitoring the condition, we explicitly call `restart()` at the end of the event's action.

12.3.4   The inventory policy

With the `Sales` event in place, sales start to happen. The 400 items produced at the beginning of the simulation disappear in about a week. The `Users` stock increases up to almost 400; it then slowly starts to decrease according to our limited lifetime assumption. And, since we have not yet implemented our inventory policy, no new items are produced. This is the last missing piece of the model. We will include the inventory policy in the same `Sales` event; the inventory level will be checked after each sale.

We will modify the `Action` of the `Sales` event this way:

```
while ( Demand >= 1 && ProductStock.size () >= 1) {
  //execute sale
  ProductStock.removefirst (); //remove a product unit from the
  stock (DE)
  Demand--; //remove a waiting client from Demand stock (SD)
  Users++; //add a (happy) client to the Users stock (SD)
}
//apply inventory policy
int inventory = //calculate inventory
  ProductStock.size () + //in stock
  Production.size () + //in production
  RawMaterialStock.size () + //raw product inventory
  Delivery.size () + //raw product being delivered
  SupplierStock.size (); //supplier's stock
if ( inventory < ReorderPoint) //QR policy
  Supply.inject ( OrderQuantity);
//continue monitoring
Sales.restart ();
```

Now, the supply chain starts to work as planned, see Figure 12.9. (Here the inventory chart and the market charts are combined: one was dragged onto the top of the other, and the labels were put on different sides.)

During the early adoption phase the supply chain performs adequately, but as the majority of the market starts buying, the supply chain cannot keep up with the market. In the middle of the new product adoption (days 40–100), even though the supply chain works at its maximum throughput, the number of waiting clients still remains high. As the market becomes saturated, the sales rate reduces to the replacement purchases rate, which equals the `Discard` rate in the completely saturated market,

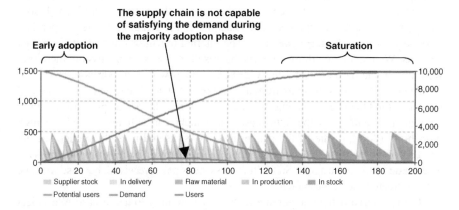

Figure 12.9   The supply chain dynamics pattern changes as the market gets saturated.

that is, `TotalMarket/ProductLifetime` = 16.7 sales per day. The supply chain handles that easily.

An interesting exercise would be to make the supply chain adaptive. You can try to minimise the order backlog and at the same time minimise the inventory by adding the feedback from the market model to the supply chain model.

12.4   Example: Epidemic and clinic

We will create a simple AB epidemic model and link it to a simple DE clinic model. When a patient discovers symptoms, he will ask for treatment at the clinic, which has limited capacity. We will explore how the capacity of the clinic affects the disease dynamics. This model was suggested in 2012 by Scott Hebert, a consultant at AnyLogic North America.

12.4.1   The epidemic model

We will add the `Agent Population` to the editor of `Main` (the object that is created by default with the new model). The agents will be our `patients`; their initial number is 2000. We will place the agents into a rectangular area of 650 by 200 miles (1040 by 320 km) and create a distance-based network; two patients are connected if they live at most 30 miles (48 km) away.

The next step is to define the behaviour of our patient. In the `Patient` object we will create a statechart. The statechart (see Figure 12.10) is similar to the classical SEIR statechart (Wikipedia, 2013). The patient is initially in the `Susceptible` state, where he can be infected. Disease transmission is modelled by the message `"Infection"` sent from one patient to another. Having received such a message, the patient transitions to the state `Exposed`, where he is already infectious, but does not have symptoms. After a random incubation period, the patient discovers symptoms and proceeds to the `Infected` state. We distinguish between the `Exposed` and `Infected` states because the contact behaviour of the patient is different before and after he discovers symptoms: the contact rate in the `Infected` state is 1 per day, as opposed to 5 in the `Exposed` state. The internal transitions in both states model contacts. We model only those contacts that result in disease transmission; therefore, we multiply the base contact rate by `Infectivity`, which, in our case, is 7%.

There are two possible exits from the `Infected` state. The patient can be treated in a clinic (and then he is guaranteed to recover), or the illness may progress naturally without intervention. In the latter case, the patient can still recover with a high probability (90%), or die. If the patient dies, he is deleted from the model, see the `Action` of the `Dead` state. The completion of treatment is modelled by the message `"Treated"` sent to the agent. So far, this message is never received, because we have not yet created the clinic model.

The recovered patient acquired a temporary immunity to the disease. We reflect this in the model by having the state `Recovered`, where the patient does not react to the message `"Infection"` that may possibly arrive. At the end of the immunity

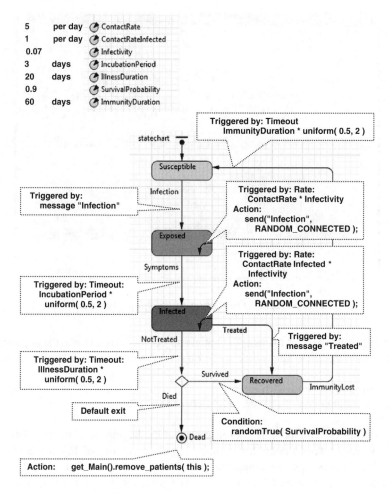

Figure 12.10   The patient behaviour statechart.

period the transition **ImmunityLost** takes the patient back to the **Susceptible** state.

Note that as long as we have defined the parameters *inside the agent*, we can make their values different for different agents. In our model, however, for simplicity they are the same throughout the whole population.

For animation purposes, we will paint the patient differently, depending on his state. In the **Entry action** field of each state, we will type the code that changes the colour of the patient animation into the colour of the state, for example, in the **Entry action** of the state **Exposed** we will type: **person.setFillColor (darkOrange);**.

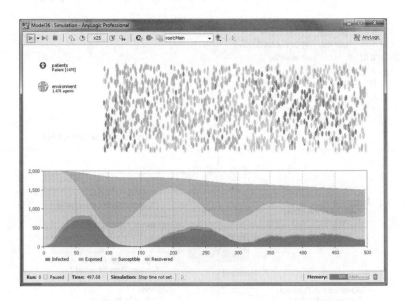

Figure 12.11 Oscillation in the epidemic model.

And finally, we need to create the initial entry of the infection into the population. In the `StartUp code` field of the `Main` object we will type the following code:

```
for( int i=0; i<5; i++)
    patients.random().receive( "Infection");
```

This will infect five randomly chosen people at the beginning of the simulation.

Now we can run the epidemic model. Look at the top of Figure 12.11. The contagious disease spreads around the initially infected agents (remember that our network of contacts is based on distance). The epidemic does not end after the first wave, because the immunity period is not long enough. We will add a chart to view the type of SD.

AnyLogic supports the collection of statistics on agent populations. We will define four statistics in the patient population. The first one will be called `NSusceptible` and will count the patients for which the condition `item.statechart.isStateActive-`
`(item.Susceptible)` evaluates to true. Similarly, we will count all exposed, infected, and recovered patients. We will use the AnyLogic `Time stack chart` with a time window of 500 days to display the statistics. At the bottom of Figure 12.11 you can see the oscillation and gradual decrease of the total population due to deaths.

12.4.2 The clinic model and the integration of methods

The next step is to add the clinic, and let the patients be treated there. Our clinic will be modelled via a very simple DE model: the `Queue` for the patients waiting to be treated and the `Delay` modelling the actual treatment.

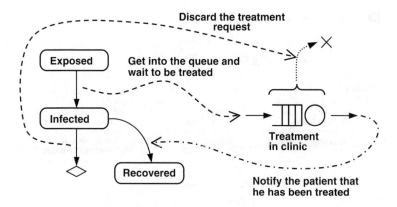

Figure 12.12   Interface between the agent-based and the discrete-event parts of the model.

We will put the process flowchart (see Figure 12.13A) in the `Main` object. Unlike in classical DE models, however, the entities in this process are not generated by a `Source` object, but are injected by the agents via an `Enter` object. The communication scheme between the patients–agents and the clinic process is shown in Figure 12.12.

Once the patient discovers symptoms, he creates an entity – let's call it "treatment request" – and injects the entity into the clinic process. Once the treatment is completed, the entity notifies the patient by sending him a message "Treated" that causes the patient to transition to the `Recovered` state. If, however, the patent is cured or dies before the treatment is completed, he will discard his treatment request by removing it from whatever stage it is at in the process. On the technical side, we need:

- The entity that will carry the reference to the patient.

- The ability to remove the entity originated by a particular patient from the process.

We will create a custom entity class `TreatmentRequest` that has a field `patient` – this will be the reference to the patient (agent) who originated the treatment request. In the process flowchart (Figure 12.13A), we identified that the entities passing through the `wait`, `treatment`, and `finished` objects are not of the generic `Entity` class, but of its subclass `TreatmentRequest`. This is necessary because we plan to use the `patient` field of those entities. For example, when the treatment is finished, the `finished` object sends a message "Treated" to the patient referenced by the entity before disposing of the entity.

We also specified that the queue has infinite capacity, that the treatment takes exactly seven days, and that there are only 20 beds in the clinic, so only 20 patients can be treated simultaneously.

Also, we prepared the function `cancelTreatmentRequest()` that will be called by patients who got well on their own or died before getting the chance to be treated.

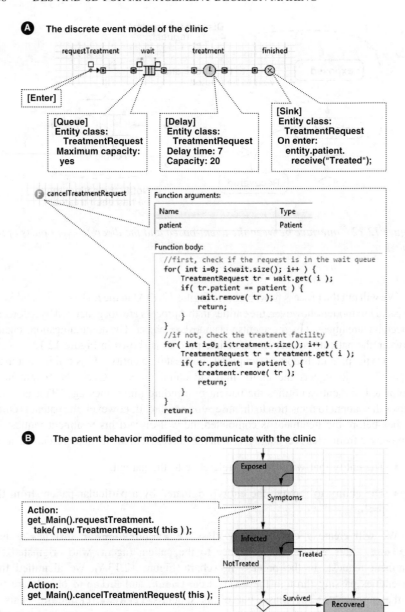

Figure 12.13   The discrete-event model of the clinic and integration with the agent-based model.

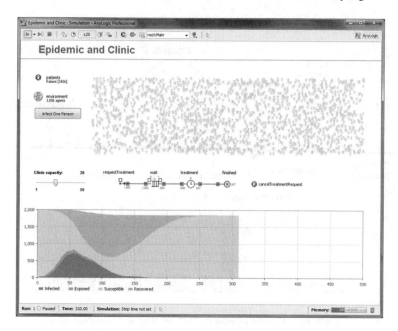

Figure 12.14   The behaviour of the integrated model.

That function uses the API (Application Programming Interface) of the `Queue` and `Delay` objects to search for a particular entity in them and remove it.

The remaining task is to modify the behaviour of `Patient` to link it to the model of clinic. We will add the actions to the statechart transitions `Symptoms` and `Not-Treated` as shown in Figure 12.13B. Remember that since the clinic process is located one level above the patient's statechart, in the `Main` object, the clinic objects and functions should be preceded by the prefix `get_Main()`. The Java word `"this"` references the object to which the code belongs, in this case the patient. The model is complete.

Now the model shows a different dynamic or, to be more precise, a different range of dynamics. The oscillations are still possible, but a possibility also exists that the epidemic will end after the first wave, as you can see in Figure 12.14. You may experiment with different clinic capacities to figure out the number of beds needed in order to treat everybody on time and prevent further waves of the epidemic.

12.5   Example: Product portfolio and investment policy

A company develops and sells consumer products with a fairly short lifecycle. After the product has been successfully launched, its revenue peaks, and then falls, as shown in Figure 12.15. To keep the business going the company has to maintain a

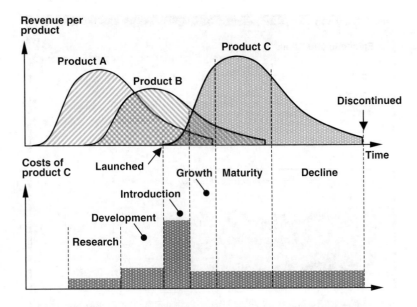

Figure 12.15 The product lifecycle and associated costs.

continuous process of new product research and development. Part of the company's revenue is therefore reinvested in R&D, and another significant part is spent on introducing new products. We will investigate how the investment policy affects the business.

12.5.1 Assumptions

We will make some simplifying assumptions, as follows.
The product lifecycle:

- The duration of the research phase of a product is uniformly distributed between 1.5 and 6 years. During this phase, each product is assigned a random "success factor" which is uniformly distributed between 0 and 1. At the end of the research phase, the product is killed if its success factor is less than 0.5; otherwise, it proceeds to the development phase.

- The duration of the development phase is also uniformly distributed between 1 and 3 years. During the development phase, the initial success factor is modified by adding a random number uniformly distributed between −0.3 and 0.3. At the end of the development phase, the project is killed if its success factor is less than 0.5; otherwise, it is released to the market.

- When the product is launched, its success factor is once again modified by adding a random number between −0.3 and 0.3. This value of the success factor then stays the same. As you can see, the value is between 0.2 and 1.6.

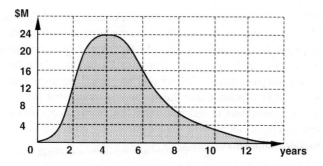

Figure 12.16   The base revenue curve of a product.

Revenue and cost:

- While in the market, all products have the same curve of base revenue over time (see Figure 12.16), and the actual revenue equals base times the "success factor".

- The first two years of the product in the market are considered as an "introduction period". At any time after the introduction period the product is discontinued if its annual revenue falls below $5M.

- The annual cost of research is $0.5M per product and is the same for all products. The annual cost of development is $1M per year. The cost of introducing a new product is $10M, which is spent evenly during the two years. In addition there is a one-time fixed cost of $0.5M for starting a new R&D project.

- After the introduction period we will assume no cost per product in market, in other words, we will treat the revenue as revenue after production and distribution costs.

Investment policy:

- A fraction of the company revenue goes into "investment capital." All R&D costs are paid from the investment capital.

- The company has a limited R&D capacity and cannot perform more than 100 projects concurrently.

- Once the company determines that the accumulated investment capital is greater than (the number of ongoing R&D projects + 1) times the "average project cost" ($3M is assumed), a new project is started.

- The invested fraction of the revenue is determined as follows. If the accumulated investment capital is greater than the R&D capacity times the "average project cost," no money is invested. Otherwise, 20% of the revenue goes into the investment capital stock.

- The remaining part of the revenue goes into the "main capital" stock, and product introduction costs are paid from there.

Assumptions, as you can see, are quite strong. For example, R&D projects may be killed at only two points, at the end of the research phase and at the end of the development phase, but not halfway through. The money spent on introducing the new product does not vary from product to product, the product lifecycles are similar, and there are no complete market failures and no great, long-lasting successes. All these things can be incorporated into the simulation model, but for the purpose of demonstrating the interaction of different modelling methods a simpler model will work just fine. Of course, all numeric values previously given are not fixed and will become the model parameters.

12.5.2   The model architecture

We will model each product individually as an agent (as you know, agents in AB models are not necessarily people – they can be anything), so the product portfolio will be a population of agents. The company finances and investment policy will be a SD component. The overall architecture of the model is shown in Figure 12.17. The

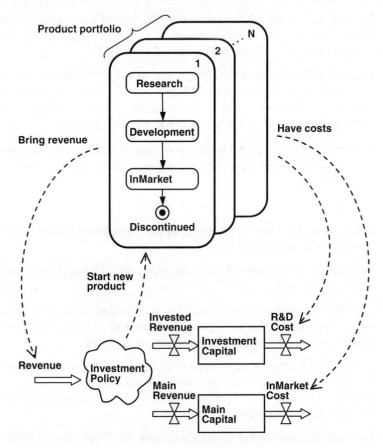

Figure 12.17   Architecture of the product portfolio and investment policy model.

product lifecycle is naturally represented as a statechart that starts in the `Research` state and is deleted from the model when it is discontinued or killed. The implementation of the interface between the products–agents and the SD part will be clear from the step-by-step description that follows.

12.5.3   The agent product and agent population portfolio

In a new model, we will create an agent population portfolio with agent class *Product*. The initial population size will be 100. We will use a circle (`bubble`) as the animation of product. Later we will create a bubble chart, and as the product progresses through the lifecycle phases, its bubble will move.

Product behaviour will be defined in the form of a statechart, see Figure 12.18. The statechart structure straightforwardly reflects the product lifecycle.

Also, the product will have the parameters and variables shown in Figure 12.19. As you can see, the numeric values in the problem statement have become parameters in the model.

For example, the duration of the development phase was originally specified as uniformly distributed between 1 and 3 years. In the model, the parameter `DevelopmentTime` has an initial value of 2, and in the timeout expression in

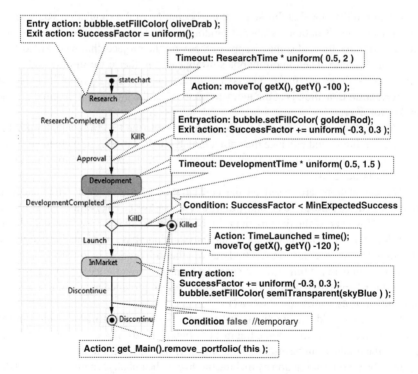

Figure 12.18   Statechart of the agent Product.

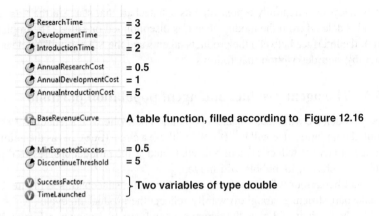

Figure 12.19   Variables, and parameters of the agent Product.

the `DevelopmentCompleted` transition, it is multiplied by a random coefficient taking values between 0.5 and 1.5, which gives us the distribution we need. Should we decide to change the average value of the development duration, the resulting interval will correctly follow it.

If you run the model at this stage, at the beginning of the simulation you will see 100 olive-coloured bubbles scattered randomly. After a while, some bubbles turn brown and move up, and some disappear; these are the projects that were killed after the research phase. Then, more bubbles disappear, and the rest turn blue and move further up – these are the products that go to market. As the condition of the `Discontinue` transition is at the moment set to false, the products will remain in the market for ever.

We will now add the cost and revenue calculation to the `Product` agent, fill in the missing condition, and enhance the animation. The calculation is implemented in the form of three functions at the agent level, see Figure 12.20 (the return type of all functions is `double`).

The calculations are based on the current state of the product. The statechart functions `isStateActive()` and `getActiveSimpleState()` are used to obtain the state. However, while the product is in the market, there is one special state that is not reflected in the statechart, namely, the introduction phase that lasts for two years, according to our specification. To find out whether the product is in the introduction phase, we compare the time from the product launch (`time() - TimeLaunched`) and `IntroductionTime`. The time from the launch is also provided as an argument to the table function `BaseRevenueCurve()`.

Now we can add the condition that triggers the `Discontinue` transition in the product statechart, see the middle fragment of Figure 12.20. The condition reflects the fact that the product can be discontinued only after the introduction phase. We can also enhance the product animation by making the size of the bubble dynamically reflect the revenue brought by the product, see the dynamic expression of the bubble radius.

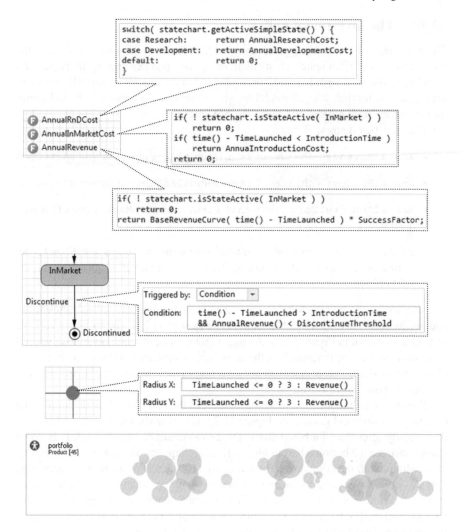

```
switch( statechart.getActiveSimpleState() ) {
case Research:      return AnnualResearchCost;
case Development:   return AnnualDevelopmentCost;
default:            return 0;
}
```

AnnualRnDCost
AnnualInMarketCost
AnnualRevenue

```
if( ! statechart.isStateActive( InMarket ) )
    return 0;
if( time() - TimeLaunched < IntroductionTime )
    return AnnuaIntroductionCost;
return 0;
```

```
if( ! statechart.isStateActive( InMarket ) )
    return 0;
return BaseRevenueCurve( time() - TimeLaunched ) * SuccessFactor;
```

InMarket

Discontinue

Discontinued

Triggered by: Condition ▾

```
Condition:    time() - TimeLaunched > IntroductionTime
              && AnnualRevenue() < DiscontinueThreshold
```

Radius X: `TimeLaunched <= 0 ? 3 : Revenue()`

Radius Y: `TimeLaunched <= 0 ? 3 : Revenue()`

portfolio
Product [45]

Figure 12.20 Functions calculating cost and revenue. Updated statechart and animation.

Now, if you run the model, you can see that the bubbles of the products in the market change their sizes dynamically as the revenue rises and then falls. Eventually, the bubbles seem to disappear.

In fact, the bubbles (and the corresponding products) do not disappear completely, as nobody is telling the agent to recalculate the condition of the **Discontinue** transition. This will be done by the SD part of the model that we will build next.

12.5.4   The investment policy

The next step is to model the company investment policy. This will be done at the **Main** level. We will create statistical items in the **portfolio** agent population calculating the total revenue and costs, and use the statistics in the SD model of investment. After that, we will model the start of new R&D projects, which depends on the money accumulated in one of the SD stocks.

The statistics items will be:

- **AnnualRevenue** – the sum of **item.AnnualRevenue()** across all products

- **AnnualRnDCost** – the sum of **item.AnnualRnDCost()** across all products

- **AnnualInMarketCost** – the sum of **item.AnnualInMarketCost()** across all products

- **NinRnD** – the count of **! item.statechart.isStateActive(item .InMarket)** across all products. This is the number of products in the R&D phase.

As a first iteration of the SD part, we will add just one dynamic variable, **Revenue**, and set its formula to **portfolio.AnnualRevenue()**. The chart of that variable over time is shown in Figure 12.21. As one might expect, the total revenue of several products launched approximately at the same time is similar to the base revenue curve of a single product. The company is not investing in new product R&D, and in about 15 years it goes out of business.

Now we will draw the first meaningful draft of the stock and flow diagram of the company's investment policy, see Figure 12.22. If we run the model, we will observe that, shortly after the simulation starts, the **InvestmentCapital** stock falls down below zero, see the bottom chart in Figure 12.22. It happens because, at the beginning of the simulation, the company starts too many (100) R&D projects simultaneously; they consume money, but bring in no revenue. As the products are launched in the market, the stock goes back up and then follows the S-shaped curve typical of systems with saturation. The **MainCapital** stock has a slight depression during the products' introduction period and then follows the same S-shaped curve.

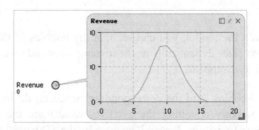

Figure 12.21   Total revenue of several products launched at approximately same time.

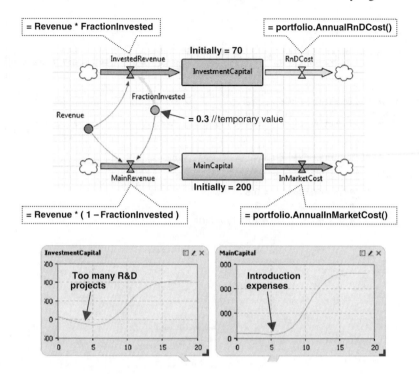

Figure 12.22   The first draft of the SD model of investment policy.

12.5.5   Closing the loop and implementing launch of new products

In the next step, we will close the loop and implement the rule for starting new products, depending on the available investment money. This will be done by an event `StartNewProject` at the level of `Main`.

See Figure 12.23. The parameter `AverageProjectCost` is an estimation of how much money will be required (per project) to finish all the ongoing projects plus a new one. The event `StartNewProject` is constantly monitoring the `InvestmentCapital` stock and, when it detects room for one more project, starts it. The one-time project setup cost is immediately subtracted from the stock, and a new `Product` agent is created in the `portfolio` population. The last statement in the event action (the call of the `restart()` function) tells the event to resume monitoring the condition after each occurrence. Also, because the new products are now created automatically, we will set the initial number of products to 0.

Now the company acts safely and is profitable. The continuous R&D process ensures that the company always has new products to offer. The revenue stream grows during the first three decades up to approximately $800M per year and then starts

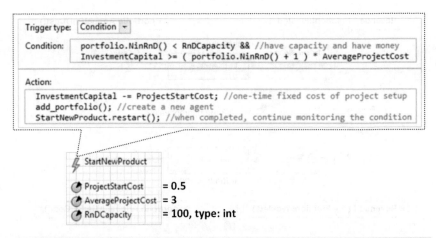

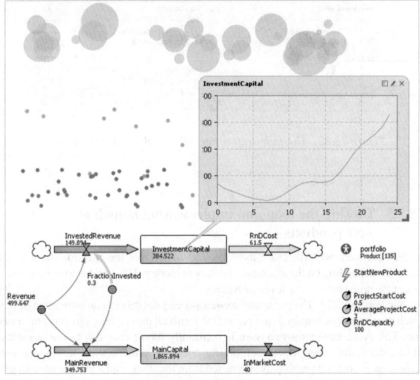

Figure 12.23   Condition event that creates new products. The new dynamics of the model.

oscillating irregularly around that value. Further revenue growth is limited by the R&D capacity of the company.

12.5.6 Completing the investment policy

Under the current model setup, 30% of the company's gross revenue always goes into the investment capital stock, which continues to build up and remains largely unused. In the last step, we will implement the remaining part of the investment policy, that is, we will make the invested fraction of the revenue a variable that depends on the accumulated resources. This is done purely at the SD level by changing the formula of `FractionInvested` to

```
InvestmentCapital > RnDCapacity * AverageProjectCost ? 0 :
MaxFraction,
```

where `MaxFraction` is a new model parameter with a default value of 0.2.

The picture is different now (Figure 12.24). The `InvestmentCapital` stock reaches the value of $300M and stops growing. The revenue oscillates around $800M per year, which, as we know already, is the upper limit with the given R&D capacity.

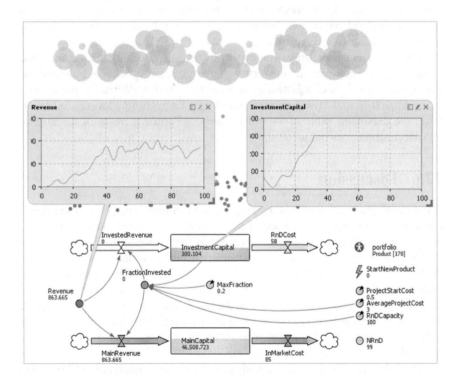

Figure 12.24   Total company dynamics under the fully implemented investment policy.

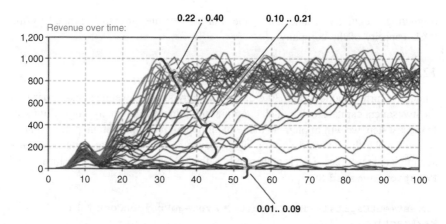

Figure 12.25   Sensitivity analysis: revenue dynamics under different invested fraction values.

We can use this model to optimize the investment policy. For example, we can investigate how sensitive the company dynamics are to the parameters of the investment policy, say, to `MaxFraction`. We will compare the curves of revenue over time obtained in different simulation runs. The results of the sensitivity analysis experiment is shown in Figure 12.25.

The results are interesting. The curves of revenue over time cluster into three groups, as in the figure. In the first one the company goes out business; this corresponds to values of `MaxFraction` from 0.01 to 0.09. When the parameter is in the range 0.10 to 0.21, the revenue climbs up – the higher the `MaxFraction`, the faster the maximum value of $800M is reached. Further increase of the invested revenue fraction does not affect the growth.

12.6   Discussion

When developing a DE model of a supply chain, IT infrastructure, or a contact centre, a modeller would typically ask the client to provide the arrival rates of the orders, transactions, or phone calls. The modeller would then be happy to get some constant values, periodic patterns, or trends, and treat arrival rates as variables exogenous to the model. In reality, however, those variables are outputs of another dynamic system, such as a market, a user base. Moreover, this other system can, in turn, be affected by the system being modelled. For example, the supply chain cycle time, which depends on the order rate, can affect the satisfaction level of the client, which impacts repeated orders and, through word of mouth, new orders from other customers. The *choice of the model boundary* therefore is very important.

The only methodology that explicitly talks about the problem of model boundary is the SD one (Sterman, 2000). However, the SD modelling language is limited by its high level of abstraction, and many problems cannot be modelled with the necessary accuracy. With multi-method modelling one can choose the best-fitting method and language for each component of the model and combine those components while staying on one platform.

For those who would like to continue learning the exciting discipline of multi-method modelling, we can suggest the following exercise:

A telecom company is about to introduce a new type of service, say, HDTV or high-speed Internet access, and is planning the additional network infrastructure, the tariff policy, and the marketing campaign. Model the adoption of the new technology by the users in the loop with the network infrastructure performance. Consider potential dissatisfaction effects, incremental growth of the infrastructure capacity, and ROI.

All models considered in this chapter are available with AnyLogic software and also at the online simulation portal www.RunTheModel.com.

References

Bass, F. (1969) A new product growth model for consumer durables. *Management Science*, **15**(5), 215–227.

Sterman, J.D. (2000) *Business Dynamics: Systems Thinking and Modelling for a Complex World*, McGraw-Hill, New York.

Wikipedia (2013) Compartmental models in epidemiology, http://en.wikipedia.org/wiki/Compartmental_models_in_epidemiology.

13

Multiscale modelling for public health management: A practical guide

Rosemarie Sadsad[1,2,3] and Geoff McDonnell[3,4]
[1]*Centre for Infectious Diseases and Microbiology – Public Health, Westmead Hospital, Sydney, New South Wales, Australia*
[2]*Sydney Medical School, Westmead, The University of Sydney, New South Wales, Australia*
[3]*Centre for Health Informatics, Australian Institute of Health Innovation, University of New South Wales, Sydney, New South Wales, Australia*
[4]*Adaptive Care Systems, Sydney, New South Wales, Australia*

13.1 Introduction

System modelling and simulation is increasingly applied to manage chronic persistent problems, including public health. Simulation models of complex and multilevel organisational systems like health care are often abstracted at one level of interest. There are many challenges with developing simulation models of systems that span multiple organisational levels and physical scales. We describe several theoretical and conceptual frameworks for multilevel system analysis and present an approach for developing multiscale and multimethod simulation models to aid management decisions. We use simulation models to illustrate how management actions, informed by patterns in stock levels, govern discrete events and entities, which, collectively, change the flow mechanism that controls stock levels.

Discrete-Event Simulation and System Dynamics for Management Decision Making, First Edition.
Edited by Sally Brailsford, Leonid Churilov and Brian Dangerfield.
© 2014 John Wiley & Sons, Ltd. Published 2014 by John Wiley & Sons, Ltd.

13.2   Background

Public health management is complex and inherently multilevel. The determinants of public health are highly interrelated and context dependent (WHO, 2011). These determinants exist at different organisational levels, affect a range of population groups, and are effective at different times or for different lengths of time (Winsberg, 2010). Context refers to the conditions or circumstances surrounding an event or action. Context may include biological or individual capacities, interpersonal relationships, institutional settings (resources, culture, and leadership) or the wider infrastructure system (the physical, cultural and regulatory environment) (Pawson, 2006; Kaplan *et al.*, 2010). Interventions are more effective and sustainable when these complex and multilevel aspects are understood and considered.

13.3   Multilevel system theories and methodologies

Few approaches support comparative studies across contexts (McPake and Mills, 2000; Mills, 2012) or the development of multilevel causal theories. Theories and methodologies that adopt systems approaches recognise multilevel and dynamic determinants (Leischow and Milstein, 2006; Homer and Hirsch, 2006; Galea, Hall and Kaplan, 2009).

Systems thinking, based on general system theory (von Bertalanffy, 1968), perceives the world as a system composed of interrelated parts that are coherently organised in a way that serves a purpose (von Bertalanffy, 1968; Meadows, 2008; de Savigny and Adam, 2009). Hierarchical system theory (Simon, 1962; von Bertalanffy, 1968) describes a multilevel system as nested systems and parts. Changes to the set of systems and parts, how they are connected, or the complex and sometimes circular relationships between them, can cause outcomes to respond in a nonlinear and often unpredictable manner over time (Sterman, 2000; Meadows, 2008). Koestler (1978) expands on the concepts of 'part' and 'system' for hierarchical systems and introduces the concept of a 'holon'. A holon is a partially independent unit that is self-regulated by its internal parts and subsystems. Each holon is embedded within, serves parts of and is influenced by a larger holon. A holon is both a part and a subsystem. Holons interact with internal and external parts and can span multiple levels of a system.

Problem abstraction involves the development of a simpler, but adequate, representation of a real-world complex system. The complexity of a system can be reduced by the principle that complex systems are partially decomposable (Simon, 1962). The links between structures within a holon are stronger (or more tightly coupled) than the links between holons (see Figure 13.1). This allows holons to be separated at weaker (or loosely coupled) links, while acknowledging their interconnectedness, for a simpler analysis. This may help resolve boundary definitions between hierarchical or organisational levels that may be hard to specify in absolute terms, a commonly encountered problem (Rousseau, 1985).

The concept of decomposability aligns with the design rules of Baldwin and Clark (2000) for modular technology development. A module can be separated from the larger system within which it functions. Modular design, when applied to

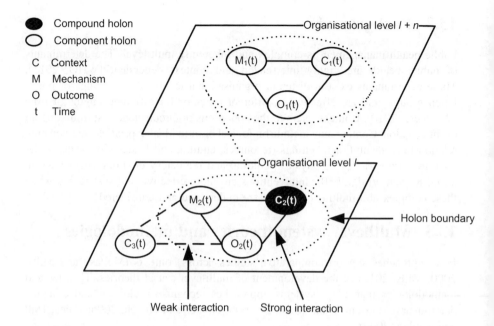

Figure 13.1   Multilevel context, mechanisms and outcomes. Adapted from Pawson and Tilley (1997) and Ratzé et al. (2007).

technology development, can produce increasingly complex technology. This technology evolves and improves in an unpredictable, yet coordinated way through a decentralised value-seeking process. With independent module improvement (or evolution), the technology as a whole evolves; consequently, the ecosystem (Adomavicius *et al.*, 2007) also evolves. Holling and Gunderson (2002) use the term 'panarchy' to describe complex evolving hierarchical systems. Panarchy is a conceptual framework that describes systems to be interlinked in ongoing adaptive cycles of growth, conservation, release and reorganisation. These cycles of change occur over a range of spatial and temporal scales. These concepts are applicable to an adaptive approach to management. By understanding these cycles and the scales in which they operate, leverage points for change that promote resilience and sustainability may be identified.

An organised view of a system can be achieved by plotting holons, events or actions and their interrelationships on a map of hierarchical organisation. This conveys key organisational levels, timescales, or spatial scales that may be involved in the public health problem and cross-level (or cross-scale) information. This can help determine appropriate levels of governance, key multidisciplinary stakeholders and actions, and shared responsibilities. An example of such a map is a Stommel diagram, a tool used in physical sciences that presents on a graph the range of levels (or physical scales such as time or space) that may be involved in the phenomena (Stommel, 1963).

The Realist approach can help identify key determinants of public health outcomes and their management. Regularities observed in public health outcomes are conceptualised as being generated by particular mechanisms acting within a context (Pawson and Tilley, 1997). Mechanisms are how outcomes are hypothesised to be produced. Context is the conditions or circumstances that trigger or control the operation of the mechanisms. These causal relationships are articulated with context–mechanism–outcome (CMO) configurations (Pawson and Tilley, 1997). CMO configurations are a list of mechanisms and contextual factors, and the outcomes that result from this interaction. Context, mechanisms and consequently outcomes each change over time. The original CMO framework, as described by Pawson and Tilley (1997), is extended to highlight the dynamic nature of this interaction (see Figure 13.1).

Contextual factors, mechanisms and outcomes are conceptualised as holons. Their state may change over time (t) and across levels of organisation (l). Holons are components and form part of a compound holon. Holons are connected by weak or strong links.

According to a Realist approach, an intervening (or management) action can change an outcome by creating a new mechanism, by modifying or disabling the operation of existing mechanisms, or by modifying the governing context (Pawson and Tilley, 1997; Kazi, 2003). Causal loop diagrams, a tool from the system dynamics approach (Sterman, 2000), can describe the direction of causation between CMO elements and whether changes in one CMO element drive a similar or opposite change in another CMO element (polarity). The concept of stocks and flows from the system dynamics approach (Sterman, 2000) can also be incorporated. With stock and flow diagrams, changes in the level of a stock or quantity is conceptualised to occur by adjusting the rate at which the stock is emptied or filled. Management actions can change flow rates to control stock levels. Analogous to stock and flow diagrams which represent changes to the accumulation of quantities or entities, Unified Modeling Language (UML) state charts (OMG, 2012) represent event-triggered transitions between finite states for a single entity, such as an individual. Management actions can govern the occurrence of these events so that the collective response of entities may impact outcomes. We describe a synthesis of these theories and methodologies for analysing complex and multilevel management outcomes. In Chapter 14, we apply this approach to frame the problem of MRSA endemicity in hospitals and explore alternative hospital infection control policies for its management.

13.4 Multiscale simulation modelling and management

Conceptual models for framing problems and planning and evaluating public health actions can convey multilevel determinants and show their interrelationships, but they may not capture the magnitude of their impact or the way these determinants and relationships change with time (Sterman, 2000). A simulation provides a platform to rigorously test single-level or multilevel hypotheses in a single experimental framework. It can capture and explore the temporal dynamics of multilevel manage-ment problems.

Single-level or single-scale simulation, where only one level of abstraction of a problem is modelled and simulated, has been extensively used to study many management problems and actions, including that of public health (Fone *et al.*, 2003; Brailsford *et al.*, 2009; Forsberg *et al.*, 2011; Sobolev, Sanchez and Vasilakis, 2011). The management model of Forrester (1961, 1992), conceptualised with stocks and flows, provides a high-level view of the impact of policy decisions. Patterns seen in stock levels over time inform management of the current state of the system. Management actions aim to control the flows into and from stocks so that the desired stock-level patterns are produced. While this model shows what is intended to happen, how this happens is not explicit at this level of abstraction. This model is extended to include details of how individuals respond to the policy. The problem and its management are also abstracted at the individual level. This describes how high-level policies shape particular individual discrete events and decisions, and how the stream of collective corrective actions subsequently changes the flow mechanism that controls outcomes (see Figure 13.2).

A high-level system structure for decision making is adapted from Forrester (1992) and Borshchev and Filippov (2004) to portray the role of individual discrete events and decisions in changing system outcomes. The multilevel management model can be developed using multiple methods. For example, high-abstraction methods, such as system dynamics, can model high-level policies and structures; low-abstraction methods, such as agent-based or discrete-event modelling methods, can

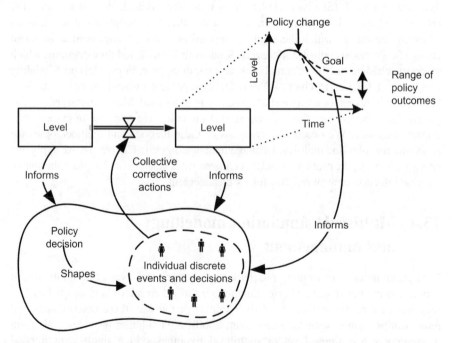

Figure 13.2 The multilevel system structure of decision making. Adapted from Forrester (1992) and Borshchev and Filippov (2004).

model discrete events and individuals interacting and acting within context and guided by policies.

A multiscale simulation[1] models multiple levels of abstraction for different time or spatial scales and can link between these levels and scales (Bassingthwaighte, Chizeck and Atlas, 2006; Meier-Schellersheim, Fraser and Klauschen, 2009; Sloot and Hoekstra, 2010). It can also switch between views at different levels and scales during real-time simulation (Bassingthwaighte, Chizeck and Atlas, 2006). This flexibility enables the presentation of both broad and specific views of the problem and its solution to multidisciplinary stakeholders and may promote consensus and collaboration (Costanza and Ruth, 1998; Etienne, Le Page and Cohen, 2003; NCI, 2012).

Multiscale simulation is primarily applied in biology (Schlessinger and Eddy, 2002; Eddy and Schlessinger, 2003; Mitha et al., 2008; Dada and Mendes, 2011), environmental sciences (Millennium Ecosystem Assessment, 2005) and physical sciences (Horstemeyer, 2010). Barriers to its application in public health include the poor availability, quality and consistency of data and theory that span multiple levels of the health care system. This is being addressed with the advancement of electronic health records and the development of frameworks for managing multilevel data collection and analyses (Eddy, 2007; IOM, 2010). Often management problems involve variables that are difficult to observe and measure. Such latent variables have been shown in single-level or single-scale models to be quantifiable (Richmond, Peterson and Vescuso, 1987) and therefore computable (Brailsford and Schmidt, 2003). Given this and the successful application of multiscale simulation in other fields, there is potential for multiscale simulation to be applied to health care and other organisational management problems.

Our multiscale simulation modelling process is composed of the following steps:

1. Develop a multilevel conceptual model of management actions.

2. Decompose the conceptual model by levels of abstraction.

3. Develop single-scale models for each abstraction.

4. Integrate the single-scale models to form one multiscale model.

5. Calibrate and validate the multiscale model.

6. Develop an interface to the model for use as an interactive learning tool, decision support tool or experimental framework.

Multilevel conceptual models, such as that described in Figure 13.2, can be broken down into smaller parts, with each part determined by considering differences in how they could be abstracted. These differences include:

---

[1] Note that the phrase multilevel modelling is often used to describe regression models that computationally account for data that varies at more than one level (Galea, Hall and Kaplan, 2009). As such, the phrase multiscale simulation is used in this chapter to avoid confusion.

1. Changes in scale or level (Bar-Yam, 2006).

2. The core focus, whether this may be patterns in aggregate quantities, key processes or interaction between individuals.

3. The treatment of time, continuous or discrete (Brennan, Chick and Davies, 2006).

4. The treatment of entities or quantities of interest, continuous or discrete (Brennan, Chick and Davies, 2006).

5. The importance of heterogeneity.

6. The importance of variability in outcomes and random events (Brennan, Chick and Davies, 2006).

7. The importance of dynamic structure (Ratzé et al., 2007).

Each smaller conceptual model informs the structure of a corresponding single-scale simulation model. Each single-scale model can be developed and validated using well-documented model development frameworks (Richmond, Peterson and Vescuso, 1987; Sterman, 2000; Bossel, 2007; Morecroft, 2007). These frameworks comprise three stages that are often repeated for ongoing refinement of the model: problem conceptualisation, model formulation and testing, and simulation analysis. Modelling methods used to develop single-scale models, such as system dynamics (Forrester, 1961; Sterman, 2000), discrete-event (Jun, Jacobson and Swisher, 1999; Fone et al., 2003; Banks et al., 2009; Gunal and Pidd, 2010) or agent-based (Bonabeau, 2002; Epstein and Axtell, 1996; Axelrod and Tesfatsion, 2011) modelling, should be selected appropriately. Several frameworks can guide this selection process (Koopman, Jacquez and Chick, 2001; Borshchev and Filippov, 2004; Brennan, Chick and Davies, 2006; RIGHT, 2009). Each framework matches characteristics of the data, theory and abstraction of the problem (i.e. the conceptual model) with characteristics of the modelling method. The RIGHT framework (2009) also considers resources such as time, money and expert knowledge to build the model in the selection process.

The independently developed and validated single-scale models are then integrated to form one multiscale and often multimethod (Koopman, Jacquez and Chick, 2001; Mingers, 1997) model. There are five ways to link single-scale models so that information can propagate across multiple levels and scales within one multiscale model (Pantelides, 2001; Ingram, Cameron and Hangos, 2004) (see Figure 13.3).

1. Serial method: The models operate sequentially. A model on one scale first generates information. Another model on a different scale then uses this information.

2. Simultaneous method: Lower scale models simulate the entire system. A higher scale model samples the information produced by the lower scale

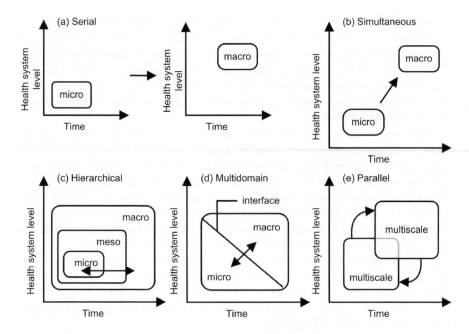

Figure 13.3   Frameworks for multiscale model construction. Adapted from Pantelides (2001) and Ingram, Cameron and Hangos (2004).

models and aggregates the information to provide a high-level summary. All the models operate simultaneously.

3. Hierarchical method: Lower scale models are embedded within higher scale models and allow information to be exchanged directly. The models operate simultaneously.

4. Multidomain method: Information is exchanged between lower and higher scale models using a common interface placed between the models.

5. Parallel method: Several multiscale models are integrated to form one model. Each model describes phenomena occurring over a range of scales with some of these scales overlapping.

The Pantelides (2001) and Ingram, Cameron and Hangos (2004) frameworks for multiscale model construction are adapted and presented as Stommel diagrams (Stommel, 1963). These multiscale models are described for two types of scales: time and health system; organisational level, however, can be described for any number of level or scale types.

The challenge when combining models is how information of different scales and levels of abstractions is bridged (Koopman, Jacquez and Chick, 2001). The 'bridging mechanism' performs either a direct translation between physical scales (Ewert *et al.*, 2006; Tsafnat and Coiera, 2009; Winsberg, 2010; Seck and Job Honig, 2012) or an approximation if no theory exists for mapping between levels of abstraction or conceptualisation. The bridging mechanism ensures information from one model at one scale is sampled at an appropriate frequency to capture important information. These samples can be aggregated or disaggregated over the time period for which they are sampled (depending on whether the sampled data is used by higher or lower scale models). Care must be taken to avoid inaccurate generalisations or variability among individual-level information.

The bridged models form the multiscale model. The multiscale model must be validated to ensure accurate and plausible results. The model is first calibrated with empirical data or plausible estimates where data is unavailable. There are a number of methods for estimating unknown parameters, such as a Least-Squares approach, Maximum Likelihood estimation and Bayesian approaches including Markov Chain Monte Carlo approaches. Parameter variation and sensitivity tests are conducted to describe the uncertainty in simulated outcomes caused by estimated parameters, and guide the interpretation of results (Granger Morgan and Henrion, 1990).

Once the model is calibrated its structure and simulated public health outcomes are validated (Barlas, 1996). There are many frameworks for validating models (Forrester and Senge, 1980; Carley, 1996; Barlas, 1996; Balci, 2007; Sargent, 2010; Gurcan, Dikenelli and Bernon, 2011). Several validation tests included in these frameworks are listed in Table 13.1.

The calibrated and validated multiscale model can then be used as an experimental framework. By performing simulation analyses, the problem can be investigated, responses and sensitivity to variations in a range of parameters can be examined, and relevant scenarios can be explored.

Table 13.1   Validation or confidence building tests of simulation models. Adapted from Forrester and Senge (1980), Barlas (1996), Carley (1996), Balci (2007) and Sargent (2010).

Tests of model structure
Compares the model equations, program code and parameters with empirical or
established theoretical relationships and values

1. Structure verification
2. Parameter verification
3. Problem scope or boundary adequacy
4. Dimensional consistency

Tests of model behaviour
Compares the range of simulated patterns in outcomes with that observed
empirically, produced by other models ('docking'), desired or expected

1. Behaviour at extreme conditions
2. The reproduction or prediction of behavioural patterns, points, distributions or values
3. Behaviour anomaly
4. Behaviour sensitivity

Multiscale simulation models can be used as learning and decision support tools. An interface can be designed that enables interactive learning, communicates and presents results simply, provides immediate feedback, and encourages thinking. To structure the design of the visual interface, the analytical design principles of Tufte (2006) can be followed: show comparisons, causality and multiple variables, use integrated text and figures, thoroughly describe the data and its sources, and use credible content. These principles are derived from analytical thinking and aim to aid the cognitive task of making sense of evidence.

The visual interface can present multiple 'viewpoints'. Multidisciplinary stake-holders can view and understand their role and the role of other stakeholders towards addressing the problem, thus promoting consensus towards management actions (Costanza and Ruth, 1998; Etienne, Le Page and Cohen, 2003; NCI, 2012). This is similar to the Enterprise Conformance and Compliance Framework (NCI, 2012) where the presentation of information is customised for the foci of stakeholder groups from a multidisciplinary team.

Interfaces can encourage learning by allowing learners to make policy or practical decisions during the simulation and observe the effect of their decisions on outcomes (Forsberg *et al.*, 2011). By providing immediate feedback to the learners, critical reflection on their decisions is encouraged and, through group interaction, can promote consensus for practice (Crichton, 1997).

13.5 Discussion

The analysis and virtual evaluation of competing public health actions are a part of a larger iterative process for effective public health management. Surrounding these steps are the gathering of knowledge, evidence and theory to be synthesised and analysed, and the communication and interpretation of the results to inform real-world decisions for action (see Figure 13.4).

While this multilevel systems approach is capable of complex analysis and virtual experimentation, the uncertainty surrounding the findings and conclusions is strongly linked to the quality of the evidence and underlying theories. The limited availability of knowledge, data and theory from multiple levels of the health system is therefore a major challenge with adopting this approach. In addition, available information must be consistent and coherent for synthesis. The difficulty in collecting this information is recognised, with progress made towards establishing a framework to assist with collecting multilevel evidence (IOM, 2010). Koopman (2004) proposes that parame-ter sensitivity experiments be conducted with mathematical or simulation models to challenge inferences and subsequently inform empirical studies. Gaps in evidence may be addressed by using techniques either to estimate missing evidence or to represent it in a different way. Further research into methods for the management of missing evidence, knowledge or theory would be of great benefit.

There are opportunities to evaluate the use of multiscale simulations by a team of multidisciplinary stakeholders and gain insights into the capability of the approach for building consensus and encouraging collaborative action.

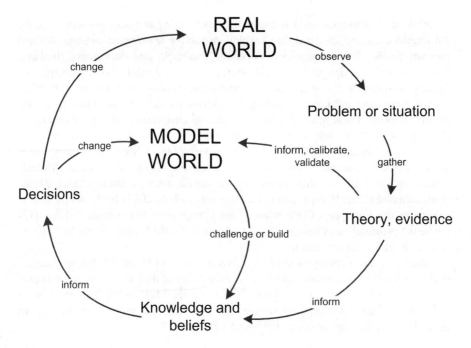

Figure 13.4   Models to inform decisions for action. Adapted from Sterman's (2000) idealised learning process.

13.6   Conclusion

We present a systematic and structured approach for analysing multilevel and systemic public health problems with system modelling and simulation. The approach explicitly considers the role of context when designing and evaluating public health actions. The approach presented extends current analytical and experimental methods and has the potential to encourage more collaborative and multidisciplinary effort towards effective public health management.

References

Adomavicius, G., Bockstedt, J.C., Gupta, A. and Kauffman, R.J. (2007) Technology roles and paths of influence in an ecosystem model of technology evolution. *Information Technology and Management*, **8**(2), 185–202.

Axelrod, R. and Tesfatsion, L. (2011) On-line guide for newcomers to agent-based modeling in the social sciences, http://www2.econ.iastate.edu/tesfatsi/abmread.htm (accessed 12 November 2012).

Balci, O. (2007) Verification, validation and testing, in *Handbook of Simulation: Principles, Methodology, Advances, Applications and Practice* (ed. J. Banks), John Wiley & Sons, Inc., Hoboken, NJ, pp. 335–398.

Baldwin, C.Y. and Clark, K.B. (2000) *Design Rules, Volume 1: The Power of Modularity*, MIT Press, Cambridge, MA.

Banks, J., Carson, J.S.II, Nelson, B.L. *et al.* (2009) *Discrete-event System Simulation*, Prentice Hall, Englewood Cliffs, NJ.

Barlas, Y. (1996) Formal aspects of model validity and validation in system dynamics. *System Dynamics Review*, **12**(3), 183–210.

Bar-Yam, Y. (2006) Improving the effectiveness of health care and public health: a multiscale complex systems analysis. *American Journal of Public Health*, **96**(3), 459–466.

Bassingthwaighte, J.B., Chizeck, H.J. and Atlas, L.E. (2006) Strategies and tactics in multiscale modeling of cell-to-organ systems. *Proceedings of the IEEE*, **94**(4), 819–830.

Bonabeau, E. (2002) Agent-based modeling: methods and techniques for simulating human systems. *Proceedings of the National Academy of Sciences*, **99**(3), 7280–7287.

Borshchev, A. and Filippov, A. (2004) From system dynamics and discrete event to practical agent based modeling: reasons, techniques, tools. Proceedings of the 22nd International Conference of the System Dynamics Society, Oxford, England.

Bossel, H. (2007) *Systems and Models: Complexity, Dynamics, Evolution, Sustainability*, Books on Demand, Norderstedt.

Brailsford, S.C., Harper, P.R., Patel, B. *et al.* (2009) An analysis of the academic literature on simulation and modelling in health care. *Journal of Simulation*, **3**, 130–140.

Brailsford, S. and Schmidt, B. (2003) Towards incorporating human behaviour in models of health care systems: an approach using discrete event simulation. *European Journal of Operational Research*, **150**(1), 19–31.

Brennan, A., Chick, S.E. and Davies, R. (2006) A taxonomy of model structures for economic evaluation of health technologies. *Health Economics*, **15**(12), 1295–1310.

Carley, K.M. (1996) Validating computational models, http://www.casos.cs.cmu.edu/publications/papers/howtoanalyze.pdf (accessed 12 November 2012).

Costanza, R. and Ruth, M. (1998) Using dynamic modeling to scope environmental problems and build consensus. *Environmental Management*, **22**(2), 183–195.

Crichton, S. (1997) Learning environments online: a case study of actual practice. PhD thesis. University of Sydney.

Dada, J.O. and Mendes, P. (2011) Multi-scale modelling and simulation in systems biology. *Integrative Biology*, **3**(2), 86–96.

de Savigny, D. and Adam, T. (eds) (2009) *Systems Thinking for Health Systems Strengthening*, Alliance for Health Policy and Systems Research, World Health Organization, Geneva, http://www.who.int/alliance-hpsr/resources/9789241563895/en/index.html (accessed 12 November 2012).

Eddy, D.M. (2007) Linking electronic medical records to large-scale simulation models: can we put rapid learning on turbo? *Health Affairs*, **26**(2), w125–w136.

Eddy, D.M. and Schlessinger, L. (2003) Archimedes: a trial-validated model of diabetes. *Diabetes Care*, **26**(11), 3093–3101.

Epstein, J.M. and Axtell, R. (1996) *Growing Artificial Societies: Social Science from the Bottom Up*, Brookings Institution, Washington, DC.

Etienne, M., Le Page, C. and Cohen, M. (2003) A step-by-step approach to building land management scenarios based on multiple viewpoints on multi-agent system simulations. *Journal of Artificial Societies and Social Simulation*, **6**(2), 257–262.

Ewert, F., van Keulen, H., van Ittersum, M.K. *et al.* (2006) Multi-scale analysis and modelling of natural resource management options, in *Proceedings of the iEMSs Third Biennial Meeting Summit on Environmental Modelling and Software* (eds A. Voinov, A.J. Jakeman and A.E. Rizzoli), International Environmental Modelling and Software Society, Manno, Switzerland.

Fone, D., Hollinghurst, S., Temple, M. *et al.* (2003) Systematic review of the use and value of computer simulation modelling in population health and health care delivery. *Journal of Public Health Medicine*, **25**(4), 325–335.

Forrester, J.W. (1961) *Industrial Dynamics*, MIT Press, Cambridge, MA.

Forrester, J.W. (1992) Policies, decisions and information sources for modeling. *European Journal of Operational Research*, **59**(1), 42–63.

Forrester, J.W. and Senge, P.M. (1980) Tests for building confidence in system dynamics models, in *System Dynamics*, Studies in the Management Sciences, vol. 14 (eds A.A. Legasto, J.W. Forrester and J.M. Lyneis), Elsevier, Amsterdam, pp. 209–228.

Forsberg, H.H., Aronsson, H., Keller, C. *et al.* (2011) Managing health care decisions and improvement through simulation modeling. *Quality Management in Health Care*, **20**(1), 15–29.

Galea, S., Hall, C. and Kaplan, G.A. (2009) Social epidemiology and complex system dynamic modelling as applied to health behaviour and drug use research. *International Journal of Drug Policy*, **20**(3), 209–216.

Granger Morgan, M. and Henrion, M. (1990) *Uncertainty: A Guide to Dealing with Uncertainty in Quantitative Risk and Policy Analysis*, Cambridge University Press, Cambridge.

Gunal, M.M. and Pidd, M. (2010) Discrete event simulation for performance modelling in healthcare: a review of the literature. *Journal of Simulation*, **4**, 42–51.

Gurcan, O., Dikenelli, O. and Bernon, C. (2011) Towards a generic testing framework for agent-based simulation models. Proceedings of the Federated Conference on Computer Science and Information Systems, IEEE, pp. 635–642.

Holling, C.S. and Gunderson, L.H. (2002) Resilience and adaptive cycles, in *Panarchy: Understanding Transformations in Human and Natural Systems* (eds L.H. Gunderson and C.S. Holling), Island Press, Washington, DC, pp. 25–62.

Homer, J.B. and Hirsch, G.B. (2006) System dynamics modeling for public health: background and opportunities. *American Journal of Public Health*, **96**(3), 452–458.

Horstemeyer, M.F. (2010) Multiscale modelling: a review, in *Practical Aspects of Computational Chemistry: Methods, Concepts and Applications* (eds J. Leszczynski and M.K. Shukla), Springer, New York, pp. 87–135.

Ingram, G.D., Cameron, I.T. and Hangos, K.M. (2004) Classification and analysis of integrating frameworks in multiscale modelling. *Chemical Engineering Science*, **59**(11), 2171–2187.

IOM (Institute of Medicine) (2010) *Bridging the Evidence Gap in Obesity Prevention: A framework to inform decision making*, The National Academies Press, Washington, DC.

Jun, J.B., Jacobson, S.H. and Swisher, J.R. (1999) Application of discrete-event simulation in health care clinics: a survey. *Journal of Operations Research Society*, **50**(2), 109–123.

Kaplan, H.C., Brady, P.W., Dritz, M.C. *et al.* (2010) The influence of context on quality improvement success in health care: a systematic review of the literature. *Milbank Quarterly*, **88**(4), 500–559.

Kazi, M.A.F. (2003) *Realist Evaluation in Practice: Health and Social Work*, Sage, London.

Koestler, A. (1978) *Janus: A Summing Up*, Random House, New York.

Koopman, J.S. (2004) Infection transmission through networks, in *Biological Networks* (ed. F. Kepes), World Scientific, Singapore, pp. 449–505.

Koopman, J.S., Jacquez, G. and Chick, S.E. (2001) New data and tools for integrating discrete and continuous population modeling strategies. *Annals of the New York Academy of Sciences*, **954**, 268–294.

Leischow, S.J. and Milstein, B. (2006) Systems thinking and modeling for public health practice. *American Journal of Public Health*, **96**(3), 403–405.

McPake, B. and Mills, A. (2000) What can we learn from international comparisons of health systems and health system reform? *Bulletin of the World Health Organization*, **78**(6), 811–820.

Meadows, D.H. (2008) *Thinking in Systems: A Primer*, Chelsea Green, Burlington, VT.

Meier-Schellersheim, M., Fraser, I.D. and Klauschen, F. (2009) Multiscale modeling for biologists. *Wiley Interdisciplinary Reviews: Systems Biology and Medicine*, **1**(1), 4–14.

Millennium Ecosystem Assessment (2005) *Ecosystems and Human Well-Being: Multiscale Assessments*, Island Press, Washington, DC.

Mills, A. (2012) Health policy and systems research: defining the terrain; identifying the methods. *Health Policy and Planning*, **27**(1), 1–7.

Mingers, J. (1997) Multi-paradigm multimethodology, in *Multimethodology: Theory and Practice of Combining Management Science Methodologies* (eds J. Mingers and A. Gill), John Wiley & Sons, Ltd, Chichester, pp. 1–20.

Mitha, F., Lucas, T.A., Feng, F. *et al.* (2008) The multiscale systems immunology project: software for cell-based immunological simulation. *Source Code for Biology and Medicine*, **3**(6). doi 10.1186/1751-0473-3

Morecroft, J. (2007) *Strategic Modelling and Business Dynamics: A Feedback Systems Approach*, John Wiley & Sons, Ltd, Chichester.

NCI (National Cancer Institute) (2012) SAIF interoperability reviews. White Paper – Introduction to SAIF and ECCF, https://wiki.nci.nih.gov/display/VCDE/Introduction+to+SAIF+and+ECCF (accessed 12 November 2012).

OMG (Object Management Group) (2012) ISO/IEC 19505-2. Information Technology – Object Management Group Unified Modeling Language (OMG UML) – Part 2: Super-structure, http://www.omg.org/spec/UML/ISO/19505-1/PDF (accessed 12 November 2012).

Pantelides, C.C. (2001) New challenges and opportunities for process modelling. *Computer Aided Chemical Engineering*, **9**, 15–26.

Pawson, R. (2006) *Evidence-based Policy: A Realist Perspective*, Sage, London.

Pawson, R. and Tilley, N. (1997) *Realistic Evaluation*, Sage, London.

Ratzé, C., Gillet, F., Muller, J.P. *et al.* (2007) Simulation modelling of ecological hierarchies in constructive dynamical systems. *Ecological Complexity*, **4**(1–2), 13–25.

Richmond, B., Peterson, S. and Vescuso, P. (1987) *An academic user's guide to Stella Software*, High Performance Systems, Inc., Lyme, NH.

RIGHT (Research Into Global Healthcare Tools) (2009) *Modelling and Simulation Techniques for Supporting Healthcare Decision Making – A Selection Framework*, Engineering Design Centre, University of Cambridge, Cambridge.

Rousseau, D.M. (1985) Issues of level in organisational research: multi-level and cross-level perspectives. *Research in Organizational Behaviour*, **7**, 1–37.

Sargent, R.G. (2010) Verification and validation of simulation models, in *Proceedings of the 2010 Winter Simulation Conference* (eds B. Johansson *et al.*), IEEE Press, Piscataway, NJ, pp. 166–183.

Schlessinger, L. and Eddy, D.M. (2002) Archimedes: a new model for simulating health care systems – the mathematical formulation. *Journal of Biomedical Informatics*, **35**(1), 37–50.

Seck, M.D. and Job Honig, H. (2012) Multi-perspective modelling of complex phenomena. *Computational & Mathematical Organization Theory*, **18**(1), 128–144.

Simon, H.A. (1962) The architecture of complexity. *Proceedings of the American Philosophical Society*, **106**(6), 467–482.

Sloot, P.M.A. and Hoekstra, A.G. (2010) Multi-scale modelling in computational biomedicine. *Briefings in Bioinformatics*, **11**(1), 142–152.

Sobolev, B.G., Sanchez, V. and Vasilakis, C. (2011) Systematic review of the use of computer simulation modeling of patient flow in surgical care. *Journal of Medical Systems*, **35**(1), 1–16.

Sterman, J.D. (2000) *Business Dynamics: Systems Thinking and Modeling for a Complex World*, Irwin/McGraw-Hill, New York.

Stommel, H. (1963) Varieties of oceanographic experience. *Science*, **139**(3555), 572–576.

Tsafnat, G. and Coiera, E.W. (2009) Computational reasoning across multiple models. *Journal of the American Medical Informatics Association*, **16**(6), 768–774.

Tufte, E.R. (2006) *Beautiful Evidence*, Graphics Press, Cheshire, CT.

von Bertalanffy, L. (1968) *General System Theory: Foundations, Development, Applications*, George Braziller, New York.

WHO (World Health Organization) (2011) The determinants of health, http://www.who.int/entity/hia/evidence/doh/en/ (accessed 12 November 2011).

Winsberg, E.B. (2010) *Science in the Age of Computer Simulation*, The University of Chicago Press, Chicago.

14

Hybrid modelling case studies

Rosemarie Sadsad,[1,2,3] Geoff Mcdonnell,[3,4] Joe Viana,[5]
Shivam M. Desai,[5] Paul Harper[6] and Sally Brailsford[5]
[1]Centre for Infectious Diseases and Microbiology – Public Health, Westmead
Hospital, Sydney, New South Wales, Australia
[2]Sydney Medical School, Westmead, The University of Sydney,
New South Wales, Australia
[3]Centre for Health Informatics, Australian Institute of Health Innovation,
University of New South Wales, Sydney, New South Wales, Australia
[4]Adaptive Care Systems, Sydney, New South Wales, Australia
[5]Southampton Business School, University of Southampton, UK
[6]School of Mathematics, Cardiff University, UK

14.1 Introduction

The objective of this chapter is to demonstrate the use of hybrid modelling for management decision support. The chapter presents three separate case studies that are unified both by the common theme of using different modelling techniques in a hybrid manner and by the health and social care context.

The first case study combines the use of system dynamics and agent-based modelling to better understand and control the spread of a specific drug-resistant pathogen in hospitals; the second case study demonstrates how a hybrid system dynamics and discrete-event simulation model can support clinical and management decision making in the context of sexual health services; and the third case study is dedicated to the investigation of a hybrid model that consists of a system-dynamics-inspired cell-based population model and a discrete-event simulation model, and is used to explore the performance of the contact centre for long-term care for people aged 65 and over.

Discrete-Event Simulation and System Dynamics for Management Decision Making, First Edition.
Edited by Sally Brailsford, Leonid Churilov and Brian Dangerfield.
© 2014 John Wiley & Sons, Ltd. Published 2014 by John Wiley & Sons, Ltd.

14.2   A multilevel model of MRSA endemicity and its control in hospitals

14.2.1   Introduction

A health-care-associated infection is an infection acquired by patients while receiving medical treatment in a health care facility (CDC, 2011; WHO, 2011). Health-care-associated infections are the most common complication to affect patients in hospitals; most cases are preventable (WHO, 2011). Many infections are caused by endemic and often drug-resistant pathogens such as Methicillin-resistant *Staphylococcus aureus* (MRSA) (Gould, 2005; Grundmann *et al.*, 2006; Köck *et al.*, 2010). MRSA acquired during a hospital stay is associated with significant morbidity and mortality and places a large burden on health care resources. Worldwide, hospital infection control has had mixed success for reducing hospital-acquired MRSA infections, highlighting the challenge of identifying the most effective infection control policy for a given setting and patient mix. In this case study a conceptual model is developed that provides a multilevel view of MRSA endemicity in hospitals. The model illustrates how factors from different levels of the health system affect MRSA endemicity in hospitals. Key stakeholders and responsibilities towards MRSA management and the areas they may target are identified. The process for how a multiscale simulation model is developed from the conceptual model is described.

14.2.2   Method

The conceptual model was developed using several multilevel system theories and frameworks. The merits of the theories and frameworks used are described in detail in Chapter 13. Several determinants of MRSA endemicity in hospitals were described using the context–mechanism–outcome (CMO) framework of the Realist approach (Pawson and Tilley, 1997). Two public health outcomes relating of MRSA endemicity in hospitals were of interest: the prevalence of MRSA and the incidence rate of MRSA acquired in a hospital. These outcomes are monitored at the health care facility and regional level. The prevalence of MRSA is the total number of people colonised or infected with MRSA at one point in time. The incidence rate of MRSA is the total number of people newly colonised or infected with MRSA in hospital over a period of time. The determinants of MRSA endemicity in hospitals were categorised as either mechanisms for the prevalence and/or incidence of MRSA, or contextual factors that could affect the operation of these mechanisms. The levels of the health system, within which the determinants or their relationships exist, were identified to indicate the levels of management and responsibilities involved. These levels were presented as a Stommel diagram (Stommel, 1963) to form a Health System Framework (Ferlie and Shortell, 2001; Glass and McAtee, 2006). The relationships between determinants that existed at the individual level, either within the individual or between individuals, were described with Unified Modeling Language (UML) statecharts (OMG, 2012). The relationships between determinants that exist at the health care facility and community level of the health care system were described using stock and flow diagrams of the system dynamics approach (Forrester, 1961; Sterman, 2000). The UML statecharts and stock and flow diagrams were embedded within the Health System Framework, forming a multilevel conceptual model. As discussed in Chapter 13, flows in the stock and flow diagrams and the occurrence of events in UML diagrams are points of leverage health care management can target to change outcomes. Contextual factors may also be targeted or introduced by health care management to control the flow and event mechanisms.

14.2.3   Results

Our conceptual model (see Figure 14.1) is based on classic models of infectious disease (Hamer, 1906; Kermack and McKendrick, 1927). In our model, the mechanisms for the prevalence of MRSA are the net admission (patient admissions less discharges and deaths) of patients already colonised or infected with MRSA and the incidence of MRSA acquired in hospital. The incidence is conceptualised as both an outcome and mechanism. When the prevalence of MRSA is the outcome of interest, the incidence of MRSA is a mechanism; however, it is also a separate outcome of interest with its own mechanisms and context. Variables may be perceived as a contextual factor, mechanism or outcome based on the perspective from the outcome of interest. The numbers of people susceptible and infectious (colonised or infected) with MRSA are represented as stocks. People entering and leaving the hospital, and new MRSA cases, are represented as flows into, out of and between these stocks, respectively. An individual in the hospital is described with a UML statechart. An individual is in one of two states: susceptible to acquiring MRSA or is infectious, being colonised or infected with MRSA. An individual transitions from the susceptible state to being infectious in the event of successful transmission of MRSA upon contact with another individual colonised or infected with MRSA or their environment. This is the mechanism for a new case of MRSA acquired in hospital in our model. All other variables interact and, together, can control the flow rates and consequently the number of people in the stocks. They can also control the occurrence of MRSA transmission events.

Several contextual factors are shown that affect the mechanisms for the prevalence of MRSA and the incidence of MRSA acquired in hospitals. The contextual factors also reflect the multilevel nature of the problem. At the community level, the prevalence of MRSA among those living in the community may affect the prevalence of MRSA among those admitted to the hospital. At the facility level, bed capacity can limit the number of people admitted to the hospital. At the individual level, high pathogen transmissibility or an individual's infectivity or susceptibility can increase the chance of MRSA transmission upon contact.

The effect of several hospital infection control policies was explored with our conceptual model. Our model illustrates that hand hygiene, where MRSA pathogens are removed or reduced from the contaminated hands of people using antimicrobial hand wash, reduces the likelihood of MRSA transmission. All other policies investigated are shown to reduce the number of contacts made between susceptible and infectious people. These policies include: cohorting, where either staff are assigned to particular patients, or infectious patients are allocated to a designated area or isolated in single-bed

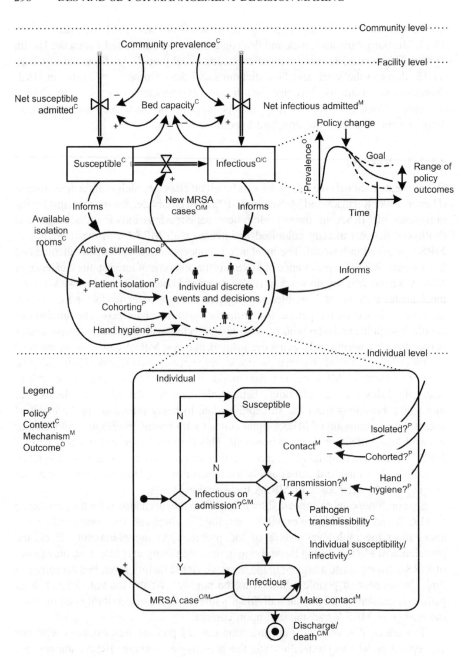

Figure 14.1   A multilevel conceptual model of MRSA endemicity and its control in hospitals. The model was developed by applying the CMO framework of the Realist approach (Pawson and Tilley, 1997), stock and flow concepts of the system dynamics approach (Sterman, 2000) and UML statecharts (OMG, 2012). The levels of the health system involved in its management are shown.

rooms; and active surveillance for MRSA, for example routine screening of patients for MRSA. Single-bed room capacity is also shown to limit the number of patients that can be isolated. Our conceptual model portrays the response from individuals to hospital infection control policies. The policies target both the mechanisms and context for the incidence of MRSA, but do not address the prevalence of MRSA directly. Our model suggests community-level strategies, reduce the prevalence of MRSA in the community, as well as facility-level strategies, could reduce the prevalence of MRSA in the hospital and, indirectly, further reduce the incidence of MRSA in the hospital.

The conceptual model provides insight into the determinants of MRSA endemicity in hospitals and the relationships between these determinants. To quantify the magnitude of their impact and describe how these determinants and their relationships change over time, the conceptual model can be translated into a multiscale simulation model. This multiscale simulation model provides a framework to explore alternative scenarios and evaluate infection control policies that may not be feasible to conduct in the real world.

We followed the six-step process for developing multiscale simulation models described in Chapter 13. The conceptual model (see Figure 14.1) shows the management of hospital acquired MRSA at the health care facility and individual level. Hospital-wide infection control policies differ in their implementation across ward specialties. As such, phenomena were modelled at the individual, ward, and hospital level.

Three single-scale models were developed individually and then combined using a hierarchical and parallel approach to form the multiscale model (see Figure 14.2):

1. The individual-based model simulates contact events between patients and health care workers and the chance of MRSA transmission based on individual attributes and the infection control precautions taken. This is an agent-based model.

2. The ward model simulates ward-specific settings, infection control policies, patient flow through the ward and patient bed allocation. This is a multimethod model. System dynamics was used to model patient flow. Patient bed

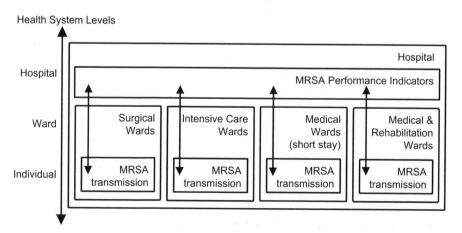

Figure 14.2   The structure of a multiscale model of MRSA endemicity and its control in hospitals.

(a) Hospital management view

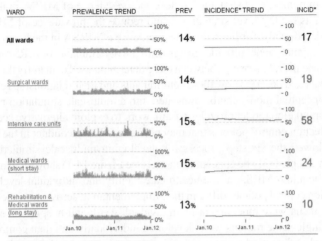

* MRSA incidence rate per 10,000 overnight bed days

(b) Ward management view

* MRSA incidence rate per 10,000 overnight bed days

Figure 14.3   The hospital and ward management view of our model of MRSA endemicity and its control in hospitals.

allocation to single-bed rooms and shared rooms was modelled using agent-based modelling methods.

3. The hospital model encapsulates four replicated ward models that have been calibrated to reflect surgical, intensive care, medical, and rehabilitation ward specialties. This model simulates patient transfers between ward specialties. The average MRSA prevalence and incidence rates in each ward speciality and in the hospital overall are collected to describe the current state of MRSA in the hospital.

The interactive visual interface was designed to assist ward- and hospital-level management and to improve understanding of MRSA transmission and its control. The hospital management view provides a direct comparison of the average daily MRSA prevalence and incidence rate for each ward speciality, and shows changes in these levels over time (see Figure 14.3a). The ward management view (see Figure 14.3b) also displays how beds in the ward (circles) are allocated to MRSA colonised and susceptible patients. It highlights when a patient acquires MRSA during their stay and whether there are empty beds. Beds that are in shared rooms are differentiated from beds in single-bed rooms (within the box). These views visually describe the state of MRSA in each ward speciality and for the hospital as a whole.

Decision makers and learners can simulate alternative scenarios and generate different MRSA outcomes by modifying several infection control, ward and disease variables (see Figure 14.4). These variables can be modified before the simulation begins or while the simulation is running. The effect of changing these variables on

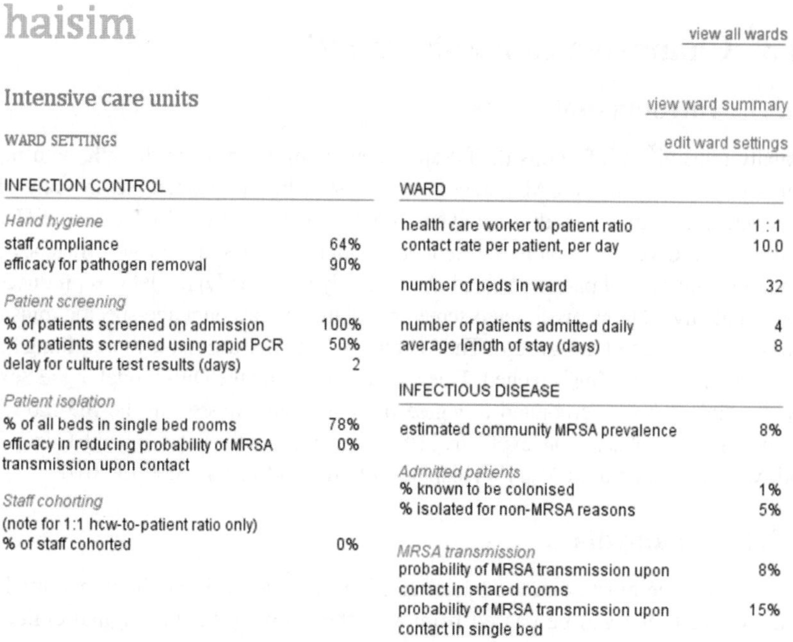

haisim   view all wards

Intensive care units   view ward summary

WARD SETTINGS   edit ward settings

INFECTION CONTROL		WARD	
Hand hygiene		health care worker to patient ratio	1 : 1
staff compliance	64%	contact rate per patient, per day	10.0
efficacy for pathogen removal	90%		
		number of beds in ward	32
Patient screening			
% of patients screened on admission	100%	number of patients admitted daily	4
% of patients screened using rapid PCR	50%	average length of stay (days)	8
delay for culture test results (days)	2		
		INFECTIOUS DISEASE	
Patient isolation			
% of all beds in single bed rooms	78%	estimated community MRSA prevalence	8%
efficacy in reducing probability of MRSA	0%		
transmission upon contact		*Admitted patients*	
		% known to be colonised	1%
Staff cohorting		% isolated for non-MRSA reasons	5%
(note for 1:1 hcw-to-patient ratio only)			
% of staff cohorted	0%	*MRSA transmission*	
		probability of MRSA transmission upon contact in shared rooms	8%
		probability of MRSA transmission upon contact in single bed	15%

Figure 14.4   The context and decision parameters for a ward in our model of MRSA endemicity and its control in hospitals.

the prevalence and incidence rate of MRSA for different ward specialties can be explored. Computer simulations provide immediate feedback to the learner, encourage critical reflection on decisions, and can promote consensus for practice. This simulation model was used to evaluate active surveillance for MRSA, ward staffing level, staff contact cohorting, isolation of MRSA patients in single-bed rooms, and staff hand hygiene policies in surgical, intensive care, and medical wards of a tertiary public hospital in Sydney, Australia (Sadsad et al. 2013). The model can be accessed at http://www.runthemodel.com/models/1055/

14.2.4 Conclusion

We demonstrate that the multilevel systems approach of Chapter 13 can be applied to develop a hybrid, multiscale simulation model of MRSA endemicity in hospitals and its control. Using this approach, key mechanisms of MRSA endemicity in hospitals and their controlling contextual factors were identified across different levels of the health system and discussed as potential areas for infection control strategies to target. The multilevel conceptual model indicates that the infection control policies studied address the incidence of MRSA in hospitals directly. There is opportunity to address the prevalence of MRSA directly as a complementary strategy. The multiscale simulation model enables ward- and hospital-level infection control strategies to be employed and explored for different ward settings. Immediate feedback in response to decisions promotes learning and multilevel perspectives encourage a collaborative response to address MRSA endemicity in hospitals.

14.3   Chlamydia composite model

14.3.1   Introduction

Sexually transmitted infections (STIs) are a continuing problem in the UK with many genito-urinary medicine (GUM) clinics being overstretched due to increases in many STIs, in particular chlamydia. A discrete-event simulation (DES) model of a Sexual Health Clinic was produced, to explore different clinic configurations to improve efficiency and increase the number of patients treated. A system dynamics (SD) model was produced to capture chlamydia transmission and screening taking into account age-specific transmission rates and different risk groups. The SD model contained a variable representing those treated as a result of being screened. This was replaced with the DES model of the Sexual Health Clinic. This incorporated a degree of detail (randomness) in the SD model of chlamydia and feedback more explicitly in the DES model of the Sexual Health Clinic. The work was conducted at St Mary's Hospital in Portsmouth from 2006 to 2010.

14.3.2   Chlamydia

Chlamydia is the most common bacterial STI in the world. It can be transmitted via sexual contact, and can be passed from a mother to baby during vaginal childbirth (NCSP, 2011). Chlamydia is in many cases asymptomatic, presenting with no symptoms in women (75%) and men (50%) (FPA, 2006). Chlamydia can result in

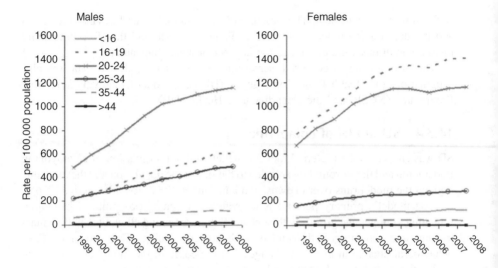

Figure 14.5   Rates of diagnoses of uncomplicated genital chlamydial infection by sex and age group, GUM departments, UK: 1999–2008 (HPA, 2007).

serious consequences (sequelae), including: pelvic inflammatory disease (PID) for women, which can result in infertility and ectopic pregnancy which is potentially fatal; and, for men, possibly infertility and inflammation of the testicles. It is estimated that the costs that result from the development of these sequelae cost the National Health Service (NHS) at least £100 million per year.

Figure 14.5 illustrates that the rates of diagnoses of uncomplicated genital chlamydial infection have been increasing in the UK since 1999. As can be seen, the rate of chlamydial diagnoses has increased disproportionately among the younger age groups (HPA, 2007).

This makes early identification and prevention vital in maintaining viable reproductive capabilities. Prevention of PID and the protection of fertility were key factors in the development of the government's sexual health strategy (Department of Health, 2001). Sexual health was also a key component of many white papers: for example, 'Choosing health: our health, our care, our say' (Department of Health, 2006) and 'Choosing health: making healthy choices easier' (Department of Health, 2004). These papers recommended the use of screening for chlamydia with a particular focus on the young (those under 25) who carried a burden of the infection.

14.3.3   DES model of a GUM department

DES was chosen as the preferred tool for modelling the walk-in clinic at St Mary's Hospital. DES has been used successfully in many other models of outpatient departments (Jun, Jacobson and Swisher, 1999; Fone *et al.*, 2003).

The model allowed the department to assess different configurations that it was interested in, for example changing the number of staff and rooms, time to complete tasks, opening hours, and so on. The model was produced in collaboration with the staff within the department. The structure of the system, key performance indicators and the data for the model were agreed. Patients missed and therefore potentially missed conditions were chosen as the key performance indicators.

Data was manually collected, screening data was analysed, and data was also sought from the Health Protection Agency (HPA), National Chlamydia Screening Programme (NCSP) and the Department of Health (DH).

14.3.4   SD model of chlamydia

SD was chosen as the preferred tool for modelling the transmission of chlamydia in Portsmouth and the surrounding area due to the size of the population and the long time horizon the model runs over (Townshend and Turner, 2000; Evenden *et al.*, 2006).

The model's primary purpose is to evaluate screening programmes including: screening younger age groups with contact tracing; screening all women with contact tracing; and screening both genders with contact tracing. This model captures behavioural change in terms of changes as the people in the model age, but there is also the probability of changing behaviour following treatment.

Data for the SD model was obtained from multiple sources: the population (Office of National Statistics); probability of infection (Turner *et al.*, 2006); contact rate (Mercer *et al.*, 2009; Townshend and Turner, 2000); mean time to recovery (Evenden *et al.*, 2006; NCSP, 2011); about the high risk group (Zenilman *et al.*, 1999).

14.3.5   Why combine the models

A composite model was constructed for the following reasons:

1. Health care models tend to focus on a specific area and do not take into account the wider system in terms of both upstream and downstream effects. The wider system can be incorporated in the composite model.

2. Models of different levels can interact and have been constructed in other industries specifically to do this. Why can the principle not be applied to health care?

Figure 14.6 gives a simple overview of the approach chosen for this work.

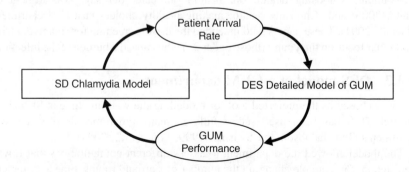

Figure 14.6   Overview of composite approach.

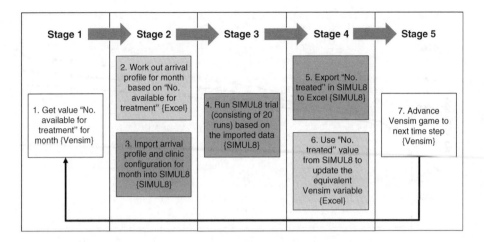

Figure 14.7   Stages involved in linking the SD and DES models.

14.3.6   How the models were combined

The models were constructed in commercially available packages SIMUL8 (DES) and Vensim (SD). It is possible to create interfaces and to communicate with both of these packages with VBA, through Microsoft Excel. How the models communicate and interact with one another is shown in Figure 14.7.

14.3.7   Experiments with the composite model

When the models are combined this dramatically increases the size of the scenario landscape.

The composite experiments are shown in Table 14.1. The staff rows in the table relate to full time equivalents (FTEs). The example row means that the department would have one member of staff working each day the department is open (without taking any breaks). The waiting area relates to the number of physical resources, waiting area chairs (capacity), etc.

To demonstrate the results from the composite model, two graphs have been chosen. Figure 14.8 represents the cumulative number of people infected over time

Table 14.1   DES parameters for the three experiments.

	Staff					Waiting areas				
	Reception	Female	Male	Blood	Lab	Reception	Standing	Female	Male	Shared
Example	5.00	5.00	5.00	5.00	5.00					
Base	14.91	10.50	7.73	4.75	4.75	4.00	10.00	10.00	10.00	No
Optimal	28.48	14.24	14.24	4.75	4.75	4.00	10.00	10.00	10.00	Yes
Max	47.46	47.46	47.46	4.75	4.75	4.00	10.00	10.00	10.00	Yes

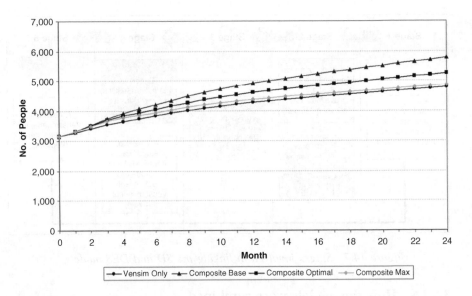

Figure 14.8   Cumulative number of people infected by month comparing composite and SD only models.

(results from the SD model), while Figure 14.9 represents the number of people not seen at the GUM department over time (results from the DES).

Figure 14.8 illustrates that by combining the DES and SD models there is an impact on the number of people who are infected over time. This is due to the

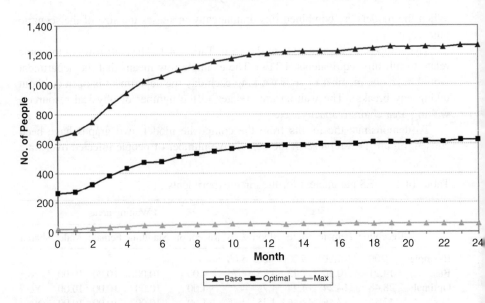

Figure 14.9   DES composite results, number of patients not seen by GUM.

configuration of the GUM department. If the GUM department is limited due to its resources then more people become infected irrespective of the level of screening, as people are unable to be treated. This is illustrated further in Figure 14.9 which clearly demonstrates that the base configuration of the GUM department results in significantly more people not being seen and subsequently potentially returning to the wider community to further spread the infection.

14.3.8   Conclusions

The key outcomes derived from the composite model are as follows:

- The composite model illustrates that decisions made at the GUM level have consequences in the population that the GUM department serves and these can be quantified based on the assumptions made.

- By making savings at the GUM department configurations can still be devised to make effective use of the resources available, but this has an effect on the prevalence of chlamydia in the population, which can now be accounted for.

- By combining the models in this way feedback normally associated with SD modelling is incorporated in the DES model.

- The SD model benefits by incorporating the detail usually associated with DES modelling.

Many have spoken about and suggested the use of hybrid simulation in various industries. This is one of a few working models in health care which does not simply mention the possibility of combining the models, but actually works in practice. The communication between the two modelling approaches is fully automated, but to make interaction easier the new model has been semi-automated to enable the DES model to be updated in between the time steps of the SD model.

There are potentially many ways in which SD and DES can be combined. The amount of and how frequently the information is exchanged are dependent upon the purpose of the combined models produced and what outcomes are required. It is important to note that the composite model produced here was not created as a replacement for existing techniques. The technique was devised to complement existing modelling approaches and to be used when a problem does not lend itself solely to one particular modelling approach.

Through the development of the composite model presented in this chapter, potential avenues for further work include:

- exploration of the timings and mechanisms through which the DES and SD models communicate and exchange information;

- evaluation of the combination of other forms of modelling including: agent-based simulation (ABS), SD, DES and other forms of model;

- further evaluation of the composite modelling approach in relation to health care and other settings; and

- comparing the development of combined models from scratch compared with the integration of software packages.

14.4   A hybrid model for social care services operations

14.4.1   Introduction

The ageing population presents many significant challenges for social care services at both a national and local level, one of which is to meet the demand for long-term care. It has been predicted that the population aged over 65 in the UK will continue to grow for some time. Social care consists of help with the activities of daily living provided to anyone with a chronic condition, that is an ongoing condition or disease that limits a person's ability to carry out everyday tasks. The concern for the government at both the national and local level is whether there will be enough resources in place to handle the expected strain on the system in the future. A study was undertaken with a local authority in the south of England, namely Hampshire County Council (HCC), to assist them with their planning. People aged over 65 are a major client group for the Council and it was this client group that was the concern of the research.

HCC have begun addressing the issue through a number of initiatives, one of which is through a modernisation programme. Part of this has been the establishment of a contact centre called Hantsdirect which will help improve efficiency so that HCC can take advantage of economies of scale. The contact centre plays an important role in the provision of social care. The service has an important role as an information provider to those in need of care (not only for publicly founded services) but acts as one of the main gatekeepers to social care. It provides an initial assessment to determine if a person is eligible for care. The design, staffing and planning of the contact centre is an important concern of the Council.

Two models were built as part of the project. First, a cell-based population model was built to forecast the demand for long-term care in Hampshire from people aged 65 and over for the period 2009 to 2026. The model uses the same logic as an SD model. DES was used to model the contact centre. The model was designed with data from HCC as well as data collected by the researcher.

The two models were combined as a hybrid model to explore the short- and long-term performance of the contact centre in the light of demographic change. The benefits and insights – of which there were many – of combining the two models were also explored in this study.

14.4.2   Population model

The cell-based population modelled the home and community system only. Projections for long-term care were made for the period 2009 to 2026 for people aged over 65 and living in the household population of Hampshire. It is important that HCC have a useful set of data to help plan services. This will potentially allow appropriate funding and resources to be allocated to meet the increases in demand. The results of this model can potentially aid the decision-making process.

The model was designed at a local level and populated with actual data from a wide range of sources including population data from the Department for Communities and Local Government (2012) and the National Statistics Subnational Population Projections Unit (2012). An important part of the model was the creation of a disability index using data from the 2001 General Household Survey (Economic and Social Data Service, 2001). This was supplemented with data from HCC.

The cell-based model uses the same logic as an SD model in terms of moving a population through different compartments. Each compartment represents an age group which is further broken down by level of disability and main service receipt.

The disability groups explored were *low, moderate, substantial, critical*, and the service providers modelled were *formal care (HCC), informal care, no care, private care*.

Due to the lack of relevant data, it was decided not to include feedback effects but to treat the model as a population projection model. This is not a weakness of the model, as the fundamental objective of projecting the demand for long-term care is still achieved. The inclusion of feedback would have relied on assumptions based solely on expert judgement. This would have been a potential weakness of the model and would have added no additional value.

14.4.3   Model construction

The model was constructed as a spreadsheet model as the features that were needed could be constructed within Microsoft Excel (2010) and hence the model could be easily shared with HCC. The model can be described by the two equations below. The population in the model for the end of each calendar year was calculated using equation (14.1). With

h_{kgt}: the household population of age group k (65–69,70–74,75–79,80–84,85+), gender g (*male, female*) at calendar year t,

u_{kgt}: the numbers of age group k, gender g, who have reached the highest year of the age group at year t,

n_{kgt}^1: the net population change for age group k, gender g, through migration and death, for year t, which was derived from n_{agt},

d_{kge}: the disability rate for disability category e (*low, moderate, substantial, critical*), age group k, gender g, and

s_{kgef}: the service rate for service category f (*formal care, informal care, no care, private care*), disability category e, age group k, gender g,

then the population of age group k, gender g at calendar year t is given by

$$h_{kgt} = h_{kgt-1} + u_{k-1gt-1} - u_{k\,gt-1} + n_{kgt}^1 \qquad (14.1)$$

The household population (h_{kgt}) at the end of the calendar year for any age group is calculated by taking the household population for the same age group from the previous year (h_{kgt-1}) and accounting for population changes. The first population change is the number ageing from the younger age group ($u_{k-1gt-1}$). For example, for the age group 70–74 in 2010, $u_{k-1gt-1}$ is the number of people aged 69 in 2009. The second population change is to account for the number of people ageing to the next age group (u_{kgt-1}). For example, for the age group 70–74 in 2010, the number of people aged 74 from 2009 are removed from their current age group and are now added to the age group 75–79 in 2010. The final population change to account for is the net population change for the age group (n_{kgt}^1) *accounting* for all the ages that are represented in that age group. For example, the net population change for the age group 65–69 accounts for migration and mortality changes for the ages 65, 66, 67, 68 and 69.

Using the results of equation (14.1), the following expansions are applied in the cell-based model. First, the disability rates for each age group and gender are multiplied by the household population in the equivalent age group and gender. The population in disability category e, age group k, gender g at time t is given by $d_{kge}h_{kgt}$. Second, the proportions of service receipt for each disability category are multiplied by the number of disabled people in each age group by gender. Thus, the population in service category f, disability category e, age group k, gender g at time t, p_{kgeft}, is given by

$$p_{kgeft} = s_{kgef}d_{kge}h_{kgt} \qquad (14.2)$$

This allowed the number of people with a disability to be predicted by age group, gender and the likely source of help.

14.4.4   Contact centre model

The contact centre was modelled using DES, which is a proven and successful approach to modelling call centres in order to test various changes and observe the impact upon performance. It allows for all the key features of Hantsdirect to be modelled. The software used was SIMUL8 (SIMUL8, 2012).

Hantsdirect has many complex features, which suggests that DES would be an appropriate approach. These features include different types of staff, time-dependent arrival times, different types of call which require different call handling times, complex routing rules, various forms of contact (telephone, e-mail, fax), outbound and abandoned calls. The purpose of modelling HCC's call centre was to be able to produce performance outputs under various demand levels. Many different performance measures can be produced from a simulation model. However, measures of particular interest in this study are:

- Distribution of time in the system (total processing time)

- Staff utilisation rates

- Number of abandoned calls

- Percentage of calls answered

- Percentage of calls answered within 20 seconds.

The simulation model was used to test a variety of different staffing scenarios but, most importantly, it was used as part of the hybrid framework.

14.4.5   Hybrid model

The combination of the two models allowed the performance of the contact centre to be explored over a 10-year period (2010–2020). In addition to standard call centre statistics, the cell-based model was used to predict the increase in demand in terms of the number of calls received due to demographic change. Based on this demand, the monthly performance of the contact centre was explored through the simulation model. This performance (in terms of abandoned calls) was then fed back into subsequent runs of the DES model, by assuming that a fraction of abandoned calls would be followed by increased demand in future months.

For the purposes of the hybrid model, the cell-based model was extended to include the rest of the population, as the contact centre receives calls from all age groups. This was only in terms of population projections and not a breakdown by various levels of disability. An additional extension included accounting for people aged 65 and over and who are not members of the household population; that is, are living in an institution. Calls can be received from people in institutional care.

This hybrid framework is useful from both a theoretical and a practical perspective. The following questions are addressed by the hybrid framework:

1. How could a detailed tactical model for the contact centre benefit from the additional use of a long-term dynamic demographic model for population change?

2. What benefits and insights would result from a combined approach based on these two different models?

Figure 14.10 illustrates the hybrid framework. The cell-based model generates the number of initial contacts for the contact centre. These initial contacts are fed into the

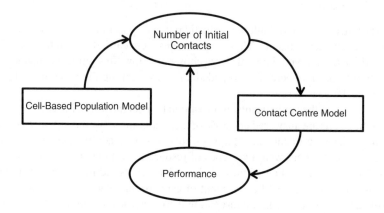

Figure 14.10   Hybrid model.

contact centre model to complete the top half of the loop shown in Figure 14.10. Each month a number of key performance indicators (KPIs) are reported from the DES contact centre model.

One assumption that potentially could be made is that a proportion of callers who abandoned their calls would subsequently experience deterioration in their condition. This proportion is unknown but it has important implications for both the contact centre and the welfare of the people concerned. The hybrid framework accounts for the people whose situation has deteriorated and who call back later requesting help, by including them in the new initial number of client profiles. This completes the bottom half of the loop illustrated in Figure 14.10. These people are added in a month later. However, as a consequence of the delay in getting the support they need, the model assumes that disability level for people aged 65 and over will be altered. People with a slight disability will have their disability status altered to moderate, and people with a moderate disability will have their disability status altered to severe.

The two models have been combined manually using a spreadsheet as an interface to transfer data and transform data between the two models. Four main experiments have been run as part of the research.

- *Experiment 1:* the model is run with no inclusion of people who abandoned their calls previously. The aim was to provide a baseline against which a number of scenarios could be compared.

Since the true proportion of people who abandon their calls and then have their situation deteriorate is unknown, the following experiments were carried out:

- *Experiment 2:* 2% of callers have their situation deteriorate and call back after one month.

- *Experiment 3:* 5% of callers have their situation deteriorate and call back after one month.

- *Experiment 4:* 10% of callers have their situation deteriorate and call back after one month.

Experiments 2, 3 and 4 allow a realistic range of possibilities to be explored and provide a useful set of results. These experiments were carried out for the years 2010, 2015 and 2020. Five-year intervals were chosen to allow for significant changes in the population to occur. There was no added value for running the experiment for each year of the 10-year period.

At the time of the study, Hantsdirect aspired to meet the following two targets: that 95% of their calls are answered and 80% of their calls are answered within 20 seconds. If in any of the experiments these targets were not met, an additional two experiments were carried out. There were two types of resources working at Hantsdirect: *agents* and *advisors*. Advisors had additional skills that agents did not have, including the ability to carry out an initial assessment of care for an elderly person potentially in need of state support. The first intervention is to increase the number of agents to the point where the KPIs are achieved and the second is to increase the number of advisors to the point where the KPIs are once again achieved. This gives a total of nine experiments (Experiments 2, 3 and 4, each for the years 2010, 2015 and 2020). This number will increase if any staffing interventions are required.

The hybrid process involves a monthly interaction between the cell-based model and the simulation model. The call profile for the first month is used as an input for the simulation model. The call centre simulation is run with the new call arrival configuration. The simulation is run for one month and the outputs are reported after 40 trials. The simulation is a terminating simulation.

For Experiment 1, for all three of the experimental years, the same process is carried out for each of the remaining 12 months. The call profiles vary each month due to changes in the number of people in the Hampshire population.

Experiments 2, 3 and 4 allow for the inclusion of abandoned callers redialling the service. The call profile for the first month is calculated in the same way as Experiment 1. After the first month, the process changes for the remaining months. The call profiles have been altered to include the impact of abandoned callers.

For example, in month 2 of Experiments 2, 3 and 4 the call profiles include a certain percentage of additional calls. For example, in Experiment 2, 2% of callers who abandoned their call in month 1 call back, for Experiment 3 it is 5% and for Experiment 4 it is 10%. The process is replicated for the remaining 10 months. This is the same for all three experimental years. Therefore, each month the call profile varies due to changes in the population and as a result of including abandoned call feedback. Table 14.2 illustrates the results of the hybrid framework and where staffing interventions were required.

When the contact centre conditions are kept the same, the model predicts that the number of staff needs to be increased in both 2015 and 2020. Surprisingly enough, there is not predicted to be much difference between the impact of both agent and advisor interventions upon performance.

Feedback does have a slight impact upon the system for all three years and the impact increases as the number of call arrivals increases. The true impact upon performance over the test period is likely to be due to changes in the population of Hampshire.

The hybrid framework has shown to be a useful toolkit in examining the impact of increasing demand upon the performance of the contact centre. It should be noted that the results for 2015 and 2020 are highly theoretical. Much can change over the next 10 years. There are many unknown future parameters. This in no way should undervalue the importance of the experiments. It is a means of validating the decisions of the Hantsdirect manager's current staffing policies. It also highlights the importance of modelling other issues such as the impact of feedback from abandoned calls.

14.4.6   Conclusions and lessons learnt

The hybrid framework has had a practical impact for HCC. This research not only looks at the changing demographics of Hampshire, but also models the impact it has had on the contact centre. It allows for a more sophisticated demand profile to be created instead of simply inflating call arrival rates by allowing for the inclusion of

Table 14.2 Summary of simulation results.

Experiment number	Year	Feedback	Intervention	Average number of abandoned calls (95% CI)	Average % of calls answered (95% CI)	Average % of calls answered within 20 seconds (95% CI)
1	2010	0%	—	630 (614 to 646)	95.37% (95.26 to 95.49%)	92.48% (92.27 to 92.68%)
2	2010	2%	—	630 (613 to 647)	95.38% (95.26 to 95.50%)	92.41% (92.19 to 92.62%)
3	2010	5%	—	634 (617 to 651)	95.36% (95.24 to 95.47%)	92.36% (92.14 to 92.59%)
4	2010	10%	—	639 (621 to 656)	95.33% (95.21 to 95.45%)	92.34% (92.12 to 92.57%)
5	2015	0%	—	802 (783 to 822)	94.46% (94.33 to 94.58%)	90.62% (90.39 to 90.85%)
6	2015	2%	—	805 (786 to 824)	94.44% (94.32 to 94.57%)	90.56% (90.33 to 90.78%)
7	2015	5%	—	813 (794 to 832)	94.40% (94.28 to 94.52%)	90.49% (90.26 to 90.71%)
8	2015	10%	—	819 (800 to 838)	94.37% (94.25 to 94.50%)	90.46% (90.23 to 90.69%)
9	2015	0%	One more agent	688 (670 to 705)	95.25% (95.14 to 95.36%)	92.14% (91.89 to 92.39%)
10	2015	2%	One more agent	693 (675 to 710)	95.22% (95.10 to 95.33%)	92.06% (91.84 to 92.27%)
11	2015	5%	One more agent	696 (679 to 714)	95.20% (95.09 to 95.32%)	92.01% (91.80 to 92.22%)
12	2015	10%	One more agent	702 (684 to 721)	95.17% (95.05 to 95.29%)	91.94% (91.74 to 92.15%)
13	2015	0%	One more advisor	673 (655 to 690)	95.35% (95.24 to 95.47%)	92.44% (92.23 to 92.65%)
14	2015	2%	One more advisor	677 (660 to 695)	95.32% (95.21 to 95.44%)	92.37% (92.16 to 92.58%)
15	2015	5%	One more advisor	679 (662 to 696)	95.32% (95.20 to 95.43%)	92.35% (92.14 to 92.56%)
16	2015	10%	One more advisor	684 (667 to 702)	95.30% (95.18 to 95.41%)	92.30% (92.10 to 92.50%)
17	2020	0%	—	1125 (1101 to 1148)	92.84% (92.70 to 92.98%)	87.48% (87.23 to 87.73%)
18	2020	2%	—	1131 (1108 to 1154)	92.80% (92.66 to 92.94%)	87.39% (87.14 to 87.63%)
19	2020	5%	—	1141 (1118 to 1165)	92.76% (92.61 to 92.90%)	87.33% (87.10 to 87.57%)
20	2020	10%	—	1158 (1136 to 1180)	92.67% (92.54 to 92.81%)	87.20% (86.96 to 87.44%)
21	2020	0%	Three more agents	741 (723 to 759)	95.28% (95.17 to 95.39%)	92.10% (91.89 to 92.31%)
22	2020	2%	Three more agents	741 (723 to 759)	95.28% (95.17 to 95.39%)	92.12% (91.90 to 92.33%)
23	2020	5%	Three more agents	744 (727 to 762)	95.27% (95.16 to 95.38%)	92.05% (91.83 to 92.26%)
24	2020	10%	Three more agents	754 (736 to 771)	95.22% (95.12 to 95.33%)	91.99% (91.77 to 92.21%)
25	2020	0%	Two/three more advisors[a]	703 (686 to 721)	95.52% (95.41 to 95.62%)	92.79% (92.59 to 92.98%)
26	2020	2%	Two/three more advisors[a]	708 (690 to 726)	95.50% (95.38 to 95.61%)	92.76% (92.55 to 92.97%)
27	2020	5%	Two/three more advisors[a]	694 (676 to 711)	95.59% (95.48 to 95.70%)	92.96% (92.76 to 93.15%)
28	2020	10%	Two/three more advisors[a]	710 (693 to 728)	95.49% (95.39 to 95.60%)	92.77% (92.57 to 92.98%)

[a] ... 1 staff were needed for most of the year and three for the rest

feedback created from abandoned calls from the previous month. Many existing simulation models of contact centres only look at short-term issues. Such models typically do not account for factors in the external environment. The combination of the two models permitted the exploration of performance issues as a result of long-term demographic change which is one of the main advantages of the framework.

Cell-based models (SD) and DES models alone have been proven to work successfully across many application areas, but this research shows that by having a combined framework a more robust modelling technique can be formed. The new framework has all the strengths of the original methodologies and also allows the researchers to address new questions that could not have been answered with one model alone. In this research, a DES model study of the contact centre would have only looked at the performance of the contact centre today and would not have studied the impact of changing population demographics and increased numbers of disabled people in the Hampshire population.

The two standalone models can potentially be useful for HCC. The Adult Services Department has taken ownership of the cell-based model and is using it as an additional piece of evidence in the planning of future services. The data provided the Council with a set of information they did not have before and the results of the model have the potential to improve decisions in both the short and long term.

The cell-based model was an ideal choice for modelling the Hampshire population to project future levels of disability and service receipt. The model is able to produce results almost instantaneously at the population level. This model can easily be used by HCC, as it was built in Excel, software which HCC already own and with which HCC staff are comfortable and familiar. This allows the Council to run their own experiments and also increases user acceptance of the model. It would not have been useful to provide HCC with the hybrid model: in addition to requiring the purchase of the SIMUL8 software and training in its use, the hybrid model requires considerable technical expertise to run.

DES models are ideal for modelling queuing systems (such as call centres) in order to understand performance under varying conditions. The cell-based model would not have been able to model the contact centre in sufficient detail and capture the stochastic aspects of the real-world system. Since the DES model was developed as a research tool and was not intended to be used by the Council, there was no need to develop a user interface or train HCC staff on how to use the SIMUL8 software. The model was used to explore various staffing scenarios that were not part of the hybrid framework. Hantsdirect requested these additional experiments over and above the modelling that was carried out to address the research questions in this section. Since the DES model was separate from the hybrid framework, these experiments were rapidly and easily carried out. The results can be potentially useful for resource planning.

As the two models have been shown to be useful on their own, it was not considered important to explore the possibility of building the model in one environment. This would have been a time-consuming programming task, and only of value if HCC had requested a user-friendly software tool which did this. Moreover, by validating the models separately, we did not need to worry about validating the combined model. The two individual models were validated using standard validation and verification approaches. Much time would be needed to ensure that the single environment model was fully validated before the results could be relied on.

The hybrid framework potentially helps decision making in a number of issues which would not have been possible through the use of either one of these models on their own, through its ability to address both strategic and operational issues, and provides a more robust approach.

References

CDC (Centers for Disease Control and Prevention) (2011) Healthcare-associated infections (HAI), http://www.cdc.gov/hai/index.html (accessed 12 November 2012).

Department for Communities and Local Government (2012) http://www.communities.gov.uk/corporate/ (accessed 27 May 2012).

Department of Health (2001) National strategy for sexual health and HIV, www.dh.gov.uk.

Department of Health (2004) Choosing health: making healthy choices easier, www.dh.gov.uk.

Department of Health (2006) Choosing health: our health, our care, our say, www.dh.gov.uk.

Economic and Social Data Service (2001) www.esds.ac.uk/ (accessed 27 May 2012).

Evenden, D., Harper, P.R., Brailsford, S.C. and Harindra, V. (2006) Improving the cost-effectiveness of Chlamydia screening with targeted screening strategies. *Journal of the Operational Research Society*, **57**(12), 1400–1412.

Ferlie, E.B. and Shortell, S.M. (2001) Improving the quality of health care in the United Kingdom and the United States: a framework for change. *Milbank Quarterly*, **79**(2), 281–315.

Fone, D., Hollinghurst, S., Temple, M. *et al.* (2003) Systematic review of the use and value of computer simulation modelling in population health and health care delivery. *Journal of Public Health*, **25**(4), 325–335.

Forrester, J.W. (1961) *Industrial Dynamics*, MIT Press, Cambridge, MA.

FPA (2006) Chlamydia: looking after your sexual health, Pamphlet.

Glass, T.A. and McAtee, M.J. (2006) Behavioral science at the crossroads in public health: extending horizons, envisioning the future. *Social Science Medicine*, **62**(7), 1650–1671.

Gould, I.M. (2005) The clinical significance of Methicillin-resistant *Staphylococcus aureus*. *Journal of Hospital Infection*, **61**(4), 277–282.

Grundmann, H., Aires-de-Sousa, M., Boyce, J. *et al.* (2006) Emergence and resurgence of Methicillin-resistant *Staphylococcus aureus* as a public-health threat. *The Lancet*, **368** (9538), 874–885.

Hamer, W.H. (1906) The Milroy lectures on: Epidemic disease in England – the evidence of variability and of persistency of type. *The Lancet*, **167**(4307), 733–739.

HPA (Health Protection Agency) (2007) All new STI episodes seen at GUM clinics in the UK 1998–2007, www.hpa.org.uk/STIannualdatatables (accessed 20 September 2008).

Jun, J.B., Jacobson, S.H. and Swisher, J.R. (1999) Application of discrete-event simulation in health care clinics: a survey. *Journal of the Operational Research Society*, **50**(2), 109–123.

Kermack, W.O. and McKendrick, A.G. (1927) A contribution to the mathematical theory of epidemics. *Proceedings of the Royal Society*, **115**, 700–721.

Köck, R., Becker, K., Cookson, B. *et al.* (2010) Methicillin-resistant *Staphylococcus aureus* (MRSA): burden of disease and control challenges in Europe. *Euro Surveillance*, **15**(41), www.eurosurveillance.org/ViewArticle.aspx?ArticleId=19688 (accessed 12 November 2012).

Mercer, C.H., Copas, A.J., Sonnenberg, P. *et al.* (2009) Who has sex with who? Characteristics of heterosexual partnerships reported in a national probability survey and implications for STI risk. *International Journal of Epidemiology*, **38**, 206–214.

National Statistics Subnational Population Projections Unit (2012) http://www.ons.gov.uk/ons/index.html (accessed 27 May 2012).

NCSP (2011) National Chlamydia Screening Programme, http://www.chlamydiascreening .nhs.uk/ps/ (accessed 5 January 2011).

Object Management Group (OMG) (2012) ISO/IEC 19505-2: Information technology – Object Management Group Unified Modeling Language (OMG UML) – Part 2: Superstructure, http://www.omg.org/spec/UML/ISO/19505-1/PDF (accessed 12 November 2012).

Pawson, R. and Tilley, N. (1997) *Realistic Evaluation*, Sage, London.

Sadsad, R., Sintchenko, V., McDonnell, G.D., Gilbert, G.L. (2013) Effectiveness of Hospital-Wide Methicillin-Resistant Staphylococcus aureus (MRSA) Infection Control Policies Differs by Ward Specialty. PLoS ONE 8(12): e83099. doi:10.1371/journal.pone.0083099-

Sadsad, R. (2013) Hospital Acquired Infections. Run the model: AnyLogic North America. http://www.runthemodel.com/models/1055/ (accessed 13 January 2014).

SIMUL8 (2012) www.simul8.com/ (accessed 27 May 2012).

Sterman, J.D. (2000) *Business Dynamics: Systems Thinking and Modeling for a Complex World*, Irwin/McGraw-Hill, New York.

Stommel, H. (1963) Varieties of oceanographic experience. *Science*, **139**(3555), 572–576.

Townshend, J.R.P. and Turner, H.S. (2000) Analysing the effectiveness of Chlamydia screening. *Journal of the Operational Research Society*, **51**(7), 812–824.

Turner, K.M.E., Adams, E.J., Gay, N. *et al.* (2006) Developing a realistic sexual network model of Chlamydia transmission in Britain. *Theoretical Biology and Medical Modelling*, **3**(3), 3.

WHO (World Health Organization) (2011) Report on the burden of endemic health care-associated infection worldwide, http://www.who.int/gpsc/country_work/burden_hcai/en/index.html (accessed 12 November 2012).

Zenilman, J., Ellish, N., Fresa, A. and Glass, G. (1999) The geography of sexual partnerships in Baltimore: applications of core theory dynamics using a geographic information system. *Sexually Transmitted Diseases*, **26**(2), 75–81.

15

The ways forward: A personal view of system dynamics and discrete-event simulation

Michael Pidd

Lancaster Business School, University of Lancaster, UK

15.1 Genesis

In a chapter that speculates on the future, it seems wise to begin by looking backwards to see what we can learn. Both system dynamics and discrete-event simulation have been used for over 50 years. System dynamics, originally christened industrial dynamics by its MIT-based, founding father Jay Forrester, was born in the late 1950s. The first publication had the title 'Industrial dynamics: a major breakthrough for decision makers' (Forrester, 1958) and later became the second chapter of the book *Industrial Dynamics* (Forrester, 1961), which presents and explains the methods and assumptions of what we now refer to as system dynamics. Given Forrester's engineering background it is hardly surprising that system dynamics treats system structure and feedback control as key to system behaviour. Forrester's argument in these early publications promoting industrial dynamics is based on a straightforward analogy between control engineering and organisational behaviour. Forrester (1961) argues that organisations include feedback structures that need to be understood and managed to achieve high performance. Forrester argued that industrial dynamics provided the tools needed to do this.

Discrete-Event Simulation and System Dynamics for Management Decision Making, First Edition.
Edited by Sally Brailsford, Leonid Churilov and Brian Dangerfield.
© 2014 John Wiley & Sons, Ltd. Published 2014 by John Wiley & Sons, Ltd.

If the approach now known as system dynamics appeared largely as the result of one man's work, doubtless supported by others, the origins of discrete-event simulation are harder to pin down. According to Hollocks (2006), what we now know as discrete-event simulation was first employed in the UK steel industry in 1957 in the GSP system. The first European text on discrete-event simulation, though it did not use that term, was Tocher (1962) and it is interesting to note that its author regarded computer simulation as a temporary solution to problems that would eventually be solved in mathematical statistics. The year 1962 also saw the publication of the first paper on the GPSS simulation language (Gordon, 1962) in the United States. Variants of GPSS are still in use. Just as Forrester's engineering background is clearly seen in the methods and assumptions of system dynamics, the influence of the discrete-event simulation pioneers is also evident today. Keith Tocher was a pioneer in both statistical methods and computing and it remains true that becoming proficient in the use of discrete-event simulation requires mastery of appropriate statistical methods and a facility with computers.

15.2   Computer simulation in management science

I joined Lancaster University as a young academic in 1979, after a spell at Aston University that was preceded by a job in industry, and was asked to teach simulation methods to undergraduates. I struggled to find a suitable book for the course, mainly because existing ones seemed to me to be out of date. They assumed inflexible mainframe computing, whereas I was using interactive time-shared computing services and personal computers and was convinced that interactive computing was the way ahead. After three years of teaching at Lancaster I was naive enough to think I could write a better and more up-to-date book than those available. I was fortunate to work with some very perceptive colleagues at Lancaster, in particular John Crookes, who argued that PCs (actually Apple II computers at that stage) would fundamentally change how we built and used simulation models. I also discovered word processing on Apple IIs and laboriously used the rather crude Applewriter software to write the text. This allowed the publishers to typeset from disk, which seemed very advanced at the time.

Hence, I wrote my own book, *Computer Simulation in Management Science* (Pidd, 1984), which stayed in print for 30 years, though heavily revised from time to time and now in its final, fifth, edition. Though it touched the bases needed for simulation on mainframe computers, it emphasised simulation models developed and run using interactive computing. It was, perhaps, rather light on the statistical aspects of discrete-event simulation, mainly because these were covered very well in books by other authors. Unusually, it included both discrete-event simulation and system dynamics in its coverage; a lead that few authors have chosen to follow.

Before taking up a university post, I had worked as an operational research practitioner for a few years, having originally trained as an engineer. Discussions with practitioners as I mapped out my simulation book confirmed my hunch that two approaches to computer simulation predominated: discrete-event simulation and system dynamics. Hence it seemed obvious that the book should devote space to these approaches and should also discuss the similarities and differences between them. At that time, users of both system dynamics and discrete-event simulation struggled with inadequate software and sometimes ended up writing their own from scratch. Thus the book took a rather detailed approach to the internal workings of both system dynamics and discrete-event simulation and included code for a discrete-event simulation system written in Basic. As it evolved, the book had three parts: an initial overview consisting of three chapters that explored the differences and similarities, a section devoted to discrete-event simulation and another devoted to system dynamics. More space was devoted to discrete-event simulation than to system dynamics because I believed, and still do, that the technical demands of its use are greater.

When I wrote the system dynamics chapters I was fortunate, once again. A new colleague, Brian Parker, joined us at Lancaster, and he was a system dynamics practitioner. Whereas I had seen system dynamics as primarily a simulation technique, he took a much broader view, seeing it as a way to explore how systems behave or might behave under different conditions. Hence, anyone reading that first edition (published in 1984) may detect a slight shift in emphasis over the three chapters of the system dynamics section, from simulation to exploration. The shift in emphasis is more marked across editions, especially comparing the fourth and fifth editions with their predecessors, and also affects the way in which discrete-event simulation is discussed in those later editions. Across all editions it is clear that I regard both discrete-event simulation and system dynamics as extremely useful and widely used, but as very different approaches.

15.3   The effect of developments in computing

Publication rates in academic journals for both system dynamics and discrete-event simulation have grown over those years. As evidence of this, Figure 15.1 shows the results of a search of *International Abstracts in Operations Research*, for five-year periods from 1985, searching separately for 'system dynamics' and 'simulation'. Though neither linear nor exponential, the growth in the academic papers published in operational research journals is obvious and is likely to be more than matched by unpublished work conducted by practitioners. Neither method can really be confined only to operational research, so the underlying growth in use is likely to be higher if other areas such as healthcare or logistics were included.

Why has this growth occurred? I suspect that the main reason is the availability of better and easier-to-use computer software. Over the more than 50 years since both approaches appeared, computers have become cheaper, more powerful and easier to use. Anyone old enough to have created programs on a remote mainframe computer will recall the utter frustration of making a small mistake and having to wait at least a day until it could be corrected. It was not unusual to spend more time bug hunting than code writing. In the same era, computer graphics were out of the question for most users and, compared even with today's domestic appliances, mainframe computers were not very powerful and offered limited memory. Hence there was a great

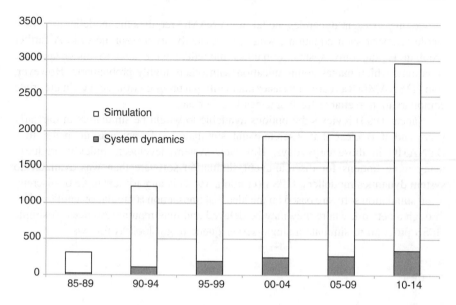

Figure 15.1   Papers published in IAOR *journals on system dynamics and discrete-event simulation.*

disincentive to use computer-based simulation and modelling approaches. As computers became more powerful and cheaper, software developers began to develop much more friendly software than previously possible and what we now think of as simulation software began to appear.

In mainframe days, developing and running a system dynamics model meant using DYNAMO (fully specified in Pugh, 1973), which requires the modeller to write equations in a tightly defined and not very intuitive format. Thus, a pair of level and rate equations might look as follows:

$$S.K = S.J + (DT)(I.JK-O.JK) \qquad (15.1)$$

$$O.KL = S.K/A \qquad (15.2)$$

where S.K is the value of a level (or stock) at time K, S.J is the value of the level at time J, and I.JK and O.JK are the input at output rates controlling the level S over the period JK, which is DT in duration. In mathematical terms, this level equation is a first-order difference equation. The rate equation shown computes the output rate over the next interval KL as the level S at time K divided by the constant A. DYNAMO computation proceeds in a defined sequence in which the level equations are computed at some defined time point K, based on their values at the previous time point J. After the level equations, the rate equations are computed for the interval KL, the next period of length DT.

Coding a system dynamics model in DYNAMO requires the modeller to write a whole series of such equations, which is inevitably an error-prone task. A further problem is that only system dynamics modellers were able to understand the equations, which makes communication with others highly problematic. However, using DYNAMO was certainly easier than writing a program from scratch in the crude compilers then available for languages like Fortran.

Nance (1993) reviews the options available to would-be simulators in the early years and it is clear that discrete-event simulation software was no better than DYNAMO in those early days. However, system developers quickly produced many more options for discrete-event simulation software than was available to system dynamics modellers. GPSS (Gordon, 1962) is an early example of discrete-event simulation software based on the idea that transactions (simulation entities) flow through a network where they may be delayed and may require resources. A simple GPSS program to simulate a single-server queue might look as follows:

```
GENERATE 6, 2
QUEUE 1
SEIZE 1
DEPART 1
ADVANCE 6, 3
RELEASE 1
TERMINATE
```

This GPSS program GENERATES a transaction every 6 ± 2 time units; that is, at intervals that are uniformly distributed between 4 and 8 time units. Once created, each transaction joins the first-in–first-out QUEUE 1. When it reaches the head of the queue it SEIZES resource 1 when this is available and keeps hold of that resource for 6 ± 3 time units. Once this time is over, the transaction RELEASES resource 1, which is then free for use by the next transaction to reach the head of QUEUE 1. The active life of the current transaction is now TERMINATED. It should be obvious that this GPSS program has much the same virtues and vices as the DYNAMO program.

Nowadays we rarely use such direct programming approaches when initially developing a system dynamics or discrete-event simulation model. Instead, we tend to use graphical tools to develop the skeleton of a model and may later resort to some programming to develop it further into a model that is valid for a defined purpose. The developers of early simulation software were well aware of the need for graphical tools to support model development and communication. However, these remained paper-based tools for many years because computer graphics devices were crude and much too expensive for widespread use. Industrial dynamics (Forrester, 1961) includes many examples of level-rate diagrams. Figure 15.2 shows a simple level-rate diagram in which the birth and death rates of a population in a particular period are based on the actual population and the fractional births and deaths (the percentage of the population that is born or dies each year). The rates themselves are shown as valves that control inflows and

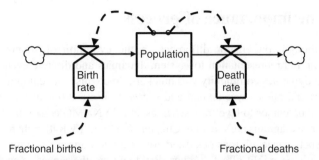

Figure 15.2   Level-rate diagram for simple population dynamics.

outflows to and from the population, represented as a tank or level. The dashed lines represent information flows; thus information about the current population level affects the birth and death rates in any time period, which hence control the population level. This is a simple example of feedback control.

Likewise in discrete-event simulation, the pioneers were aware of the power of graphical representation and devised ways to exploit this. The GPSS listing shown earlier actually corresponds to a block diagram in which each line of code is a separate block using some of the symbols shown in Ståhl *et al.* (2011, Figure 2). The GPSS simulation modeller, probably not an expert programmer, was encouraged to draw a block diagram to represent the system logic and from it to write GPSS code. Each block corresponded to a punched card which formed the program and data entry mechanism for most computers at the time. In the UK, Tocher (1962) uses what were termed *stochastic gearwheels*, to represent logical interactions between classes of simulation objects. These later evolved into the familiar activity cycle diagrams shown in Chapters 2 and 6.

As computers grew cheaper and more powerful, software developers exploited this and users soon took it for granted that they could interact graphically with simulation software. Thus the now familiar software packages appeared. Stella (isee systems, 2013) was the first widely used graphical system dynamics simulation system and exploited the superior graphics of Apple computers. Stella was followed by Vensim (see Chapter 11) and Powersim (Powersim Corporation, 2013). All three of these provide a graphics palette that allows on-screen model development by drawing diagrams based on Forrester's level-rate graphics. The discrete-event simulation field similarly realised the potential of computer graphics and experienced much more explosive growth than system dynamics. Many different graphical packages appeared, of which well-known examples include SIMUL8 (see Chapter 10), Witness (Lanner, 2013), Arena (Rockwell Automation, 2013) and Simio (Simio, 2013). In addition, various vendors developed graphical software that supports both system dynamics and discrete-event simulation. Examples include AnyLogic (see Chapter 12), which also supports agent-based modelling, and ExtendSim (ImagineThat, 2013).

15.4   The importance of process

There is, though, one major difference in the evolution of support for model development and representation for system dynamics and discrete event simulation. The system dynamics community was quicker to recognise that communication with model users and clients was crucial and started to use a different type of influence diagram as a precursor to level-rate schemes and DYNAMO equations. Examples of these influence diagrams are shown in Chapters 6, 11 and 13. Rather than attempting to distinguish between levels (or stocks) and flows (or rates), the aim is simply to represent how one variable might affect another. By doing so, the aim is to understand the feedback behaviour generated by the structure of the system being modelled, which may mean that there is no need actually to develop and run a fully fledged simulation model. Thus the approach described in Wolstenholme (1990), and called qualitative system dynamics in Chapter 6, was born.

This qualitative representation of system structure allowed writers such as Senge (1990) to popularise the idea of system archetypes, enabling the use of qualitative system dynamics in system design as well as analysis.

This brought system dynamics to the attention of the much less technical management learning community, whereas the technical demands of discrete-event simulation meant that, despite its power, its use was mainly restricted to technically trained analysts. In Chapter 10, Mark Elder sensibly argues that such analysts must also pay great attention to the process aspects of their work, using tools like SIMUL8, of which he was originator.

System dynamics and discrete-event simulation have followed similar tracks that were influenced mainly by developments in computing. In addition, system dynamics modellers were quick to realise the power of graphical conceptualisation in influence diagrams. In recent years, both communities have realised that close user engagement in conceptualisation and model development is key to successful implementation of analysis based on simulation modelling.

15.5   My own comparison of the simulation approaches

The comparisons in the previous chapters are at a conceptual level, but it is helpful, I think, to go into more detail about both system dynamics and discrete-event simulation – to look under the hood. Doing so should help our understanding of why the close integration of the two approaches, if we take them to be simulation methods, may or may not be wise.

15.5.1   Time handling

The types of system usually simulated by both system dynamics and discrete-event simulation are dynamic, rather than static. That is, things happen through time and the simulation model is intended to capture the effects of those changes, so as to understand why they occur and may be used to investigate what might be done to improve things. A very early definition of computer simulation captures this rather well: 'Simulation always involves the manipulation of a model so that it yields a motion picture of reality' (Ackoff and Sasieni, 1968, p. 97). Simulation models run on digital computers, which means, fundamentally, that the progress of time is not smooth, but proceeds in discrete steps. System dynamics and discrete-event simulation models treat this advance in simulated time in different ways. Pidd (2004, Chapter 2) provides examples to illustrate these differences.

As normally implemented, system dynamics simulations employ a *time slicing* approach in which the system is indexed through simulation time in predefined time increments. These time slices are usually known as d*t* or DT, depending on the software in use. Thus level equation (15.1) shown earlier in this chapter includes a constant DT, which represents this time slice. The level equations are updated at these regular intervals to capture the effect of the inflows and outflows over that interval. Once these level equations have been computed, the rate equations are computed to determine the rates that will apply over the next time interval of length DT. A system dynamics model that employs time slicing is actually performing an Euler–Cauchy integration of the difference equations, though this is rarely mentioned in practice. Time slicing is very simple to implement in simulation software and is easily understood.

Most discrete-event simulations model the movement of time rather differently. Instead of employing fixed time slices that are determined in advance, the simulation proceeds from event to event and the intervals between those events vary. Events are state changes that affect one or more entities in the system being simulated. Examples might be the arrival of a patient at an Emergency Department, the discharge of a patient from hospital or the start or finish of a lunch break. To move simulated time from event to event in this way requires a more complex computer program than needed for time slicing, though it is not too difficult to implement. The efficiency of a discrete-event simulation system is to a large extent dependent on the way that its executive is implemented. The executive maintains an event calendar on which future events are entered. At each event time, the executive checks the calendar to see when the next event is due and moves the simulation clock to that time, at which more future events may be entered on the calendar (see Pidd, 2004, Chapter 6, for more details). The intervals between events depend entirely on the rules coded into the entity definitions and will vary. Thus the simulation proceeds quickly when events are sparse, but slows down when events are frequent. If the nature of the system is such that all events occur at regular intervals (e.g. once each day) then a discrete-event simulation automatically defaults to time slicing.

Any attempt to integrate system dynamics and discrete-event models must cope with these different approaches to time handling in the implementation and execution of a simulation. In essence this means that either both models must briefly cease operation in order to exchange data, or the discrete-event model must schedule some regular, time-sliced events so that interaction can occur with the system dynamics model at those regular events. In the accident and emergency model discussed in Chapter 13, the system dynamics and discrete-event models are both stopped every simulated hour to enable this data exchange. A similar approach is taken in the chlamydia integrated model discussed in Chapter 14. The AnyLogic software, which allows system dynamics, discrete-event models and agent-based approaches, manages its integration by forcing system dynamics time points to become discrete events.

15.5.2   Stochastic and deterministic elements

Many dynamic systems include non-deterministic elements. Thus we may not be sure exactly how many patients will arrive at an Emergency Department or how long it will take a nurse to treat a patient. These are examples of stochastic elements, a term used to denote something that varies probabilistically through time. Stochastic elements are usually represented in a simulation model by appropriate probability distributions. These may be mathematically defined distributions such as Normal, Poisson and gamma or may be histograms directly based on data. Simulations with stochastic elements employ sampling routines in which random numbers are used to generate samples from the distributions (see Pidd, 2004, Chapter 10, for details).

Discrete-event simulations are usually employed on systems with stochastic elements, whereas most system dynamics models are wholly deterministic. The use of random sampling in a discrete-event simulation means that each simulation run is, in effect, a sampling experiment and each such run will produce different results. How different they are will depend on the extent of the stochastic variation in the model. To allow for this sampling variation, discrete-event modellers employ careful analysis methods to ensure that any conclusions drawn from a set of simulation runs are statistically sound (see Pidd, 2004, Chapter 11, for details). This stochastic variation means that it is usually unwise to draw conclusions based on a single run of a discrete-event simulation.

By contrast, system dynamics models are based on an assumption that system structure leads to system behaviour and that any observed variation in that behaviour is a result of interactions between system substructures. Thus, there is no need to run a system dynamics model many times, unless the values assigned to system parameters vary. For example, in Figure 15.2 we may be uncertain of the percentage of deaths expected each year. This is normally handled by running the model several times, each with different values that represent the likely range of variation in that parameter. This obviously needs careful thought if there are multiple parameters with values that are uncertain. Some system dynamics software allows the modeller to specify that the value of a parameter is sampled from a probability distribution, but this provision feels like an added extra.

Fundamentally, then, system dynamics models are deterministic and discrete-event models are usually stochastic. This means that, in addition to the issues caused by their different approaches to time handling, great care needs to be taken in interpreting the results from an integrated model or those from a pair of cooperating models. In essence the two different types of variation can lead to a factorial explosion when attempting to understand the results of such simulations. As yet, there is no sign that this is an issue to which researchers have given much attention.

15.5.3   Discrete entities versus continuous variables

The basic building blocks of the two approaches also differ. In a discrete-event simulation, the concept of a simulation entity is fundamental. A simulation entity is an object that is included in the simulation model because it changes state and is believed to contribute to the behaviour of the system. Examples might include doctors, nurses and patients in a healthcare simulation or equipment such as imaging machines. The behaviour of some of the entities included in a discrete-event model is likely to be stochastic, which means that the events scheduled for their state changes will not occur at regular, time-sliced intervals. If a discrete-event simulation includes multiple entities of the same type that need to be separately represented, these are usually organised into classes, for example of patients. The entities included in a class need not be identical, but must have enough similarity to allow general statements to be made about their behaviour. For example, unless arriving in an ambulance, patients register their arrival at the clinic before they are seen by a clinician. Fundamentally, discrete-event models (and agent-based models, too) are atomistic, in that system behaviour is a result of interactions occurring between individual entities and the resources they use while in the system being simulated.

By contrast, the fundamental building blocks of system dynamics models are two types of variable: levels (or stocks) and rates (or flows). Levels are accumulations within the system that will persist even if activity ceases elsewhere in the system. Rates represent the activity within a system and control the levels, like inflows and outflows from a tank of liquid. Rates are constant over a time interval DT and levels are computed as a result of those rates at the instants between each time slice. This means that system dynamics variables are quasi-continuous as shown in Figure 15.3, because the rates and levels only change at discrete, predefined points of time and are held constant between the points. That is, a system dynamics model is a quasi-continuous simulation, based on straightforward difference equations in which time moves between discrete, predefined points. The focus of a system dynamics model is the way that the variables vary through simulated time and not on individual items counted within those aggregated variables. Thus system dynamics models are not atomistic, but quasi-continuous aggregations in which variation in behaviour occurs as a result of relationships between the variables included.

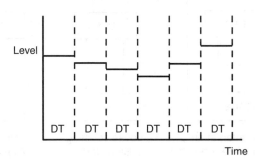

Figure 15.3   Quasi-continuous representation of levels in system dynamics.

As with the differences in time handling and the use of random sampling in a discrete-event model, the emphasis on discrete entities or continuous variables presents issues to be faced when attempting to integrate or cooperate the two approaches. It is relatively straightforward to use the value taken by a continuous variable to fire a discrete event. For example, if the number of patients seen in a clinic reaches some prescribed value, the simulation may ensure that extra staff are drafted in. That is, the value of a continuous variable may affect the population of entities in a discrete-event simulation. Chapter 12, which describes the operation of the AnyLogic simulation software, provides a thorough discussion of the different ways that discrete entities and continuous variables may interact in a simulation that contains both types of object.

We might note in passing that simulation time in a discrete-event simulation is actually a quasi-continuous variable, albeit one that changes its value at non-constant intervals. Since most discrete-event software implements efficient mechanisms for managing this quasi-continuous variable via an event calendar, it is worth speculating on whether a similar approach could be used to manage interactions between a discrete-event simulation and a system dynamics model. This might be a fertile research area to pursue.

15.6   Linking system dynamics and discrete-event simulation

Any attempt to link system dynamics and discrete-event simulation must take account of their different ways of handling simulated time, their assumptions about the causes of variations in system behaviour and the different degrees of aggregation. None of these is insurmountable, as is evidenced by software such as AnyLogic and ExtendSim, which allow both approaches to coexist within the same model.

Broadly speaking, there are three ways in which the two approaches may be linked and these are summarised in Figure 15.4:

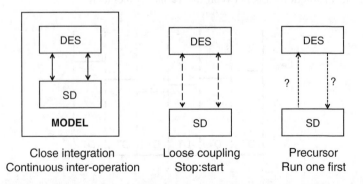

Figure 15.4   Ways to combine system dynamics and discrete-event simulation.

1. The close integration of different approaches within a single model: examples of this might be the use of AnyLogic or ExtendSim to embed both discrete-event and system dynamics elements within the same model.

2. The loose coupling of different approaches: examples might be situations in which a discrete-event simulation model runs separately from a system dynamics model and the two periodically communicate and exchange data. This is described as parallel interaction in Chapter 13 and as combined modelling in the description of the chlamydia model in Chapter 14.

3. The use of one approach as a precursor to the other. For example, values produced from a discrete-event simulation might be used as parameters in a system dynamics model, or vice versa. This is described as sequential interaction in Chapter 13.

Whether it is always wise to have both elements within the same model is another question altogether and asserting that a real-world system contains both continuous and discrete elements may not be cause enough. It is important to remember that models are always simplifications. Pidd (2009, pp. 8–10) defines a model as 'an external and explicit representation of part of reality as seen by the people who wish to use that model to understand, to change, to manage and to control that part of reality'. The aim of a systems modelling exercise is to find some way to represent the factors that are believed to contribute to the aspects of the behaviour of the system in which we are interested. That is, choices have to be made and some factors will be included and others excluded. Even if a factor is included, it may be possible to represent in an approximate way that avoids the need to operate a combined model.

This book does not include a detailed discussion of model validation, but this is an important part of any modelling exercise if modellers and model users are to have confidence in the results produced by the model. It may be sufficient to represent continuous variables by discrete elements and build a wholly discrete-event simulation. Alternatively, it may be sufficient to represent discrete elements by suitable equation parameters in a system dynamics model. No modelling exercise is perfect and no model is 100% accurate. Models need to be fit for purpose and clarity about that purpose is vital. Thus embedding both system dynamics and discrete-event elements in a model is possible, but may not be necessary – depending on the intended use of the model.

15.7 The importance of intended model use

Though there is never a guarantee that a model will be used for its intended purpose, or indeed whether it will ever be used at all, the intended model use should always be a major determinant of what is included in the model and how it is represented. Pidd (2009, p. 66) discusses model use and includes the figure shown simplified here as Figure 15.5. This is a spectrum of model use that includes the four points discussed below. The first two positions (decision automation and routine decision support) are

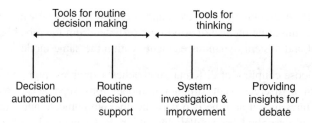

Figure 15.5   A spectrum of model use.

shown as requiring tools for routine decision making. The other two (system investigation and providing insight) are shown as requiring tools for thinking. Clearly these four points are archetypes and in-between positions are possible.

15.7.1   Decision automation

As used here, the term decision automation refers to situations in which a model is developed to replace routine, human decision making. Many of us use such models regularly but are unaware of it. For example, hotel rooms and airline tickets are sold through web sites, behind which sit one or more automated systems of revenue management. The aim of revenue management systems is to ensure that seats and hotel rooms are offered at prices that are neither too high nor too low. If prices are too high, seats and rooms will be left empty and profit forgone; if they are too low, revenue will be depressed. Revenue management systems rely on automated forecasting models to predict sales at different prices, which cooperate with optimisation models that allocate seats and rooms to price bands through time. From time to time, the parameters of the models will need to be updated to reflect changes that occur in the environment in which the models are employed, but day-to-day routine use is fully automatic.

For successful decision automation, great attention must be paid to model and data accuracy and to speed and efficiency of execution, and it is usually crucial that the model is properly linked to other corporate systems. Since an organisation is trusting the model to take decisions on its behalf and to commit resources, fidelity and detail are crucial, which means that model testing and validation loom large and may require considerable time and resources.

It would, I believe, be most unwise to use either system dynamics or discrete-event simulation, unless the model is wholly deterministic, in a fully automated decision-making process. Once a discrete-event simulation includes stochastic elements, the results of the simulation will vary from run to run, depending on the random sampling processes employed.

This induced variation means that human intervention is needed to interpret the results of the simulation. It would also be unwise to build an automated decision-making system around a system dynamics model because such models are approximate and very aggregated representations that are unlikely to be fine grained enough for routine decision making. Both system dynamics and discrete-event models are best regarded as adjuncts to human decision making.

15.7.2   Routine decision support

Models that underpin routine decision support are also used regularly, though less frequently, than those employed in decision automation. The aim here is not to replace human decision makers but to support them. Ensuring that they have the right crews and aircraft in the right place at the right time is a major concern for airlines. Commercial aircraft are very expensive and do not make money when on the ground, so need to be flying full of passengers as often as possible. This requires the correct number of flight crew to be available as well as having the aircraft ready. Having enough aircraft and crew available is easy if the airline employs an excessive number of both, but doing so costs money. Hence the need to get the levels right, which is a complex problem for airlines that operate many flights from many airports.

Considerable effort has gone into developing models to support human decision makers as they plan rosters and schedules, and the accuracy and validity of the models improve each year. However, the models are not used to make decisions, but rather to propose rosters and schedules that a human decision maker can alter because of circumstances of which he or she is aware that are not part of the model. Without the model, the decision maker would face a daunting task of developing rosters and schedules from scratch and is likely to rely on heuristic approaches that may or may not be valid. With a decision support model, the decision maker can greatly reduce his or her work and focus on the areas of uncertainty.

Though, as with models intended for use in decision automation, fidelity and detail are important, there is no requirement for the model to be wholly complete, though this does not mean that sloppy modelling is allowed. Rather, because the model is an adjunct to decision making by humans, there is a recognition that the model and human decision maker working together form the decision-making system. Since the model is to be used regularly, ease of use looms large and there may also be a requirement that it can interface with corporate systems.

Both discrete-event and system dynamics models could be used in routine decision support; however, there are few, if any, reports of their actual use in this way. This may be because discrete-event models are not transparent to the user and because the nature of the results from system dynamics models can seem rather imprecise. Perhaps this may be another fertile area of research for the future?

15.7.3   System investigation and improvement

This type of model use is perhaps the most common way in which both system dynamics and discrete-event simulations are employed, particularly the latter. Typically, an organisation or individual wishes to gain better understanding of how a system operates, or to find better ways for it to do so. In the latter case, the model becomes a vehicle for experimentation; an artificial world in which options can be compared and explored before any real implementation. Most simulation case studies reported at conferences or in journals are examples of such work. In this book, most of the examples in the preceding chapters discuss the development and use of discrete-event, system dynamics, linked or hybrid models for such system investigation and improvement.

Earlier in this chapter I argued that system dynamics modellers realised the importance of the process of interactive model building much earlier than did their discrete-event counterparts. When seeking system improvement a triad of modeller, decision maker and model constitute the decision-making, or learning, system. Indeed, when a model is developed for use in system investigation and improvement, the processes of model conceptualisation, development and validation become a learning process for all involved. It is sometimes the case that creating the model provides enough insights for people to act with confidence without actually running the model. In this regard, visualisation, whether paper-based or on-screen, becomes very important, to help participants strive for common understanding of the system being simulated.

Given all this, it should be no surprise that both discrete-event and system dynamics approaches can be used, and are used, to develop models to support system investigation and improvement. Indeed, most books on these subjects assume that this is why the model is to be developed.

15.7.4   Providing insights for debate

The final point on the spectrum of Figure 15.5 is labelled as providing insights for debate. Whereas decision automation may be appropriate for operational decisions, here we are concerned with strategy and policy issues in which people may hold fundamentally different positions. Strategic decisions are rarely made solely on the basis of the results from running a simulation model and this is especially true in public policy, as discussed in Thissen and Walker (2013). Instead, people argue their case and it can be helpful to understand and demonstrate the effect of their beliefs about how a system does or might operate. There is an underlying assumption in the other three positions that the model is representing something 'out there', but models developed to provide insights for debate are more often concerned with explicating mental models.

The explication of mental models has long been a concern in the system dynamics community. As pointed out by Doyle and Ford (1998), mental models were a major concern of the *Industrial Dynamics* book (Forrester, 1961), and Forrester (1971) elaborates further on this theme. Forrester's main argument is that everyone has mental models of how systems operate or could operate and the job of a modeller is to tease these out and to represent them explicitly. Once explicated, they are available for clear discussion and debate.

Thissen and Walker (2013) argue that rational analysis has a place in public policy development and suggest that this explication is an important contribution from the analytical community. However, it should be noted that the political arena is characterised by ambiguity and sometime this is essential if accommodation is to be achieved between warring parties (Noordegraaf and Abma, 2003). This does not mean that modelling has no place in policy analysis, but it does suggest that its use may require political skill and a willingness to live with ambiguity rather than to insist on its removal.

Although discrete-event simulations could be used in policy debate, they are more often used later, after agreement has been reached about the broad shape of a policy that now needs tuning. By contrast, system dynamics is an approach that could be used to support some types of policy development, particularly for the exploration of scenarios, in which ambiguity plays an important role.

15.8   The future?

Attempting to forecast the future more than a short period ahead is a risky business. It was, after all, Thomas Watson, Chairman of IBM, who stated in 1943 that the world market for computers would be no more than five. One of the lessons of longer term business forecasting is that flexible responses to changes in the environment may be more important than apparent forecasting accuracy. Hence, in this final section I will attempt to examine what might happen to system dynamics and discrete-event simulation in the future. Inevitably, this exploration is based on my own prejudices and experience, as well as evidence from the past.

15.8.1   Use of both methods will continue to grow

Figure 15.1 gives some idea of the growth in the use of both approaches over the last 30 years. What might cause such growth to stall or to be reversed? Two possibilities come to mind. The first is that better and more appropriate methods might appear and become popular. Perhaps Keith Tocher was right and developments in mathematical statistics will mean it is unnecessary to use discrete-event simulation? It is very likely that new methods will be developed and used that will make the use of simulation unnecessary in some projects. However, little has happened in the last 50 years that makes such a major decline very likely, so it seems that the need for discrete-event simulation will remain. System dynamics likewise offers an appealingly simple formalism for thinking about the behaviour of dynamic systems.

For both approaches, new or expanding application areas have appeared and will continue to do so. For example, though discrete-event simulation has been used in healthcare for many years (Tunnicliffe-Wilson, 1981) its use has grown greatly in recent years. For example, the 2000 Winter Simulation Conference, which has a main focus on discrete-event simulation, had only a partial stream of seven papers devoted to healthcare. In 2006, the conference introduced a full Healthcare track, which included nine papers. By 2012 this track had grown to 32 papers. It is harder to establish the similar growth in system dynamics, as the organisation of the Proceedings of the International Conferences of the System Dynamics Society is harder to disentangle in its earlier years; however, the other chapters of this book suggest that this is currently a major area of activity. Why has this growth occurred? Because healthcare is expensive and there is a need to increase its efficiency and effectiveness and both system dynamics and discrete-event simulation are a great help in this. At some point, growth in modelling work in healthcare will slow down and may even decline, but it seems likely that new application areas will appear because they face similar issues and require the use of similar approaches.

15.8.2   Developments in computing will continue to have an effect

I argued earlier that much of the previous growth in the use of both approaches was due to developments in computing technology. In the very early years of discrete-event simulation, there was two-way traffic between the worlds of simulation and computing. For example, the basic ideas of object orientation were first implemented in Simula, a much underused simulation programming language. However, in the last 20 years or so simulation software providers have been quick to exploit developments in computing and it seems likely that this trend will continue. What none of us know is what these developments will be, but the following few seem likely.

Most users now take mobile computing for granted and today's smartphones are much more powerful than the PCs of a decade ago. Smartphones and tablet computers have more than enough power to run simulations, especially if parts of the execution are managed over the Internet. This cooperative execution has been possible for some time with Web-based SIMUL8 (SIMUL8 Corporation, 2013), which allows a user to run part of the simulation on a PC and with the rest run on a Web-based, or intranet, server. There is no technical reason why simulations, whether system dynamics or discrete-event, cannot be run on a smartphone or tablet or via a Web-based cooperation. This does not mean, however, that such devices are ideal for model development, in which small screen size is a definite handicap.

For model development, especially when this is group-based, it seems obvious that the two communities should exploit large graphics walls with touch-based interfaces. In 2013, when I wrote this chapter, I visited my three-year-old grand-daughter's nursery and saw that she and her playmates knew how to use a smart-board. Better large display walls could enable group participation, allowing people to stand and create models on-the-fly, assuming the software is smart enough to exploit this. Since 1987, the advocates of the Strategic Choice Approach (Friend and Hickling, 1987) have argued that allowing people to walk around and to stand while considering strategic options adds greatly to the value of strategic discussions. Large display walls with touch-based interfaces running simulation software would seem to be ideal for this.

Ubiquitous computing is now ubiquitous. That is, we take for granted that the devices that we use in our mundane, daily lives are smart, and this is because they include embedded computing devices. The defence community has long been a user of 'man in the loop' simulations in which humans interact with computer simulations, sometimes for training, but also in strategy and tactical development. These approaches recognise that it is better to allow humans really to interact than to try to model those interactions. It seems possible that wearable or implantable computing devices may be of great value in further developing these approaches.

15.8.3   Process really matters

Earlier in this chapter, I argued that the system dynamics community was quick to realise that much modelling involves the explication of mental models, which requires careful process management. This emphasis continues to this day. The discrete-event simulation community came rather later to the same view, partly because its models are usually rather more complex than system dynamics models. Thus, part of Mark Elder's argument in Chapter 10 is that highly developed skills in computing and statistics are no guarantee that a modeller will be effective in helping users to employ simulation so as to develop and understand options for change. This also chimes with the recent emphasis on conceptual modelling (Robinson *et al.*, 2011) in discrete-event simulation, which stresses that conceptualisation is a process rather than an outcome. Likewise, it also links into the theme developed in Chapter 4 by Kathy Kotiadis and John Mingers that problem structuring methods can and should be combined with simulation approaches, though this needs to be done with some care. This emphasis on carefully managed processes for model use and development also seems set to continue and we may expect the appearance of further tools to support this.

References

Ackoff, R.L. and Sasieni, M.W. (1968) *Fundamentals of Operations Research*, John Wiley & Sons, Inc., New York.

Doyle, J.K. and Ford, D.N. (1998) Mental models concepts for system dynamics research. *System Dynamics Review*, **14**,(Spring), 1.

Forrester, J.W. (1958) Industrial dynamics: a major breakthrough for decision makers. *Harvard Business Review*, **36**(4), 37–66.

Forrester, J.W. (1961) *Industrial Dynamics*, MIT Press, Cambridge, MA.

Forrester, J.W. (1971) Counterintuitive behavior of social systems, in *Collected Papers of J. W. Forrester*, Wright-Allen Press, Cambridge, MA, pp. 211–244.

Friend, J.K. and Hickling, A. (1987) *Planning Under Pressure: The Strategic Choice Approach*, Pergamon Press, Oxford.

Gordon, G. (1962) A general purpose system simulator. *IBM Systems Journal*, **1**, 18–32.

Hollocks, B.W. (2006) Forty years of discrete-event simulation: a personal reflection. *Journal of the Operational Research Society*, **57**(12), 1383–1399.

ImagineThat   (2013) ExtendSim powertools for simulation, www.extendsim.com.

isee systems   (2013) Stella: systems thinking for education and research, www.iseesystems.com.

Lanner (2013) Witness 12 – Making it easier to provide answers for business critical problems, www.lanner.com.

Nance, R.E. (1993) A history of discrete event simulation programming languages. Technical Report TR-93-21, Computer Science, Virginia Polytechnic Institute and State University.

Noordegraaf, M. and Abma, T. (2003) Management by measurement? Public management practices amidst ambiguity. *Public Administration*, **81**(4), 853–873.

Pidd, M. (1984) *Computer Simulation in Management Science*, 1st edn., John Wiley & Sons, Ltd, Chichester.

Pidd, M. (2004) *Computer Simulation in Management Science*, 5th edn., John Wiley & Sons, Ltd, Chichester.

Pidd, M. (2009) *Tools for Thinking: Modelling in Management Science*, 3rd edn., John Wiley & Sons, Ltd, Chichester.

Powersim Corporation (2013) Powersim, www.powersim.com.

Pugh, A.L. (1973) *DYNAMO II User's Manual*, MIT Press, Cambridge, MA.

Robinson, S.L., Brooks, R., Kotiadis, K. and van der Zee, D.-.K. (eds) (2011) *Conceptual Modelling for Discrete-Event Simulation*, CRC Press, Boca Raton, FL.

Rockwell Automation (2013) Arena simulation software, www.arenasimulation.com.

Senge, P.M. (1990) *The Fifth Disciple: The Art and Practice of the Learning Organization*, Random House, London.

Simio (2013) Simio simulation software, www.simio.com.

SIMUL8 Corporation (2013) Web based SIMUL8, www.simul8.com/products/simul8_web. htm.

Ståhl, I., Henricksen, J.O., Born, R.G. and Herper, H. (2011) GPSS 50 years old, but still young. Proceedings of the 2011 Winter Simulation Conference, Phoenix, AZ.

Thissen, W.A.H. and Walker, W.E. (eds) (2013) Public Policy Analysis: New Developments, International Series in Operations Research & Management Science, vol. 179, Springer, New York.

Tocher, K.D. (1962) *The Art of Simulation*, English Universities Press, London.

Tunnicliffe Wilson, J.C. (1981) Implementation of computer simulation projects in health care. *Journal of the Operational Research Society*, **32**, 825–832.

Wolstenholme, E.F. (1990) *System Enquiry: A System Dynamics Approach*, John Wiley & Sons, Ltd, Chichester.

Index

Discrete-Event Simulation and System Dynamics for Management Decision Making, First Edition.
Edited by Sally Brailsford, Leonid Churilov and Brian Dangerfield.
© 2014 John Wiley & Sons, Ltd. Published 2014 by John Wiley & Sons, Ltd.